G Proteins

G Proteins

Edited by

Patrick C. Roche

Department of Experimental Pathology
Mayo Clinic
Rochester, Minnesota

ACADEMIC PRESS

San Diego   New York   Boston   London   Sydney   Tokyo   Toronto

Front cover illustration: Immunoperoxidase stain (red reaction product) for growth hormone (GH) in a pituitary adenoma. Molecular characterization of these neoplasms has shown that nearly 40% of all GH-secreting human pituitary adenomas contain an activating mutation in the α subunit of G_s (see Chapter 23). Hematoxylin counterstain; original magnification 100X.

Academic Press, Inc.
A Division of Harcourt Brace & Company
525 B Street, Suite 1900, San Diego, California 92101-4495

United Kingdom Edition published by
Academic Press Limited
24-28 Oval Road, London NW1 7DX

International Standard Serial Number: 1043-9471

International Standard Book Number: 0-12-185299-7

PRINTED IN THE UNITED STATES OF AMERICA
96 97 98 99 00 01 EB 9 8 7 6 5 4 3 2 1

Table of Contents

QP
552
,G16
G19
1996

Contributors

Article numbers are in parentheses following the names of contributors. Affiliations listed are current.

JOEL ABRAMOWITZ (6), Department of Medicine and Cell Biology, Baylor College of Medicine, Houston, Texas, 77030

JOSEPH M. ALEXANDER (23), Department of Medicine, Harvard Medical School, Boston, Massachusetts, 02114

NIVA ALMAULA (12), Dr. Arthur M. Fishberg Research Center in Neurobiology, and Department of Neurology, Mount Sinai School of Medicine, New York, New York, 10029

JA-HYUN BAIK (7), Laboratoire de Génétique Moléculaire Eucaryotes Centre National de la Recherche Scientifique, U184 Institut National de la Santé et de la Recherche Medicale Institut de Chimie Biologique, Faculté de Medecine, 67085 Strasbourg, France

MADHAV BHATIA (7), Laboratoire de Génétique Moléculaire Eucaryotes Centre National de la Recherche Scientifique, U184 Institut National de la Santé et de la Recherche Medicale Institut de Chimie Biologique, Faculté de Medecine, 67085 Strasbourg, France

MARK W. BITENSKY (18), Biophysics Group, Los Alamos National Laboratory, Los Alamos, New Mexico, 87545

EMILIANA BORRELLI (7), Laboratoire de Génétique Moléculaire Eucaryotes Centre National de la Recherche Scientifique, U184 Institut National de la Santé et de la Recherche Medicale Institut de Chimie Biologique, Faculté de Medecine, 67085 Strasbourg, France

MICHEL BOUVIER (19), Departement de Biochimie, Groupe de Recherche sur le Systeme Nerveux Autonome, Université de Montreal, Montreal, Quebec, Canada H3C 3J7

PATRICK J. CASEY (8), Department of Molecular Cancer Biology, Duke University Medical Center, Durham, North Carolina, 27710

SHERRY COULTER (5), Laboratory of Cellular and Molecular Pharmacology, National Institute of Environmental Health Sciences, Research Triangle Park, North Carolina, 27709

CHARLES W. EMALA (24), Department of Anesthesiology, The Johns Hopkins Medical Institutions, Baltimore, Maryland, 21205

Ross D. Feldman (19), Departments of Medicine and Pharmacology and Toxicology, University of Western Ontario, London, Ontario, Canada N6A 5A5

James C. Garrison (15), Department of Pharmacology, University of Virginia School of Medicine, Charlottesville, Virginia, 22908

Susan L. Gillison (11), Department of Pharmacology, University of Washington, Seattle, Washington, 98195

Stephen G. Graber (15), Department of Pharmacology and Toxicology, West Virginia University, Morgantown, West Virginia, 26506

John R. Hadcock (10), Division of American Home Products Agricultural Research Center, Molecular and Cellular Biology American Cyanamid Company, Princeton, New Jersey, 08543

Hazem Hallak (3), Department of Pharmacology, University of Pennsylvania School of Medicine, Philadelphia, Pennsylvania, 19104

Hyung-Mee Han (22), Department of Immunotoxicology, National Institute of Safety Research, Seoul, South Korea

T. Kendall Harden (17), Department of Pharmacology, University of North Carolina School of Medicine, Chapel Hill, North Carolina, 27599

Joyce B. Higgins (8), Department of Chemistry, Eastern Illinois University, Charleston, Illinois, 61920

Zhengmin Huang (13), Endocrine Research Unit, Departments of Medicine and Physiology, Veterans Affairs Medical Center, University of California, San Francisco, San Francisco, California, 94121

Saleem Jahangeer (5), Laboratory of Cellular and Molecular Pharmacology, National Institute of Environmental Health Sciences, Research Triangle Park, North Carolina, 27709

Richard S. Jope (14), Department of Psychiatry and Behavioral Neurobiology, University of Alabama at Birmingham, Birmingham, Alabama, 35294

Anjaneyulu Kowluru (20), Section of Endocrinology, Department of Medicine, University of Wisconsin School of Medicine, Madison, Wisconsin, 53792

Michael A. Levine (24), Department of Medicine, The Johns Hopkins Medical Institutions, Baltimore, Maryland, 21205

Xiaohua Li (14), Department of Psychiatry and Behavioral Neurobiology, University of Alabama at Birmingham, Birmingham, Alabama, 35294

MARGARET A. LINDORFER (15), Department of Pharmacology, University of Virginia School of Medicine, Charlottesville, Virginia, 22908

IRENE LITOSCH (4), Department of Molecular and Cellular Pharmacology, University of Miami School of Medicine, Miami, Florida, 33136

KAREN M. LOUNSBURY (3), Department of Pathology, University of Vermont Medical College, Burlington, Vermont, 05405

CRAIG C. MALBON (1), Department of Molecular Pharmacology, Diabetes and Metabolic Diseases Research Program, Health Sciences Center, State University of New York, Stony Brook, Stony Brook, New York, 11794

DAVID R. MANNING (3), Department of Pharmacology, University of Pennsylvania School of Medicine, Philadelphia, Pennsylvania, 19104

STEWART A. METZ (20), Section of Endocrinology, Department of Medicine, University of Wisconsin School of Medicine, Madison, Wisconsin, 53792

NICOLE MONS (2), Laboratoire de Neurobiologie Fonctionnelle, URA-CNRS 339, Université de Bordeaux 1, Talence, France, and Department of Pharmacology, University of Colorado Health Sciences Center, Denver, Colorado, 80262

CHRISTOPHER M. MOXHAM (1), Department of Molecular Pharmacology Diabetes and Metabolic Diseases Research Program, Health Sciences Center, State University of New York, Stony Brook, Stony Brook, New York, 11794

BIPIN G. NAIR (21), Panlabs, Inc., Bothell, Washington, 98011

ROBERT A. NISSENSON (13), Endocrine Research Unit, Departments of Medicine and Physiology, Veterans Affairs Medical Center, University of California, San Francisco, San Francisco, California, 94121

TARUN B. PATEL (21), Department of Pharmacology, University of Tennessee, Memphis, Memphis, Tennessee, 38163

ANDREW PATERSON (17), Department of Pharmacology, University of North Carolina School of Medicine, Chapel Hill, North Carolina, 27599

ROBERTO PICETTI (7), Laboratoire de Génétique Moléculaire Eucaryotes Centre National de la Recherche Scientifique, U184 Institut National de la Santé et de la Recherche Medicale Institut de Chimie Biologique, Faculté de Medecine, 67085 Strasbourg, France

HELEN POPPLETON (21), Department of Pharmacology, University of Tennessee, Memphis, Memphis, Tennessee, 38163

HANI M. RASHED (21), Department of Pharmacology, University of Tennessee, Memphis, Memphis, Tennessee, 38163

ANN E. REMMERS (9), Department of Pharmacology, The University of Michigan Medical School, Ann Arbor, Michigan, 48109

JANET D. ROBISHAW (16), Weis Center for Research, Geisinger Clinic, Danville, Pennsylvania, 17822

MARTIN RODBELL (5), Laboratory of Cellular and Molecular Pharmacology, National Institute of Environmental Health Sciences, Research Triangle Park, North Carolina, 27709

VLADIMIR RODIC (12), Dr. Arthur M. Fishberg Research Center in Neurobiology, and Department of Neurology, Mount Sinai School of Medicine, New York, New York, 10029

VITALYI RYBIN (22), College of Physicians and Surgeons, Columbia University, New York, New York, 10032

STUART C. SEALFON (12), Dr. Arthur M. Fishberg Research Center in Neurobiology, and Department of Neurology, Mount Sinai School of Medicine, New York, New York, 10029

GEOFFREY W. G. SHARP (11), Department of Pharmacology, College of Veterinary Medicine, Cornell University, Ithaca, New York, 14853

DOLORES M. SHOBACK (13), Endocrine Research Unit, Departments of Medicine and Physiology, Veterans Affairs Medical Center, University of California, San Francisco, San Francisco, California, 94121

GUNNAR SKOGLUND (11), INSERM Unit 55 Centre de Recherche, Hôpital Sainte-Antoine, F-75571 Paris, France

SUSAN F. STEINBERG (22), Departments of Medicine and Pharmacology, College of Physicians and Surgeons, Columbia University, New York, New York, 10032

JOANN STRNAD (10), Division of American Home Products Agricultural Research Center, Molecular and Cellular Biology, American Cyanamid Company, Princeton, New Jersey, 08543

HUI SUN (21), Department of Pharmacology, University of Tennessee, Memphis, Memphis, Tennessee, 38163

GRAZIELLA THIRIET (7), Laboratoire de Génétique Moléculaire Eucaryotes Centre National de la Recherche Scientifique, U184 Institut National de la Santé et de la Recherche Medicale Institut de Chimie Biologique, Faculté de Medecine, 67085 Strasbourg, France

PAUL R. TURNER (13), Endocrine Research Unit, Departments of Medicine and Physiology, Veterans Affairs Medical Center, University of California, San Francisco, San Francisco, California, 94121

HSIEN-YU WANG (1), Department of Physiology and Biophysics, State University of New York, Stony Brook, Stony Brook, New York, 11794

BARRY M. WILLARDSON (18), Biophysics Group, Los Alamos National Laboratory, Los Alamos, New Mexico, 87545

HONG XIE (4), Department of Molecular and Cellular Pharmacology, University of Miami School of Medicine, Miami, Florida, 33136

TATSURO YOSHIDA (18), Biophysics Group, Los Alamos National Laboratory, Los Alamos, New Mexico, 87545

YI-MING YU (21), Department of Pharmacology, University of Tennessee, Memphis, Memphis, Tennessee, 38163

WEI ZHOU (12), CNS Drug Exploratory, Bristol-Myers Squibb, Wallingford, Connecticut, 06492

Preface

Our knowledge of the structure, function, and diversity of G proteins has increased exponentially over the past 20 years. G proteins have been shown to play many roles in normal cellular function, including cell growth, differentiation, and neurotransmission. Likewise, aberrations in G proteins and their functions have been discovered to cause a variety of disease states, from cancer to cholera. Appropriately, the 1994 Nobel Prize in Physiology or Medicine was awarded for pioneering work in the field of G proteins.

The heterotrimeric ($\alpha\beta\gamma$) GTP-binding proteins (G proteins) are integral components of many signal transduction pathways for hormones, neurotransmitters, and autocrine/paracrine factors. G proteins transduce messages received at the cell surface to intracellular effector systems. There are hundreds of receptor types that couple to the intracellular milieu through four major families of G proteins. Each family has multiple members, and although molecular cloning studies have identified more than 20 α, 4 β, and at least 6 γ subunits, different G protein heterotrimers have classically been designated according to the identity of their α subunit. The α subunit contains the GTP-binding site and intrinsic GTPase activity and can reversibly associate with the nearly inseparable $\beta\gamma$ complex. The $\beta\gamma$ complex is required for association of G_α with the plasma membrane and for activation of G_α by a cell surface receptor. In addition to its membrane anchoring role for α subunit, the $\beta\gamma$ complex can directly modulate several intracellular effector systems.

This volume emphasizes the universal importance of G proteins and contains chapters that cover the many diverse areas of G protein investigation. Both biochemical and molecular techniques are well represented. Covalent cross-linking protocols, assays of solublized membranes, and ADP-ribosylation methods for detection and measurement of G proteins are presented. A variety of chapters contain strategies and experimental details for evaluating receptor–G protein interactions and G protein–effector system coupling. Posttranslational processing of G proteins subunits, use of baculovirus expression systems, production of specific monoclonal antibodies, and mutational analysis of G proteins are also covered in this volume.

PATRICK C. ROCHE

Methods in Neurosciences

Editor-in-Chief
P. Michael Conn

[1] G Proteins Controlling Differentiation, Growth, and Development: Analysis by Antisense RNA/DNA Technology

Christopher M. Moxham, Hsien-yu Wang,
and Craig C. Malbon*

Introduction

G Proteins: Molecular Properties

The central role of G proteins in transmembrane signaling is well-established. The actions of a wide variety of chemically diverse ligands, including hormones, neurotransmitters, and autacoids, are propagated across the lipid bilayer via G-protein-mediated signaling devices (1). Although sharing common features such as GTP binding and hydrolysis, heterotrimeric (α, β, γ) subunit structure, and, in some cases, recognition as substrates for bacterial toxins, G proteins can be distinguished by unique α subunits and functional coupling to effector units (1). Effector units displaying GTP sensitivity and/or G protein mediation include adenylylcyclase (adenylate cyclase). cGMP phosphodiesterase, phospholipases C and A_2, and K^+ as well as Ca^{2+} channels (2). The G-protein-linked receptor–effector devices control the intracellular levels of cyclic nucleotides, water-soluble inositol phosphates, diacylglycerol, arachidonic acid metabolites, and ion levels (1, 2). Alterations in these intracellular second messengers dictate cellular responses through protein phosphorylation and other mechanisms (1, 2). Olfaction, visual excitation, and other forms of sensory physiology, too, rely on G proteins to transduce information from activated receptors or photobleached pigments to membrane-bound effectors (3). Evidence is accumulating that G proteins play important, but poorly understood, roles in differentiation and development.

The heterotrimeric G proteins involved in transmembrane signaling display α subunits of M_r 39,000–52,000 (1, 2), in contradistinction to the small molecular weight G proteins, or smgs, displaying M_r ~25,000 α subunits (4), which will not be discussed further. The diversity of α-subunit structure can reflect alternative splicing, as is the case for the four forms of $G_{s\alpha}$ (two small M_r

---

* To whom correspondence should be addressed.

42,000 $G_{s\alpha\text{-}s}$ and two large M_r 52,000 $G_{s\alpha\text{-}l}$ (4). An additional form of $G_{s\alpha}$ has been identified in olfactory tissue ($G_{s\alpha\ \text{olf}}$) (5). The existence of G protein families like G_i, composed of at least three members (M_r ~41,000), $G_{i\alpha1}$, $G_{i\alpha2}$, and $G_{i\alpha3}$, exemplifies another possibility, each α subunit the product of a separate gene (6, 7). $G_{t\alpha1}$ (retinal rod), $G_{t\alpha2}$ (retinal cone), and $G_{o\alpha}$ (8) all display M_r ~39,000. Each of these subunits can act as a substrate for mono-ADP-ribosylation catalyzed by either cholera ($G_{s\alpha}$s and $G_{t\alpha}$s) or pertussis ($G_{i\alpha}$s, $G_{o\alpha}$, and $G_{t\alpha}$s) toxin. Cholera toxin catalyzes the ADP-ribosylation of Arg-201 of $G_{s\alpha}$, leading to reduced endogenous GTPase activity, and a constituitively active $G_{s\alpha}$. Although possessing an argininyl residue in a region with some conservation to the region flanking Arg-201, $G_{i\alpha}$s and $G_{o\alpha}$ are not substrates for this toxin, but rather for pertussis toxin (1, 2). Pertussis toxin catalyzes the ADP-ribosylation of the most C-terminal Cys residue of $G_{i\alpha}$s, $G_{o\alpha}$, and $G_{t\alpha}$s, stabilizing the heterotrimeric assembly, thereby inactivating these G proteins. Several other members of the G protein family have been discovered, namely, $G_{z\alpha}$ (or $G_{x\alpha}$; M_r 41,000), $G_{q\alpha}$, and $G_{\alpha11}$ (M_r 42,000), that lack the C-terminal Cys residue covalently modified by pertussis toxin (9, 10). The existence of additional forms of $G_{o\alpha}$ have been reported, although not confirmed by molecular cloning. All of these α subunits not only bind and hydrolyze GTP, but also reversibly bind a complex composed of $\beta\gamma$ subunits tightly associated in a 1:1 stoichiometry. Three β subunits have been identified (11). $G_{\beta1}$, M_r 36,000, is found in retinal rods and elsewhere. $G_{\beta2}$, M_r 35,000, is absent in rods but is otherwise ubiquitous. $G_{\beta3}$, M_r 37,200 (by deduced primary sequence), was cloned from retinal cDNA libraries and its mRNA was found in a number of human cell lines, including pheochromocytoma and neuroblastoma (11). $G_{\beta1}$ and $G_{\beta2}$ copurify as $\beta\gamma$ complexes in association with several G protein α subunits, including G_s, G_i, and G_o. The γ subunits (M_r ~8000) display similar diversity—at least seven forms have been identified (12). The functional significance of this rich diversity of subunits and possible permutations of G protein composition is only beginning to be elucidated. The functional implications of tissue-specific expression of various G protein subunits, too, have not been deduced and remain an important target for current research. Posttranslational modifications to G protein subunits include myristoylation ($G_{i\alpha}$s, $G_{t\alpha}$s, and $G_{o\alpha}$, but not $G_{s\alpha}$s) (13), farnesylation (G_γ) (14), palmitoylation ($G_{i\alpha}$ and $G_{t\alpha}$), (13) and also phosphorylation (for $G_{i\alpha2}$) (15).

G Proteins: Activation and Subunit Function

Activation of G proteins by agonist-occupied receptors or photobleached pigment (such as rhodopsin) catalyzes guanine nucleotide exchange (GDP release–GTP binding) and the generation of two species, a GTP-liganded α

subunit (G_α^*) and a dissociated $\beta\gamma$ complex (1, 2). G_α^* subunits have been identified *in vitro* using purified G proteins and GTP analogs such as GTPγS in reconstituted systems (16). $G_{s\alpha}^*$ has been identified as a product of stimulation of S49 mouse lymphoma membranes with β-adrenergic agonist. G_α^* products of other G protein activation have not been reported. Precise assignment of G protein–effector partners *in vivo* has been established in but only a few situations. Adenylylcyclase is an effector regulated by $G_{s\alpha}$ (as well as G_is and $\beta\gamma$ subunits), and cGMP phosphodiesterase is regulated by $G_{t\alpha}$. Direct addition of purified, resolved G_α^* to reconstituted systems *in vitro* has demonstrated in several instances the potential of a G protein to mediate a response (17). Unequivocal assignment of various receptors and effectors to specific G proteins *in vivo* has not been possible. $G_{i\alpha1}$, $G_{i\alpha2}$, and $G_{i\alpha3}$ expressed recombinantly, for example, all show the ability to stimulate K^+ channels in heart (18). $G_{o\alpha}$, $G_{i\alpha1}$, and $G_{i\alpha2}$, too, were shown to inhibit Ca^{2+} currents in rat dorsal root ganglion (19). Recombinantly expressed $G_{s\alpha\text{-}s}$ and $G_{s\alpha\text{-}l}$ both associate with $\beta\gamma$ subunits, are equivalent substrates for ADP-ribosylation by cholera toxin, and activate dihydropyridine-sensitive Ca^{2+} channels (17). Ascribing functional roles for G protein subunits *in vivo* remains a critical and formidable task that requires new approaches (20).

According to the α-subunit hypothesis, the $\beta\gamma$ complex plays an important but supporting role (1, 2). Evidence for a role of the $\beta\gamma$ complex in regulating ion channels both directly and indirectly has been presented (2). $\beta\gamma$ complexes have been shown to be capable of activating K^+ channels and phospholipase A_2, Cβ2, and Cβ3 under proper conditions (21). Irrespective of the hypothesis embraced, the $\beta\gamma$ complex appears to play the following roles in G-protein-mediated transmembrane signaling: (1) dampening or buffering α-subunit regulation of effector unit(s), (2) participating in interactions of α subunits with effectors and receptors, and (3) providing for $G_{s\alpha}$ perhaps some physical attachment to the membrane. Reconstitution studies and the results of G protein purification suggest that the $\beta\gamma$ complexes are largely interchangeable. Suppression of β subunits with antisense oligodeoxynucleotides and reconstitution of baculovirus-expressed $\beta\gamma$ complexes, however, suggest that the composition of the $\beta\gamma$ dimer may have important effects on G-protein-mediated pathways (1–4), including $\beta\gamma$ dimers dictating receptor specificity.

G Proteins and Cell Biology

The G-protein-linked responses regulate a wide range of physiologies and cellular processes *in vivo*. G protein involvement in the hormonal regulation of important metabolic pathways such as lipolysis, glycogenolysis, and gluconeogenesis is well known. The G proteins G_s and G_i are thought to regulate these pathways in a classical fashion by altering the activity of adenylylcyclase and hence the intracellular levels of cAMP. More recently G_s and G_i have been implicated in biological events such as oncogenesis and differentiation. Constitutively active mutants of G_s and $G_{i\alpha2}$ proteins have been identified in pituitary, thyroid, ovarian, and adrenal tumors (22–24). G_s and $G_{i\alpha2}$ have also been shown to regulate adipogenesis in mouse 3T3-L1 cells (25) and stem cell differentiation of F9 teratocarcinoma cells to primitive endoderm (26), in a manner that cannot be explained simply by changes in intracellular cAMP. Suppression of $G_{i\alpha2}$ *in vivo* provides a powerful tool to explore the roles of this subunit and the inhibitory control of adenylylcyclase in growth and metabolic regulation.

Hormonally stimulated hydrolysis of phosphatidylinositol 4,5-bisphosphate by phospholipase C (PLC) represents another major G-protein-mediated signaling pathway (27). Activation of PLC generates water-soluble inositol phosphates, which mobilize calcium, and also generates diacylglycerol, which activates protein kinase C (28). Two closely related G proteins, G_q and G_{11}, have been identified as activators of PLC-β isoforms (prominently β_1 and β_3, but also β_2) (29, 30). The $G_{q\alpha}$ family includes $G_{\alpha11}$, as well as $G_{\alpha14}$, $G_{\alpha15}$, and $G_{\alpha16}$. Both $G_{q\alpha}$ and $G_{\alpha11}$ are found in the liver, where regulation of hepatic metabolism appears to involve the PLC pathway (31). Both subunits have been found also in heart, fat, and brain (32). $\beta\gamma$ dimers also activate PLC-β, but at higher concentrations than $G_{q\alpha}$ or $G_{\alpha11}$. How suppression of or constitutive activation of the α subunits of G_q and G_{11} may modulate transmembrane signaling, growth, and metabolism *in vivo* remains a fundamental question of G protein biology. Equally intriguing, we have demonstrated that $G_{i\alpha2}$ also regulates PLC, but in an inhibitory fashion (33). Using antisense technology for cells in culture as well as to generate $G_{i\alpha2}$-deficient liver cells and adipocytes in transgenic mice, we were able to show that elimination of $G_{i\alpha2}$ was accompanied by a markedly elevated IP_3 accumulation as well as a potentiation of the stimulatory PLC pathway (33). The proposed studies shall provide the template to explore this new role of $G_{i\alpha2}$ in the context of PLC regulation.

Dominant-Negative Analysis of G Protein Function

Greater appreciation for the role of G protein diversity in transmembrane signaling will be gained once precise assignments of G protein–effector(s) coupling can be made. Reconstitution of purified G proteins with effectors such as adenylylcyclase and via patch-clamping with K^+ and Ca^{2+} channels has provided invaluable information on the potential of a specific G protein to activate an effector. As described above, however, discrimination among various members of G protein families (such as $G_{i\alpha}$ and $G_{q\alpha}$) has not been achieved. From these observations one can speculate that either no difference in effector coupling exists among these G proteins or the high degree of conservation among the members and/or the assay systems does not provide the conditions necessary for distinguishing what may be subtle, but important, differences operating *in vivo*.

Dominant-negative control of gene expression in cells has proved to be an effective means of evaluating the function of a specific gene product (34). Although first reported in 1977, use of complementary or "antisense" RNA and DNA sequences to block expression of a gene product in order to study its function is only in its infancy (35, 36). Through base pairing a DNA–RNA or RNA duplex is formed and is either rapidly degraded, or becomes an inhibitor of nuclear processing and/or translation of the target protein (36). A variety of examples exist demonstrating the utility of the antisense strategy (25, 26). Application of antisense RNA techniques to the study of the functions of G protein subunits provides a powerful new tool for establishing G protein coupling partners by blocking the expression of a specific subunit. Our laboratory has succeeded in using antisense oligodeoxynucleotides as well as stable cell transfectants expressing antisense RNA to block production of G protein subunits.

Antisense RNA/DNA Technologies

We have gained experience with the use of antisense RNA approaches. Below we summarize our work with (1) oligodeoxynucleotides antisense to the mRNA of $G_{s\alpha}$ employed to probe the roles of this subunit in adipogenesis; (2) retroviral expression vectors capable of generating constitutive expression of RNA antisense to G proteins, specifically $G_{i\alpha2}$, employed to study the role of this subunit in stem cell development to parietal and primitive endoderm; and (3) antisense vectors activated on birth in transgenic animals to examine the roles of $G_{i\alpha2}$ and $G_{q\alpha}$ in neonatal development and metabolism.

Suppression of G Proteins by Oligodeoxynucleotides Antisense to Subunit mRNA: Role of $G_{s\alpha}$ in Adipogenesis

Cellular differentiation of mouse 3T3-L1 fibroblasts is characterized by adipogenic conversion to a phenotype like that of a fat cell. Differentiation of 3T3-L1 fibroblasts to adipocytes induced by dexamethasone (DEX) and methylisobutylxanthine (MIX) occurs within 7–10 days, at which time cultures display >95% conversion to cells accumulating lipid droplets. To probe the possible role of G proteins in differentiation, the effects of bacterial toxins on the ability of fibroblasts to be induced to differentiate by DEX/MIX were explored first (25). Bacterial toxins covalently modify (ADP-ribosylate) α subunits of G proteins. Cholera toxin ADP-ribosylates and constitutively activates $G_{s\alpha}$, the stimulatory G protein coupled to adenylylcyclase. Cholera toxin was found to block differentiation. Cultures treated with toxin in combination with DEX/MIX did not differentiate. Treatment with pertussis toxin, in contrast, did not inhibit the ability of DEX/MIX to induce differentiation of 3T3-L1 fibroblasts to adipocytes. Although both agents elevate intracellular cyclic AMP levels, cholera toxin blocks differentiation, whereas pertussis toxin does not. Increasing cyclic AMP levels through activation of adenylylcyclase by forskolin or by the direct addition of dibutyryl-cAMP to the cultures, in contrast, did not block adipogenesis. The ability of cholera toxin rather than elevated cyclic AMP per se to block differentiation implicates a $G_{s\alpha}$ activity distinct from activation of adenylylcyclase. Steady-state levels of $G_{s\alpha}$ decline in 3T3-L1 cells during adipogenesis [37]. Metabolic labeling of the cultures with [^{35}S]methionine andd immune precipitation of $G_{s\alpha}$ show a frank decline in the synthesis of $G_{s\alpha}$ during differentiation, comparing days 1 and 3 after induction with DEX/MIX. Thus, $G_{s\alpha}$ synthesis declines during differentiation and constitutive activation of $G_{s\alpha}$ in cultures by cholera toxin blocks differentiation.

Would suppression of $G_{s\alpha}$ levels influence differentiation of 3T3-L1 fibroblasts to adipocytes? To explore this question we employ antisense oligodeoxynucleotides. Oligomers complementary either to the sense or the antisense strand of the 39 nucleotides of G protein α subunits immediately 5' to and including the initiator codon ATG are synthesized. Although the oligomers can be synthesized easily, the purification of the oligomers for direct use in cell culture has posed a problem to some laboratories. We typically purchase cell culture-grade oligomers from Operon (Alameda, CA). Confluent cultures are treated in chamber slides (0.1 ml volume) with 30 μM oligomers in serum-free medium. After 30 min of exposure to oligomers under these conditions, the medium is supplemented with serum (to 10%) and the incubation is continued for 24 hr. Treating 3T3-L1 fibroblasts with oligodeoxynucleotides antisense to $G_{s\alpha}$ results in a dramatic reduction (>90%) in the steady-state expression of $G_{s\alpha}$. Oligodeoxynucleotides sense as well as missense to $G_{s\alpha}$, in contrast, fail to alter the steady-state expression of this G protein subunit. The oligomers antisense to $G_{s\alpha}$ accelerate markedly the time course for differentiation induced by DEX/MIX. Few foci of adipogenic conversion are apparent in the DEX/MIX-treated as compared to nontreated cells at day 3.5. Cultures treated with DEX/MIX in combination with antisense oligomers to $G_{s\alpha}$, in sharp contrast, are fully differentiated by day 3.5. Oil red O staining of lipid droplets reveals lipid accumulation in 3.5-day

TABLE I Differentiation of Fibroblasts to Adipocytes: Effects of Oligodeoxynucleotides
Sense and Antisense for G[a]

Culture	% Differentiated at day 7	n	Culture	% Differentiated at day 3.5	n
Control untreated	1.8 ± 0.7	5	Control untreated	0	4
DEX/MIX treated	90.1 ± 1.1	5	+ DEX/MIX	11.7 ± 1.7	6
Antisense $G_{s\alpha}$ treated	$>80^b$	5	+ DEX/MIX + antisense $G_{s\alpha}$	$>90^c$	6
Sense $G_{s\alpha}$ treated	2.2 ± 0.9	5	+ DEX/MIX + sense $G_{s\alpha}$	13.5 ± 1.7	6

[a] Cells were treated with dexamethasone and methylisobutylxanthine (inducers) or oligodeoxynucleotides, or both, and observed either 3.5 or 7.0 days later. Cultures treated with DEX/MIX in combination with oligomers antisense to $G_{s\alpha}$ displayed robust accumulation of lipid and extensive adipogenic conversion by day 3.5.
[b] More than 80% of the cultures treated with oligomers antisense to $G_{s\alpha}$ alone displayed lipid accumulation.
[c] More than 90% of the cultures treated with oligomers antisense to $G_{s\alpha}$ and DEX/MIX displayed lipid accumulation.

cultures of 3T3-L1 cells induced by oligomers antisense to $G_{s\alpha}$ with DEX/MIX as compared to cells induced by DEX/MIX alone. Oligomers antisense to $G_{i\alpha1}$ and to $G_{i\alpha3}$, in contrast, do not alter the induction of adipogenic differentiation in response to DEX/MIX. Quantification of lipid accumulation in fixed, intact cells is performed following staining with oil red O (Table I). These data demonstrate that oligomers antisense to, but not sense to, $G_{s\alpha}$ enhance the rate of differentiation induced by DEX/MIX or that observed in the absence of other exogenous inducers (25). The ability of cholera toxin to block DEX/MIX-induced differentiation is progressively lost as time following treatment with oligomers antisense to $G_{s\alpha}$ progresses. We investigated next if oligomers antisense to $G_{s\alpha}$ alone, in the absence of DEX/MIX, would induce differentiation. In cultures that are 7 days postconfluence, oil red O staining revealed enhanced lipid accumulation in antisense $G_{s\alpha}$ oligomer-treated as compared to control, untreated, or sense $G_{s\alpha}$ oligomer-treated cultures. Enlargement of the fields highlighted the differential effects of the oligomers sense, as compared to antisense, to $G_{s\alpha}$ on the adipogenic conversion of the 3T3-L1 cells.

The inability of the dibutyryl analog of cyclic AMP to block differentiation argues against a role for cyclic AMP in modulating differentiation of 3T3-L1 fibroblasts. Earlier work showing that adipogenesis could be accelerated by treatment with methylxanthines, but not by elevating intracellular cyclic AMP, too, argues against adenylylcyclase functioning as the effector for $G_{s\alpha}$ in this process (3). In addition to adenylylcyclase, Mg^{2+} transport (1, 2) and Ca^{2+} channels (1, 2) have been implicated as effector units regulated via $G_{s\alpha}$. Insulin, like the oligomers antisense to $G_{s\alpha}$, accelerates differentiation of 3T3-L1 fibroblasts to adipocytes, reflecting perhaps the ability of insulin to oppose responses, such as lipolysis, mediated by $G_{s\alpha}$. These provocative observations permit us to entertain the possibility of a novel role for G proteins in complex biological responses such as cellular differentiation.

Are there other examples where G proteins have been implicated in nontraditional, more complex biological responses such as differentiation? Recent evidence localizing the G protein G_o to growth cones of neurite outgrowths in NGF-treated pheochromocytoma (PC12) cells suggests that members of the G protein family may participate in cellular functions other than transmembrane signaling (38–40). The signals that cause collapse of nerve growth cones have been reported to involve mediation by G proteins that are sensitive to pertussis toxin. G protein expression changes during differentiation of 3T3-L1 fibroblasts (37). 3T3-L1 cells differentiate to a phenotype similar to mature adipocytes, a process dramatically accelerated by insulin, dexamethasone, and methylxanthine, or growth in high serum concentrations (41). The importance of $G_{s\alpha}$ activity in a nontraditional role, i.e., modulation of differentiation, is supported by the following evidence: (1) $G_{s\alpha}$ synthesis declines early in differentiation; (2) constitutive activation of $G_{s\alpha}$ by cholera toxin blocks differentiation; (3) oligodeoxynucleotides antisense to $G_{s\alpha}$, but not those antisense to $G_{i\alpha1}$ and $G_{i\alpha3}$, or sense to $G_{s\alpha}$, accelerated differentiation, reducing the time to full adipocyte conversion from 7–10 to <3 days; and (4) oligodeoxynucleotides antisense, but not sense, to $G_{s\alpha}$ alone are capable of inducing differentiation.

Retroviral Expression Vectors Generating Constitutive Expression of RNA Antisense to G Proteins: Role of $G_{i\alpha2}$ in Early Mouse Development

The morphogen retinoic acid induces teratocarcinoma embryonic stem cells to differentiate into a primitive endoderm-like phenotype. Tissue-type plasminogen activator (tPA) production is a hallmark for the primitive endoderm phenotype, i.e., F9 stem cells produce no tPA and primitive endoderm produces and secretes significant levels of tPA. Induction of stem cells to primitive endoderm is accompanied by a marked decrease in the steady-state levels of $G_{i\alpha2}$ subunit. This decline of $G_{i\alpha2}$ in retinoic acid-induced differentiation prompted us to investigate the relationship between $G_{i\alpha2}$ and differentiation (26). $G_{i\alpha2}$ activity was reduced by antisense RNA and increased by expression of wild-type and a constitutively active mutant (Q205L) of $G_{i\alpha2}$.

Antisense RNA blocks expression of "targeted" proteins. Forty bases of the 5' noncoding region immediately upstream of, and including, the ATG translation initiation codon were used as antisense templates. This region of

$G_{i\alpha2}$ was found to have no significant homology with other sequences available from GenBank. Initial constructs employing the SV40 early promoter and calcium phosphate-mediated transfection were found inadequate and were replaced by use of the retroviral expression vector (pLNC-ASG$_{i\alpha2}$) in which the antisense RNA is driven by the cytomegalovirus (CMV) promoter (26). Following retroviral infection of F9 stem cells with pLNC-ASG$_{i\alpha1}$, pLNC-ASG$_{i\alpha2}$, or pLNCX, neomycin-resistant colonies were selected and then tested for expression of $G_{i\alpha2}$ by immunoblotting. The same vector antisense to $G_{i\alpha1}$ (pLNC-ASG$_{i\alpha1}$) was employed as a control, because F9 stem cells do not express $G_{i\alpha1}$ at either the protein or mRNA level. Infection with the retroviral vector alone (pLNCX; i.e., minus antisense sequences) provided an additional control.

Clone F9ASG$_{i\alpha2}$ demonstrated decreased expression of $G_{i\alpha2}$ (>85% reduction) compared to either the F9ASG$_{i\alpha1}$-infected cells or wild-type F9 stem cells. Staining of immunoblots with antibodies to other G protein subunits ($G_{s\alpha}$ or G_β) demonstrates that the decrease in $G_{i\alpha2}$ expression was specific for F9ASG$_{i\alpha2}$-infected cells. The inhibitory control of adenylylcyclase, believed to be mediated by $G_{i\alpha2}$ (42, 43), was examined in the F9ASG$_{i\alpha2}$ cells. Forskolin and isoproterenol stimulated adenylylcyclase in F9ASG$_{i\alpha2}$ cells to the same extent as in F9ASG$_{i\alpha1}$ and F9 stem cells. The ability of thrombin to inhibit forskolin-stimulated adenylylcyclase, in contrast, was markedly attenuated in the F9ASG$_{i\alpha2}$ cells. The suppression of $G_{i\alpha2}$ expression by antisense RNA in F9ASG$_{i\alpha2}$ clones was sufficient to decrease the inhibitory adenylylcyclase response by more than 70%. This observation confirms other recent data implicating $G_{i\alpha2}$ as the mediator of adenylylcyclase inhibition (44).

F9 stem cells in culture exhibit a simple, rounded morphology (26). When induced to differentiate by retinoic acid, the cell doubling time is prolonged and the cells assume a morphology characteristic of primitive endoderm, i.e., an extended spindle shape with "organization centers" of cell growth. F9ASG$_{i\alpha2}$ cells display a primitive endoderm-like morphology at all stages of growth, forming distinctive "organization centers" at confluence. Treating F9ASG$_{i\alpha2}$ cells with retinoic acid did not result in any further changes in morphology. F9ASG$_{i\alpha1}$ cells, in contrast, displayed normal stem cell phenotype in the absence of retinoic acid and primitive endoderm-like phenotype following 4 days treatment with the morphogen.

When F9 stem cells are treated with retinoic acid, the rate of cell growth declines as the number of cells that have terminally differentiated increases. The doubling time of F9ASG$_{i\alpha2}$, but not F9ASG$_{i\alpha1}$, cells was 28 hr in the absence of retinoic acid as compared to 16 hr for F9 stem cells (26). By the criteria of cell morphology, growth rate, and resistance to the actions of the morphogen retinoic acid, F9ASG$_{i\alpha2}$ cells have differentiated to the primitive endoderm phenotype. Conversely, expression of a constitutively active, on-cogenic mutant of $G_{i\alpha2}$ ($G_{i\alpha2}Q205L$) has been shown to decrease the doubling time of Rat 1a, Swiss 3T3 and NIH 3T3 cells (45).

The changes in doubling time and morphology of the F9ASG$_{i\alpha2}$ cells in the absence of retinoic acid prompted us to question the ability of these cells to produce tPA, a biochemical marker for the primitive endoderm phenotype. The F9ASG$_{i\alpha2}$ cells produced tPA much like the retinoic acid-induced primitive endoderm cells, albeit to a lesser extent. In addition, treatment of F9ASG$_{i\alpha2}$ cells with retinoic acid failed to induce further production of tPA.

Cyclic AMP is a mitogen for a variety of cell lines in culture (46). Basal cyclic AMP levels were found to be essentially unaltered, i.e., 1.4 ± 0.9 pmol ($n = 4$) for stem cells, $1.0 + 0.3$ pmol ($n = 5$) for F9ASG$_{i\alpha1}$ cells, and 1.9 ± 0.6 pmol ($n = 7$) cyclic AMP/10^6 cells for the F9ASG$_{i\alpha2}$ cells. A 4-day challenge with dibutyryl-cAMP (10 mM) failed to induce either the morphology characteristic of primitive endoderm or the biochemical marker tPA. These two observations demonstrate that the effects of decreased expression of $G_{i\alpha2}$ in the differentiation of F9ASG$_{i\alpha2}$ cells are independent of cyclic AMP. If decreased expression of $G_{i\alpha2}$ induces cell differentiation, overexpression of $G_{i\alpha2}$ activity might be expected to block retinoic acid-induced differentiation. To test this hypothesis we transfected F9 cells with expression vectors harboring a cDNA encoding a mutant form of $G_{i\alpha2}$, designated $G_{i\alpha2}Q205L$. $G_{i\alpha2}Q205L$ has a point mutation in the GTPase region of the molecule, rendering it constitutively active (43). This mutation has been shown to result in oncogenic expression of $G_{i\alpha2}$, *gip2* (47). Expression of $G_{i\alpha2}Q205L$ (F9G$_{i\alpha2}Q205L$) was achieved using the vectors described (16) and resulted in more than a twofold increase in immunoreactivity in blots of cell membranes stained with antibodies specific to $G_{i\alpha2}$ (26). F9 clones expressing $G_{i\alpha2}Q205L$ were found to be refractory to retinoic acid-induced differentiation. We assessed also the ability of these cells to produce tPA. F9G$_{i\alpha2}Q205L$ cells failed to produce tPA in response to retinoic acid. Thus, the increased $G_{i\alpha2}$ activity in F9G$_{i\alpha2}Q205L$ cells blocks retinoic acid-induced differentiation of F9 cells, as judged both by morphological and biochemical criteria.

These observations demonstrate the ability of $G_{i\alpha2}$ activity to regulate the differentiation of F9 stem cells to a primitive endoderm-like phenotype, a model of early mouse development. Oncogenic transformation leading to proliferation can be thought of as a failure of the cell to differentiate. Overexpression of $G_{i\alpha2}$ activity using GTPase-deficient mutants of $G_{i\alpha2}$ has been shown by others to result in transformation of fibroblasts in cell culture (45). Injection of antibodies specific for $G_{i\alpha2}$, to block $G_{i\alpha2}$ activity, inhibits serum-stimulated DNA synthesis (48). A decrease in $G_{i\alpha2}$ expression (F9ASG$_{i\alpha2}$ cells) is sufficient to induce differentiation of F9 stem cells, whereas an increase in $G_{i\alpha2}$ expression (F9G$_{i\alpha2}Q205L$ cells) is sufficient to block differentiation. These data provide a novel insight into the control of cell growth and differentiation by G proteins, highlighting a function other than inhibition of adenylylcyclase mediated by $G_{i\alpha2}$, namely, control of differentiation.

Expression of Antisense RNA in Transgenic Mice

Studies of G protein function *in vivo* are proposed. One of the most powerful approaches to define the role(s) of specific gene products is the transfer of new or altered genetic information into mouse embryos to modify stably the genetic constitution (49). We propose to probe the function of G protein subunits in transgenic mice expressing antisense RNA. Although in its infancy, application of antisense RNA in tandem with transgenic mice has been demonstrated. Conceptually it is an appealing idea that appears to be far less technically demanding than homologous gene interruption (50). Katsuki *et al.* (51) first demonstrated the feasibility of the approach, succeeding in the transfer of an antisense minigene against myelin basic proteins (MBPs) to fertilized mice zygotes. The transgenic mice developed the mutant shiverer phenotype. Antisense MBP mRNA was expressed in the mice; endogenous MBP mRNA, the gene product, and CNS myelination were reduced in the transgenic mice.

Several potential obstacles were considered in creating a strategy for suppressing the expression of a G protein subunit *in vivo*. The simpler approach of employing $G_{i\alpha2}$-specific antisense RNA was adopted over gene disruption. To enhance the accumulation of the antisense RNA the target sequence was inserted in the 5' untranslated region of the rat phosphoenolpyruvate carboxykinase (PEPCK) mRNA (33) (Fig. 1). The PEPCK gene was selected for three reasons. First, we reasoned that this 2.8-kb hybrid mRNA would be more stable than a comparatively short-lived antisense RNA oligonucleotide. Furthermore, expression of PEPCK is regulated by several hormones, including glucagon (acting via cAMP), glucocorticoids, thyroid hormone, and insulin. Insertion of the antisense sequence within the PEPCK gene thus confers regulated expression of the desired antisense sequences. cAMP plays a unique role in the control of PEPCK gene expression in that it coordinately increases the transcription rate of the gene as well as the stability of the mRNA (52). Third and most importantly, expression of the PEPCK gene is developmentally regulated with initial appearances of the mRNA only at birth (53). PEPCK mRNA is highly abundant; neonatal levels approach 0.5–1.0% of total cellular RNA. Targeted expression of the antisense RNA after birth eliminates the problem of inducing a potentially lethal outcome from the suppression of $G_{i\alpha2}$ *in utero* that might preclude transgenic pups.

The utility of the construct pPCK-ASG$_{i\alpha2}$ was evaluated first after transfec-

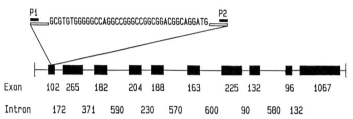

FIG. 1   Inducible antisense RNA; a Trojan horse approach to suppressing $G_{i\alpha2}$. The 39 nucleotides immediately upstream and including the translation initiation codon were chosen to serve as the antisense sequence based on the low degree of homology among the G protein α subunits within this region. This sequence did not show significant homology with any of the sequences present in the GenBank database. The $G_{i\alpha2}$ antisense sequence was excised from the vector pLNC-ASG$_{i\alpha2}$ as a 235-bp fragment and engineered into the first exon of the PEPCK gene using standard recombinant DNA techniques. Plasmids carrying the desired orientation of the insert were selected based on direct sequencing of the DNA. In order to discriminate between the endogenous PEPCK gene and the transgene, primers (P1, P2) were synthesized complementary to the flanking ends of the insert, permitting subsequent PCR amplification of only the 235-bp insert present in the pPCK-ASG$_{i\alpha2}$ construct. A physical map of the PEPCK gene was reported previously. The 5' flanking promoter sequence within the construct employed contains the responsive elements necessary for the tissue-specific, hormonal, and developmental regulation of gene expression, as observed with the endogenous PEPCK gene.

tion into FTO-2B rat hepatoma cells. These cells display cAMP-inducible PEPCK gene expression (54) and express $G_{i\alpha2}$. RNA antisense to $G_{i\alpha2}$ was detected in FTO-2B clones transfected with pPCK-ASG$_{i\alpha2}$ after reverse transcription of total cellular RNA followed by PCR amplification. FTO-2B clones transfected with pPCK-ASG$_{i\alpha2}$ displayed normal levels of $G_{i\alpha2}$ expression in the absence of cAMP, an inducer of the PEPCK gene expression. $G_{i\alpha2}$ expression declined >85% when these same cells were challenged with the cAMP analog, 8-(4-chlorophenylthio)-cAMP (CPT-cAMP) for 12 days. FTO-2B clones transfected with the vector lacking the antisense sequence to $G_{i\alpha2}$ displayed no change in $G_{i\alpha2}$ expression. In marked contrast to the suppressed expression of $G_{i\alpha2}$, the expression of $G_{s\alpha}$ and $G_{i\alpha3}$ was not changed in cells expressing the RNA antisense to $G_{i\alpha2}$, demonstrating that the antisense RNA sequence was specific for $G_{i\alpha2}$. The time elapsing between the induction of pPCK-ASG$_{i\alpha2}$ by CPT-cAMP and the decline of steady-state levels of $G_{i\alpha2}$ likely reflects the half-life of this subunit.

TABLE II BDF1 Mice Harboring pPCK-ASG$_{i\alpha2}$ Transgene Display Alterations in Tissues Targeted for G$_{i\alpha2}$ Deficiency[a]

| Tissue | Age (weeks) | | | | | |
| | 6 | | 12 | | 18 | |
	Transgenic/normal	Ratio	Transgenic/normal	Ratio	Transgenic/normal	Ratio
Brain	$0.39 \pm 0.01/0.44 \pm 0.03$	0.89	$0.42 \pm 0.01/0.43 \pm 0.01$	0.98	$0.45 \pm 0.02/0.46 \pm 0.02$	0.98
Fat[b,c]	$0.09 \pm 0.03/0.27 \pm 0.03$	0.33	$0.19 \pm 0.01/0.37 \pm 0.04$	0.49	$0.19 \pm 0.01/0.47 \pm 0.01$	0.40
Heart	$0.11 \pm 0.01/0.13 \pm 0.01$	0.85	$0.14 \pm 0.02/0.14 \pm 0.01$	1.00	$0.19 \pm 0.02/0.20 \pm 0.02$	0.95
Kidney[b]	$0.14 \pm 0.03/0.19 \pm 0.02$	0.74	$0.19 \pm 0.02/0.19 \pm 0.02$	1.00	$0.28 \pm 0.02/0.28 \pm 0.03$	1.00
Liver[b]	$1.07 \pm 0.08/1.49 \pm 0.02$	0.72	$1.08 \pm 0.04/1.49 \pm 0.03$	0.72	$1.24 \pm 0.03/1.89 \pm 0.09$	0.66
Lung	$0.13 \pm 0.01/0.15 \pm 0.03$	0.87	$0.17 \pm 0.02/0.18 \pm 0.01$	0.94	$0.18 \pm 0.02/0.19 \pm 0.04$	0.95
Skeletal muscle[d]	$0.10 \pm 0.02/0.15 \pm 0.01$	0.67	$0.14 \pm 0.02/0.14 \pm 0.02$	1.00	$0.20 \pm 0.05/0.22 \pm 0.03$	0.91

[a] Transgenic mice were produced and bred at the Stony Brook Transgenic Mouse facility. Necropsy and histology were performed by Charles River Laboratories (used with permission from Ref. 33).
[b] Denotes target tissue for pPCK-ASG$_{i\alpha2}$.
[c] Epididymal fat pad.
[d] Gastrocnemius skeletal muscle.

G$_{i\alpha2}$ is the member of the G$_i$ family most prominently implicated in mediating the inhibitory adenylylcyclase pathway. Suppression of G$_{i\alpha2}$ expression in FTO-2B cells was associated with the loss of receptor-mediated inhibition of adenylylcyclase (33, 55). Inhibition of forskolin-stimulated cAMP accumulation by either somatostatin or the A1 purinergic agonist $(-)$-N^6-(R-phenylisopropyl)adenosine (R-PIA) was nearly abolished in transfectant cells in which RNA antisense to G$_{i\alpha2}$ was first induced by CPT-cAMP for 12 days. Cells transfected with the vector lacking the antisense sequence for G$_{i\alpha2}$ displayed a normal inhibitory adenylylcyclase response following a 12-day challenge with CPT-cAMP. These data demonstrate that G$_{i\alpha2}$ mediates the hormonal inhibition of hepatic adenylylcyclase.

Having demonstrated that the pPCK-ASG$_{i\alpha2}$ construct could effectively reduce G$_{i\alpha2}$ subunit expression *in vitro*, we expanded this approach for *in vivo* studies to explore the consequences of G$_{i\alpha2}$ antisense RNA expression in transgenic mice. BDF1 mice carrying the pPCK-ASG$_{i\alpha2}$ transgene were identified by Southern analysis (55). Four founder lines were bred and characterized over several generations. Necropsy and histology of the transgenic and control mice were quite revealing. Epididymal fat mass was 0.27 ± 0.01 g in control mice and 0.09 ± 0.03 g in transgenic mice, representing a 65% decrease in fat mass by 6 weeks of neonatal growth and antisense RNA expression (Table II). By 18 weeks of age, fat mass had increased in all mice, with the transgenic mice still displaying ~60% reduction. The pPCK-ASG$_{i\alpha2}$ transgenic mice also displayed a 30% reduction in liver mass from 6

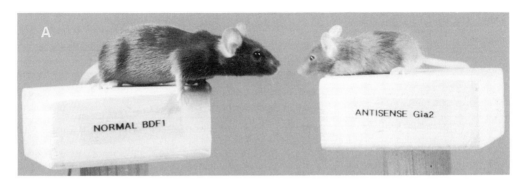

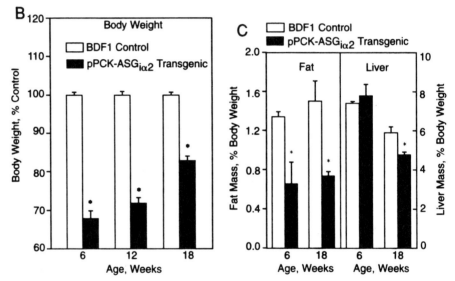

FIG. 2 Suppression of $G_{i\alpha2}$ expression by RNA antisense to $G_{i\alpha2}$ blunts neonatal growth. (A) The phenotype of the transgenic animal carrying the pPCK-ASG$_{i\alpha2}$ transgene is characterized by a marked reduction in neonatal growth. Shown are normal (left) and transgenic (right) male littermates at 6 weeks of age. The body weights of the normal and transgenic mice shown were 17 and 9.9 g, respectively. (B) Normal and transgenic mice ($n = 6$) were weighed at the times indicated after birth. Mice were then transported to the Charles River Laboratories for complete necropsy and histological examinations of targeted and nontargeted tissues. Fat mass expressed as percent body weight was significantly reduced in transgenic mice at 6 and 18 weeks of age. (C) Liver mass as percent of body weight was significantly reduced in transgenic mice at 18, but not 6, weeks of age as well. The values reported are expressed as mean ± SEM.

to 18 weeks of age (Table I). Inspection of a wider range of tissues and organs (Table I) highlights that growth has been diminished selectively in tissues targeted for pPCK-ASG$_{i\alpha2}$ gene expression (i.e., tissues that normally express PEPCK). Brain, heart, lung, and skeletal muscle growth was unaffected in transgenic mice lacking G$_{i\alpha2}$. Kidney development was unilateral in two transgenic mice, whereas prominent vacuolation localized to the proximal convoluted tubule of the cortex was observed in 30% of the other transgenic mice. Kidney mass on average, however, was not appreciably altered.

Each of the founders and their transgenic offspring displayed sharply reduced G$_{i\alpha2}$ expression in fat, liver, and in some cases kidney, target tissues for PEPCK gene expression. The level of G$_{i\alpha2}$ in fat and liver of transgenic mice in all four founder lines harboring the pPCK-ASG$_{i\alpha2}$ transgene was less than 5% of the control. For kidney, the suppression of G$_{i\alpha2}$ was more variable among transgenic animals, some displaying less than 5% of the control levels, others displaying wild-type levels of expression. The variability of G$_{i\alpha2}$ expression in the kidney may reflect epigenetic effects, i.e., differences in sites of integration.

The most striking phenotypic characteristic of the mice carrying the pPCK-ASG$_{i\alpha2}$ transgene (Fig. 2A) was a failure to thrive (33, 55). Although phenotypically unremarkable at birth, by 6 weeks of postnatal development and antisense RNA expression the transgenic mice lacking G$_{i\alpha2}$ displayed >30% reduction in body weight (Fig. 2B). Food consumption was equivalent for control and transgenic mice, yet differential loss in body weight of transgenic mice was maintained for 12 weeks, and plateaued at >20% below normal up to 24 weeks. As a percentage of body weight, fat mass is reduced in the transgenic animals lacking G$_{i\alpha2}$. Although 30% smaller in body weight at 6 weeks of age, on a percent body weight basis, transgenic mice lacking G$_{i\alpha2}$ displayed liver mass equivalent to normal mice (Fig. 2C). By 18 weeks of age, however, on a percent body weight comparison, transgenic mice displayed body weights lagging behind normal mice, but a liver mass considerably less (~50%) than that of their normal counterparts (Fig. 2C). Thus the full impact of the loss of G$_{i\alpha2}$ and its metabolic/developmental consequences was observed earlier in fat than in liver at 6 weeks of age, but was manifest in both tissues by 18 weeks of age.

Suppression of G$_{i\alpha2}$ in liver and fat clearly was associated with a marked reduction in neonatal growth. The dramatic reduction in body mass cannot be explained by the reduction in the mass of targeted organs nor by differences in food consumption, suggesting that a reduction in G$_{i\alpha2}$ expression induces a metabolic alteration adversely affecting normal, neonatal development. The marked reduction in G$_{i\alpha2}$ levels observed in the adipose tissue of pPCK-ASG$_{i\alpha2}$ transgenic mice was associated with a greater than threefold elevation in basal levels of cAMP, (4.99 $\pm$ 0.44 pmol/10^5 cells and 16.45 $\pm$ 1.37 pmol/10^5 cells for control and transgenic mice, respectively), as well as the loss of the inhibitory adenylylcyclase response in isolated fat cells (71). Thus blocked expression of a single G protein, $G_{i\alpha2}$, in a tissue-specific manner, especially for tissues critical in metabolism, affects not only the development of targeted tissues, but can also yield global metabolic consequences. The use of inducible expression of antisense sequences harbored within a functional mRNA permitted us the first insight into the role(s) of $G_{i\alpha2}$ in neonatal development *in vivo*.

Concluding Remarks

G proteins are critical elements in transmembrane signaling for a class of cell surface receptor molecules now totaling in excess of 1000. In addition to this classic role, G proteins have been demonstrated to play an important role in more complex biological processes, such as mitogenesis, oncogenesis, neonatal growth, and cellular differentiation. Antisense RNA/DNA technology provides a new strategy with which to explore the role(s) of specific subunits in both transmembrane signaling as well as the more complex biological responses. Although not without limitations and constraints for experimental use, antisense RNA/DNA technology now provides flexible, powerful tools for defining functional consequences of eliminating one or more members of the G protein family both *in vitro* and *in vivo*. Application of these technologies will permit us to address the fundamental and unresolved questions about G proteins and their functions posed above. In addition, it most certainly will become clear to the reader that this technology has applications relevant to the treatment of human diseases, much like gene therapy.

References

1. L. Birnbaumer, J. Codina, R. Mattera, A. Atani, N. Sherer, M.-J. Goro, and A. M. Brown, *Kidney Int.* **32,** S-24 (1987).
2. E. J. Neer and D. E. Clapham, *Nature (London)* **333,** 129–134 (1988).
3. G. Johnson and N. Dhanasekaran, *Endocr. Rev.* **10,** 317–331 (1989).
4. H. R. Bourne, *Cell* **53,** 669–671 (1988).
5. D. T. Jones and R. R. Reed, *Science* **244,** 790–795 (1989).
6. A. Yatani, R. Mattera, J. Codina, R. Graf, K. Okabe, E. Padrell, R. Iyengar, A. M. Brown, and L. Birnbaumer, *Nature (London)* **336,** 680–682 (1988).
7. H. Itoh, R. Toyama, T. Kozasa, T. Tsukamoto, M. Matsuoka, and Y. Kaziro, *J. Biol. Chem.* **263,** 6656–6664 (1988).
8. P. C. Sternweis and J. D. Robishaw, *J. Biol. Chem.* **259,** 13806–13811 (1984).

9. H. K. W. Fong, K. Yoshimoto, P. Eversole-Cire, and M. I. Simon, *Proc. Natl. Acad. Sci. U.S.A.* **85**, 3066–3070 (1988).

10. M. Matsuoka, H. Itoh, T. Kozasa, and Y. Kaziro, *Proc. Natl. Acad. Sci. U.S.A.* **85**, 5384–5388 (1988).

11. M. A. Levine, P. M. Smallwood, P. T. Moen, L. J. Helman, and T. G. Ahn, *Proc. Natl. Acad. Sci. U.S.A.* **87**, 2329–2333 (1990).

12. N. Gautam, M. Baetscher, R. Aebersold, and M. I. Simon, *Science* **44**, 971–974 (1989).

13. J. E. Buss, S. E. Mumby, P. J. Casey, A. G. Gilman, and B. M. Shefton, *Proc. Natl. Acad. Sci. U.S.A.* **84**, 7493–7497 (1987).

14. B. K. Fung, H. K. Yamane, I. M. Ota, and S. Clarke, *FEBS Lett.* **260**, 313–317 (1990).

15. T. Katada, A. G. Gilman, Y. Watanabe, S. Bauer, and K. H. Jakobs, *Eur. J. Biochem.* **151**, 431–437 (1985).

16. E. M. Ross, *Neuron* **3**, 141–152 (1989).

17. A. M. Brown and L. Birnbaumer, *Annu. Rev. Physiol.* **52**, 197–213 (1990).

18. A. Yatani, R. Mattera, J. Codina, R. Graf, K. Okabe, E. Padrell, R. Iyengar, A. M. Brown, and L. Birnbaumer, *Nature (London)* **336**, 680–682 (1988).

19. D. A. Ewald, I.-H. Pang, P. C. Sternweis, and R. J. Miller, *Neuron* **2**, 1185–1193 (1989).

20. P. C. Sternweis and I.-H. Pang, *Trends Neurosci.* **13**, 122–126 (1990).

21. D. E. Logothetis, D. Kim, J. K. Northup, E. J. Neer, and D. E. Clapham, *Proc. Natl. Acad. Sci. U.S.A.* **85**, 5814–5818 (1988).

22. J. Lyons, C. A. Landis, G. Harsh, L. Vallar, K. Grunewald, H. Feichtinger, Q. Y. Duh, O. H. Clark, E. Kawasaki, H. R. Bourne, and F. McCormick, *Science* **249**, 655–659 (1990).

23. C. A. Landis, S. B. Masters, A. Spada, A. M. Pace, H. R. Bourne, and L. Vallar, *Nature (London)* **340**, 692–696 (1989).

24. E. Clementi, N. Malagaretti, J. Meldolesi, and R. Taramelli, *Oncogene* **5**, 1059–1061 (1990).

25. H.-Y. Wang, D. C. Watkins, and C. C. Malbon, *Nature (London)* **358**, 334–337 (1992).

26. D. C. Watkins, G. L. Johnson, and C. C. Malbon, *Science* **258**, 1373–1375 (1992).

27. M. J. Berridge and R. F. Irvine, *Nature (London)* **341**, 197–205 (1989).

28. Y. Nishizuka, *Nature (London)* **334**, 661–665 (1988).

29. S. G. Rhee, P.-G. Suh, S.-H. Ryu, and S. Y. Lee, *Science* **244**, 546–550 (1989).

30. R. Kriz, L.-L. Lin, L. Sultzmann, C. Ellis, C.-H. Heldin, T. Pawson, and J. Knopf, *Ciba Found. Symp.* **150**, 112–127 (1990).

31. G. Berstein, J. L. Blank, A. V. Smrcka, T. Higashijima, P. C. Sternweis, J. H. Exton, and E. M. Ross, *J. Biol. Chem.* **267**, 8081–8088 (1992).

32. I.-H. Pang and P. C. Sternweis, *J. Biol. Chem.* **265**, 18707–18712 (1990).

33. C. M. Moxham, Y. Hod, and C. C. Malbon, *Science* **260**, 991–995 (1993).

34. D. C. Watkins, C. M. Moxham, A. J. Morris, and C. C. Malbon, *Biochem. J.,* **299**, 593–596 (1994).

35. J. Haseloff and W. L. Gerlach, *Nature (London)* **334**, 585–591 (1988).

36. J. Goodchild, "Oligodeoxynucleotides—Antisense Inhibitors of Gene Expression." CRC Press, Cleveland, Ohio, 1989.

37. D. C. Watkins, J. K. Northup, and C. C. Malbon, *J. Biol. Chem.* **264,** 4186–4194 (1989a).
38. S. M. Strittmatter, D. Valenzuela, T. E. Kennedy, E. J. Neer, and M. C. Fishman, *Nature (London)* **344,** 836–841 (1990).
39. M. Igarashi, S. M. Strittmatter, T. Vartanian, and M. C. Fishman, *Science* **259,** 77–79 (1993).
40. C. E. Bandtlow, M. F. Schmidt, T. D. Hassinger, M. E. Schwab, and S. B. Kater, *Science* **259,** 80–83 (1993).
41. C. S. Rubin, A. Hirsch, C. Fung, and O. M. Rosen, *J. Biol. Chem.* **253,** 7570–7578 (1978).
42. L. Birnbaumer, J. Abramowitz, and A. M. Brown, *Biochim. Biophys. Acta* **1031,** 163–224 (1990).
43. J. M. Lowndes, S. K. Gupta, S. Osawa, and G. L. Johnson, *J. Biol. Chem.* **266,** 14193–14197 (1991).
44. Y. H. Wong, A. Federman, A. M. Pace, I. Zachary, T. Evans, J. Pouyssegur, and H. R. Bourne, *Nature (London)* **351,** 63–65 (1991).
45. S. K. Gupta, C. Gallego, J. M. Lowndes, C. M. Pleiman, C. Sable, B. J. Eisfelder, and G. L. Johnson, *Mol. Cell Biol.* **12,** 190–197 (1992).
46. R. J. Klebe, T. M. Overfelt, V. L. Magnuson, B. Steffensen, D. Chen, and G. Zardeneta, *Proc. Natl. Acad. Sci. U.S.A.* **88,** 9588–9592 (1991).
47. J. Lyons, C. A. Landis, G. Harsh, L. Vallar, K. Grunewald, H. Feichtinger, Q. Y. Duh, O. H. Clark, E. Kawasaki, H. R. Bourne, and F. McCormick, *Science* **249,** 655–659 (1990).
48. V. J. LaMorte, P. K. Goldsmith, A. M. Spiegel, J. L. Meinkoth, and J. R. Feramisco, *J. Biol. Chem.* **267,** 691–694 (1992).
49. D. Hanahan, *Science* **246,** 1265–1275 (1989).
50. M. R. Capecchi, *Science* **244,** 1288–1292 (1989).
51. M. Katsuki, M. Sato, M. Kimura, M. Yokoyama, K. Kobayashi, and T. Nomura, *Science* **241,** 593–595 (1988).
52. J. Liu, E. A. Park, A. L. Gurney, W. J. Roesler, and R. W. Hanson, *J. Biol. Chem.* **266,** 19095–19102 (1991).
53. J. P. Garcia Ruiz, R. Ingram, and R. W. Hanson, *Proc. Natl. Acad. Sci.* **75,** 4189 (1978).
54. Y. Hod and R. W. Hanson, *J. Biol. Chem.* **263,** 7747–7752 (1988).
55. C. M. Moxham, Y. Hod, and C. C. Malbon, *Develop. Genetics* **14,** 266–273 (1993).

[2] Analysis of Adenylylcyclase Subspecies Gene Expression in Brain by *in Situ* Hybridization Histochemistry

Nicole Mons

Introduction

Eight isoforms of adenylylcyclase have been identified in mammalian tissues (1–15). Although these isoforms share sequence homology and common structural topology, they are divided into three major subclasses, based on the degrees of divergence in their amino acid sequence and regulation by $\beta\gamma$ subunits of G proteins, protein kinase C, and Ca^{2+} (for reviews, see Refs. 16–17). Types 1, 3, and 8 are stimulated by Ca^{2+}/calmodulin; types 2, 4, and 7 are not affected by Ca^{2+}; and types 5 and 6 are inhibited by submicromolar Ca^{2+} concentrations independently of calmodulin. In regard to their tissue distribution, each isoform has specific tissue expression, from broadly distributed (types 2 and 4) to brain specific (types 1 and 8). Adenylylcyclase is present in high concentration in neuronal tissues and the brain is the source of the greatest variety of isoforms expressed. Although Northern blotting and polymerase chain reaction (PCR) analysis demonstrate that brain is the only tissue expressing all eight isoforms (3, 6–10, 13, 18–19), *in situ* hybridization studies demonstrate that messages for several isoforms have a remarkably discrete distribution in brain (13, 16, 20–25). Thus, to gain insights into the physiological role played by individual isoforms in brain, we previously used specific rat antisense oligodeoxynucleotides to localize expression of seven isoforms (types 1–6 and 8) in rat brain (13, 16, 21, 23, 24). These and other studies suggest that, by their restricted distribution, many subspecies may play specific roles in the regulation of physiological functions in the nervous system, as in the case of the involvement of type 1 in learning and memory processes (20, 21) and type 5 in striatal dopaminergic regulation (12, 23). In this chapter we present a detailed description of procedures that provide specific signals for the eight isoforms in areas of the rat brain and in individual neuronal cells and, thus, allow a fair estimation of mRNA levels for each adenylylcyclase isoform.

In Situ Hybridization Procedures

Fixation, Embedding, and Sectioning

Fixation and sectioning are critical steps for the success of *in situ* hybridization. Paraformaldehyde fixation preserves good tissue architecture and, therefore, good association of mRNA with cellular structures, without interfering with the accessibility of short, synthetic oligonucleotide probes to the target mRNA. Paraformaldehyde fixation, either on sections from fresh frozen brain or by prior perfusion of the animal, has yielded comparable hybridization signals. All solutions must be RNase free and thus are prepared with sterile diethyl pyrocarbonate (DEPC)-treated water (see later, *Protocols*).

Perfusion of Adult Rat Brain

Rats are deeply anesthetized (pentobarbital, 40 mg/kg ip). Intracardiac perfusion is performed first with saline solution (NaCl, 9 g/liter), followed by 250–300 ml of a freshly prepared fixative solution containing 2% (w/v) paraformaldehyde. Afterward, the brain is removed, cut into blocks, and placed in the same fixative solution for an additional 3 hr at 4°C. The blocks are then washed in cold sterile phosphate-buffered saline (PBS) and infused with 20% (w/v) sucrose, dissolved in PBS, overnight at 4°C. Blocks are frozen by immersion in isopentane (chilled with liquid nitrogen) and stored at −80°C before sectioning.

Fixation of Cryosections from Fresh Brain

Adult rats are decapitated and the brain is removed immediately, placed on aluminum foil-covered powdered dry ice, frozen, and stored at −80°C before sectioning. For fixation, cryostat sections are collected, quickly dried, fixed with 2% paraformaldehyde for 20 min at room temperature, rinsed two times in PBS for 5 min, dehydrated for 2 min with a series of graded ethanol (30, 60, 80, and 100%), dried, and stored with desiccant at −80°C.

Tissue Sectioning

Cryostat sections must be obtained under RNase-free conditions. Before sectioning, the brain is allowed to equilibrate in the cryostat chamber (−18° to −20°C) and then embedded in Tissue-Tek OCT (Miles Scientific, Naperville, IL). Coronal or sagittal sections of 10–16 μm are thaw-mounted on gelatin-coated slides. The sections are dried and stored at −80° until fixation.

TABLE I   Sequences (5'–3') of Oligodeoxynucleotide Probes Used
for *in Situ* Hybridization Analysis[a]

Subtype	Sequence (5'–3')
1	ATGTTCAGGTCTACTTCAGTAGCCTCAGCCACGGATGTGATGGTATCGAT
2	CAGGTTCCTGGGCGTGCTCCTGGCTGGGTATGGCACTCAGTCCCGTGGCT
3	TGGAGCTTGTAAAGCCATTGGTGTTGACATCTGGTGTGACTCCAGAAGCT
4	AGCCGAGCCATGATCTCTGCCTTCATCTCTCGGGCCAGGTAGGCAGGAAG
5	CGTTGAGGGTCATGACCAGTTCTTGGGCAGTACACTGGGATGCCAGGCTA
6	TGTGCGAGCGGCCGACCTGGTCATAGGTGCTGGCATTTAGCCCGGAGGCG
7	CATGCTAAATTCTACCAAGACTCCGATGTGCACGTGCTT
8	TGACGTTGCGGGCCAAATAGGTCTCGTTAATCCAACAACAGGTTTTGCGG

[a] Corresponding bases of each isoform gene to which they are complementary. Oligonucleotides were designed to hybridize specifically to rat (types 1–6, 8) or mouse (type 7) adenylylcyclase mRNA. The sequences shown are complementary to the cDNA encoding type 1 (bases 1141–1190 within the full-length bovine sequence), type 2 (bases 2838–2887), type 3 (bases 2949–2998), type 4 (bases 691–740), type 5 (bases 950–999), type 6 (bases 2991–3040), type 7 (bases 3025–3064), and type 8 (bases 2304–2353).

Synthesis and Labeling of Oligonucleotides

We routinely use synthetic oligodeoxynucleotides, consisting of 50-mer oligonucleotides designed to hybridize specifically to rat types 1–6 and 8 mRNA and a 39-mer oligonucleotide specific for mouse type 7 mRNA (see Table I). These probes are synthesized on an Applied Biosystem automated DNA Synthesizer. The oligonucleotides are diluted in DEPC-treated water at a final probe concentration of 2 pmol/μl and they are stored at $-20°C$ until labeling.

^{35}S Labeling of Oligonucleotide Probes

Oligomers are labeled by using [α-^{35}S]dATP and terminal deoxynucleotidyltransferase (Tdt), which catalyzes the addition of labeled oligodeoxynucleotides to the 3' end. Add in order to a sterile microfuge tube: DEPC-treated water (3 μl); 5× Tdt reaction buffer (4 μl); oligonucleotide (16 pmol), [α-^{35}S]dATP (NEN, Dupont de Nemours; specific activity 1500 Ci/mmol; 3μl), TdT enzyme (Stratagene, 32 U). Vortex tubes at low speed and centrifuge them briefly. Incubate mixture at 37°C for 1 hr and stop by cooling on ice.

Rapid Purification of Probe by Acid Precipitation

Add 5 μl of tRNA (yeast tRNA, 10 mg/ml), 200 μl 0.25 *M* ammonium acetate, and 30 μl DEPC-treated water to the reaction mixture. Add 0.5-μl aliquots of mixture to two positively charged DE-81 filters. The probe is precipitated by adding 650 μl of cold 100% ethanol, and placing the tubes at $-80°C$ for 1 hr. The supernatant is discarded after centrifugation at 12,000 g for 15 min at 4°C. The pellet is washed with 100% ethanol and dried for 5–10 min. The pellet is reconstituted with hybridization buffer and kept at $-20°C$. After 10–15 min, 0.5 μl of mixture is transferred to a scintillation vial and counted for 1 min. If precipitation is efficient, the amount of radioactivity is similar to that of the washed filter. The incorporation should be as high as 5 × 10^8 cpm/μg. The efficiency of the reaction is calculated by determining the proportion of the radioactivity that has been incorporated into the oligonucleotide. This can be achieved by differential absorption of the incorporated/ unincorporated nucleotides on DE-81 filters. Two 0.5-μl probe aliquots are added onto two DE-81 filters. One filter is stored at room temperature until it is completely dry; the second filter is successively washed in 0.5 M Na_2HPO_4 (4×), H_2O (3×), and then 70% (v/v) ethanol (twice). Unincorporated oligonucleotides that stick less tightly to the DE-81 filter are eluted by the successive washes in sodium phosphate, whereas the radioactive nucleic acids are fixed. The amount of radioactivity is measured on each filter. The proportion of radioactivity (P) that has been incorporated into oligonucleotide is calculated from the ratio of counts on the washed filter (which measures radioactivity in oligonucleotide) to counts on the unwashed filter (which measures total radioactivity). If the reaction is efficient, P is usually 50–80% of the total radioactivity.

Separation of Labeled Oligonucleotides by Column Chromatography

After incubation at 37°C, the reaction is stopped with EDTA (0.1 ml; 10 mM). The labeled oligonucleotides are separated from unincorporated labeled nucleotides by gel filtration in a 5-ml disposable pipette filled with Sephadex G-50 (see *Protocols*). The probe is placed in the column, the column is washed with 10 ml of Tris-EDTA (TE) buffer (1×), and fractions (approximately 500 μl) are collected in microcentrifuge tubes. The fractions containing radioactive probes are in 5-μl aliquots of each fraction identified by determining ^{35}S incorporation. The first peak corresponds to the labeled probe; samples are pooled and then precipitated by adding ice-cold acid precipitation solution [1 : 10× sodium acetate (3 M, pH 4.8) and 3× cold absolute ethanol]. After precipitation for 1 hr at $-70°C$ (or overnight at $-20°C$), the probe is centrifuged at 12,000 g for 15 min at 4°C. The pellet is washed with 100% ethanol and dried for 5–10 min. The pellet is reconstituted with hybridization buffer. After 10–15 min, 0.5 μl of probe is transferred to a scintillation vial. The specific activity and percentage of incorporation are determined.

Protocols

DEPC-Treated Water (0.1%)

In a fume hood, add 1 ml DEPC to 1.0 liter of nanopure water. Shake vigorously, heat at 60°C with continuous stirring overnight, and autoclave.

Paraformaldehyde (2%) in Phosphate Buffer

In a hood, slowly add 10 g of paraformaldehyde to 200 ml of DEPC-treated water that is heated at 60°C with continuous stirring. Add 1 M NaOH dropwise to the milky solution until it becomes clear. Allow the solution to cool, adjust to pH 7.2 with HCl, add distilled water to a final volume (250 ml), and filter. Cool to 4°C and add 1 volume of phosphate buffer (PB) (0.2 M, pH 7.2).

Gelatin-Coated Slides

Soak slides in absolute ethanol for 24 hr, wipe dry, place on aluminum foil, and bake in an oven at 180°C overnight. Allow to cool to room temperature. Heat sterile DEPC-treated water to 60°C, add 0.5% (w/v) gelatin–0.1% (w/v) chromium potassium sulfate, mix thoroughly, and filter. Dip slides in gelatin mixture for 4 min and dry at 60°C overnight.

Sephadex G-50 Medium

Add 10 g Sephadex G-50 medium (Pharmacia, Piscataway, NJ) to 120 ml sterile TE buffer (10 mM Tris–base, 1 mM EDTA, pH 8.0). Leave overnight at room temperature; wash several times with TE buffer. Pour off excess supernatant and store at 4°C.

Hybridization

Pretreatments

All sections are pretreated to enhance tissue permeability and/or decrease nonspecific probe labeling. Under our conditions (i.e., light paraformaldehyde fixation), tissue accessibility is good, so that pretreatments, such as incubation of sections with pronase, proteinase K, or HCl, are omitted. To reduce nonspecific signals, sections are incubated with hybridization mixture without the oligonucleotide probe and dextran sulfate for 1 hr at 37°C. The presence of dextran sulfate, which is added to the hybridization mixture to increase the viscosity of the solution, makes it difficult to decant prehybridization mixture from the slide. After prehybridization, the slides are rinsed twice with 4× SSC (see later, *Stock Solutions*) and then acetylated. To acetylate amino groups in tissues and thereby reduce nonspecific electrostatic binding of the probe, sections are incubated for 3 min in freshly prepared 0.1 *M* triethanolamine (TEA, pH 8.0, 0.9% (v/v) NaCl) at room temperature, followed by 10 min in acetic anhydride (0.25% (v/v) final concentration) added to the TEA. Following acetylation, sections are rinsed twice in 4× SSC (5 min), then they are dehydrated through a series of graded ethanol solutions (70, 80, and 95%, 1 min). After dehydration, sections are air-dried for at least 15 min at room temperature before hybridization.

Hybridization

The hybridization solution contains (see later, *Stock Solutions*) 4× SSC (from 20× SSC), 50% deionized formamide [from 100% (v/v)], 1× Denhardt's (from 50×), 10% dextran sulfate (from 50%), 0.5 mg/ml denatured salmon sperm DNA (from 10 mg/ml), 0.5 mg/ml yeast tRNA (from 20 mg/ml), 5% *N*-lauroylsarcosine (from 20%), and 20 μM dithiothreitol (DTT) (from 1 *M*). The hybridization mixture can be prepared ahead of time and stored at $-20°C$ in aliquots. DTT is thawed and added just before use. DTT, Denhardt's solution, and nucleic acid diminish the nonspecific signal. We used a final probe concentration varying between 500,000 cpm and 2×10^6 cpm/100 μl of hybridization buffer. Coverslips are placed on slides, which are incubated overnight in humidified box at 32–37°C, depending on the probe. The initial hybridization is performed at a stringency of 35°C in 50% formamide, which is below the theoretical melting temperature (T_m) for the oligonucleotide probes.

Posthybridization Treatments

The degree of nonspecific labeling and the amount of background are controlled by the stringency of posthybridization steps. The level of stringency increases by diminishing the concentration of SSC in the wash buffer and by increasing the temperature. Coverslips are removed by washing each slide with 4× SSC at room temperature. Excess hybridization buffer and unhybridized probe are removed by rinsing slides first in 2× SSC (room temperature) and then successively in 2× SSC (37°C, 1 hr), 2× SSC (45°C, 1 hr), 0.5× SSC (45°C, 1 hr), and 0.5× SSC (RT, 1 hr). Finally, sections are dehydrated in graded ethanol solutions (70, 80, and 90%). The sections are air dried for 1 hr before exposure to X-ray film.

Stock Solutions

1. SSC, 20×: This is a stock solution of 3 *M* sodium chloride and 0.3 *M* sodium citrate, added to DEPC-treated water (pH 7.0), which is sterilized by autoclaving. Dilutions of 20× SSC are made with DEPC-treated water.

2. Deionized formamide: To 5 g of ion-exchange resin (Bio-Rad, Richmond, CA, AG 501-X8) add 50 ml formamide. Stir mixture overnight at 4°C, filter through Whatman (Clifton, NJ) filter paper, dispense into 1-ml aliquots in sterile tubes, and store at −20°C.
3. Denhardt's solution, 50×: To DEPC-treated water add 0.02% (w/v) Ficoll, 0.02% (w/v) polyvinylpyrrolidone, and 0.02% (w/v) bovine serum albumin; stir until complete dissolution and store in sterile tubes at −20°C.
4. Dextran sulfate, 50% (w/v): Add 50 g dextran sulfate to 60 ml DEPC-treated water. Stir at 60–65°C in a water bath overnight. After complete dissolution, adjust volume to 100 ml with DEPC-treated water, stir, and store at 4°C.
5. N-Lauroylsarcosine, 20% (w/v): Add 5 g N-lauroylsarcosine to 25 ml DEPC-treated water. Filter and keep at 4°C.

Detection of Hybridized Probes

The examination of hybridized sections is possible at the macroscopic level by using autoradiographic films and/or at the microscopic level by dipping the slides into emulsion. All autoradiographic procedures must be carried out under safelight illumination with appropriate filters.

Film Autoradiography

The slides are directly apposed against a β-Max Hyperfilm (Amersham International). The slides are exposed for 1–7 days at room temperature. (The exposure time depends on the amount of mRNA and the specific activity of the probe.) The films are developed in a Kodak (Rochester, NY) D-19 solution for 3 min at 20°C, rinsed in water for 30 sec, fixed with Rapid-fixer (Kodak) for 5 min, washed in running tap water for at least 15 min, rinsed briefly with deionized water, and finally air-dried at room temperature.

Emulsion Autoradiography

After sections have been exposed to autoradiographic film, slides with good signals are processed for higher resolution examination. The emulsion (Kodak NTB-2) diluted 1 : 1 in 600 mM ammonium acetate is melted in a 42–45°C water bath. The slides are dipped slowly, placed vertically to remove excess emulsion, and dried for 2–3 hr at room temperature. The slides are stored in light-tight slide boxes containing desiccant (e.g., silica gel). The boxes are sealed, covered with foil, and stored at 4°C. After 15–30 days, the boxes equilibrate for 15 min at room temperature before the slides are developed for 3 min in a Kodak D-19 solution at 20°C. The development is stopped in 2% (v/v) acetic acid for 30 sec, fixed in sodium thiosulfate solution (30% w/v) for 15 min, and rinsed in running tap water for 20–30 min. The sections are then lightly stained in toluidine blue solution [0.5% (w/v) toluidine blue, 1% (w/v) sodium tetraborate dissolved in water] for 1–2 min, successively washed in distilled water for 5 min, ammonium pentamolybdate (5% w/v) for 5 min, distilled water for 5 min, running tap water for 5–10 min, and dehydrated briefly through a graded ethanol series (80, 95, and 100%). The slides are transferred in a hood and dipped in several changes of xylene. One drop of Permount (Fisher) is added to the moist slide, and coverslips are placed on slides.

Specificity Controls

The specificity of hybridization is evaluated by competition studies performed with an excess of unlabeled probe in the hybridization solution or by using an oligonucleotide designed against a distinct region of the target mRNA. The addition of a 50-fold excess of unlabeled oligomer to the hybridization mixture, which contains the labeled probe, should eliminate the specific signal. The evaluation of the nonspecific binding is done by measuring the optical density (OD) in an area where no specific mRNA is expressed. The nonspecific OD coincides with general OD background found on the film.

Regional Distribution of Adenylylcyclase Isoforms

The brain is highly enriched in adenylylcyclase, and messages for the eight isoforms are present, each probe producing a unique pattern of hybridization. The patterns of mRNA expression for types 1–6 and type 7 are shown in Figs. 1–3.

Distribution of Type 1 mRNA

Type 1 is the most predominant adenylylcyclase mRNA expressed in mammalian brain. However, the strong expression of type 1 mRNA is restricted in areas that are associated with learning and memory processes, such as dentate gyrus, neocortex, and cerebellum (16, 21) (Fig. 1A). Lower amounts of type 1 mRNA are found in the pyriform cortex and thalamic structures. A similar pattern of type 1 mRNA has been found, using a pool of two rat-specific oligonucleotide probes that differ from the probes used in this study (12). A comparison of differential patterns of expression of type 1 mRNA using either rat oligonucleotides (12, 16, 21) or bovine ribonucleotide probes (20) shows discrepancies, particularly in the hippocampus. Bovine ribonucleotide probe gives strong and homogeneous signals in pyramidal cells of CA1–CA3 fields and dentate gyrus, whereas, using rat oligonucleotide probes, the highest expression is found in CA1–CA2 and dentate gyrus. These discrepancies are probably due to problems with the long non-species-selective bovine ribonucleotide probe. The high level of type 1 mRNA in pyramidal cells of the CA1–CA2 field and dentate gyrus is consistent with the involvement of a Ca^{2+}/calmodulin-stimulatable cyclase in NMDA- and Ca^{2+}/calmodulin-dependent LTP processes occurring at postsynaptic sites in stratum radiatum (26).

Distribution of Type 2 mRNA

The type 2 isoform is also predominantly expressed in rat brain. In contrast to type 1, type 2 mRNA is not restricted to areas associated with particular neuronal functions. The high levels of expression roughly parallel those of type 1 mRNA in hippocampus, neocortex, and cerebellum. However, the expression of type 2 mRNA in hippocampal formation is distinct from that of type 1 mRNA (16, 21) (Fig. 1B). Type 2 mRNA is uniformly expressed in hippocampal pyramidal cells of CA1–CA3 fields. Densitometric analysis showed that the expression of type 2 mRNA in granular cells of dentate gyrus is at least five times less than that of type 1 mRNA (21). Both types 1 and 2 mRNA display strong signals in cerebellar granular layers. However, emulsion autoradiography reveals that only type 2 probe gives an intense signal in Purkinje cells, which contrasts with the absence of type 1 mRNA in these cells. Additional areas expressing high levels of type 2 mRNA, but not type 1 mRNA, include hypothalamic and supraoptic nuclei and substantia nigra (24). More modest levels of expression are observed in striatum and thalamus. Similarly, high levels of type 2 mRNA were found in hippocampus, cerebellum, cerebral cortex, and olfactory bulb using a pool of three rat oligonucleotide probes (12). By contrast, Furuyama et al. (22), also using a rat-specific oligonucleotide probe, found a more restricted distribution of type 2 mRNA in hippocampal pyramidal layers and cerebellar granular cell layers. Such differences may be due to either different sensitivities of the

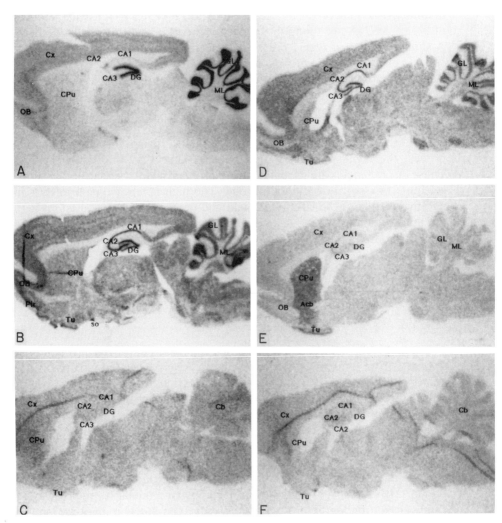

Fɪɢ. 1   Distribution of types 1–6 mRNA in sagittal sections of rat brain. The strongest
signals (corresponding to dark areas) are found for types 1 (A), 2 (B), and 5 (E).
Type 4 (D) has a lower expression and no significant signals are found for types 3
(C) and 6 (F) in brain. Acb, Nucleus accumbens; CA1–CA3, hippocampal pyramidal
cell layers; Cg, cerebellum; CPu, caudate putamen; Cx, cortex; DG, dentate gyrus;
GL, granular cell layer; ML, molecular layer; OB, olfactory bulb; Pir, pyriform
cortex; so, supraoptic nucleus; Tu, olfactory tubercle. Reprinted from *Cell. Signalling*
6; D.M.F. Cooper, N. Mons, and K. Fagan, Ca²⁺-sensitive adenylyl cyclases, 823–
840, copyright 1994, with kind permission from Elsevier Science Ltd, The Boulevard,
Langford Lane, Kidlington, OX5 1GB, UK.

probes used or different stringencies in the hybridization and washing conditions.

Distribution of Type 3 mRNA

The highest expression of type 3 mRNA is found in olfactory neuroepithelium (1). In rat brain, type 3 mRNA is homogeneously expressed, but at very low levels (16) (Fig. 1C). However, a longer exposure of autoradiographic film shows that cerebellar granular cells display a higher signal than other areas. A higher expression of type 3 mRNA in cerebellum, with much lower levels in hippocampus, cerebral cortex, and olfactory bulb, has been described also using rat oligonucleotide probes (12).

Distribution of Type 4 mRNA

Type 4 adenylylcyclase has a wide expression in rat brain, which overlaps that of type 2 mRNA but with lower levels of expression. Like type 2, higher levels of expression of type 4 mRNA are found in both hippocampus and cerebellum (16) (Fig. 1D). In the hippocampal formation, type 2 mRNA is uniformly distributed in CA1–CA3 layers and dentate gyrus, whereas no signals are detected in molecular layer.

Distribution of Type 5 mRNA

Type 5 mRNA is present at a high level in rat brain, although its expression is restricted to the striatum (12, 16, 23) (Fig. 1E). Type 5 mRNA is uniformly distributed in caudate putamen and nucleus accumbens, which receive dopaminergic inputs from substantia nigra and glutamatergic inputs from cerebral cortex. More ventrally, the olfactory tubercles also contain high levels of type 5 mRNA. Analysis of expression of type 5 mRNA at the cellular level indicates that abundant medium-sized neurons highly expressed type 5 mRNA (Fig. 2), whereas type 2 is preferentially expressed in large cholinergic interneurons (22, 23).

Distribution of Type 6 mRNA

Although the type 6 adenylylcyclase isoform is closely related to type 5, the expression of type 6 mRNA is quite different (23) (Fig. 1F). The expression of type 6 mRNA is weak and appears homogeneous throughout the rat

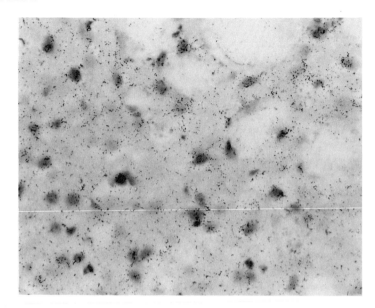

FIG. 2   Cellular localization of type 5 mRNA in the striatum. Bright-field photomicrograph of emulsion-coated sections shows that type 5 labels the majority of striatal medium-sized neurons.

brain. Even after long exposure, no distinct expression is seen over discrete brain areas.

Distribution of Type 7 mRNA

Type 7 has a high level of expression in cerebellar granular cells (Fig. 3). Other areas, including hippocampus, pyriform cortex, cortex, and striatum, contain very low amounts of type 7 mRNA. After a long exposure, type 7 is detectable in hippocampus, with higher levels in dentate gyrus than CA1–CA3 pyramidal layers.

Distribution of Type 8 mRNA

The strongest signals of type 8 mRNA are found in hippocampal pyramidal cells of layers CA1–CA3 and dentate gyrus (13). Moderate signals are present in hypothalamus and a weak signal is detected in cerebellum. Using a ribonucleotide probe derived from a partial human cDNA (13), high signals were

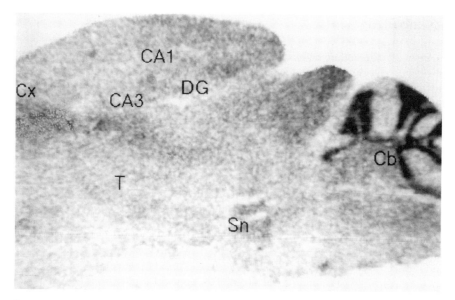

FIG. 3 Distribution of type 7 mRNA in sagittal sections of mouse brain. Note a selective expression of type 7 mRNA in the granular layer of cerebellum. CA1–CA3, Hippocampal pyramidal layers; Cb, cerebellum; Cx, cortical layers; DG, dentate gyrus; Sn, substantia nigra; T, thalamus.

previously reported not only in rat hippocampus, but also in granular cells of the cerebellum (25). The use of a non-species-specific long riboprobe probably accounts for the differential patterns observed in cerebellum.

Conclusions

In situ hybridization provides a useful tool for the identification of specific adenylylcyclase isoforms in brain and clearly demonstrates that the eight isoforms have a different pattern of expression. The restricted distributions of adenylylcyclase isoforms indicate that their expression is highly regulated for specific roles in the central nervous system. Because a number of the adenylylcyclases are rather closely related structurally, the use of species-specific oligonucleotides appears to yield less ambiguous results and more discrete signals than are obtained with long riboprobes. It seems likely that this technique can be profitably applied to future studies on pharmacologically or developmentally induced changes in adenylylcyclase mRNA expression.

Acknowledgments

The author thanks Dr. D. M. F. Cooper for careful reading of the manuscript and Dr. K. Hellevuo for providing mouse type 7 sequence. This work was supported by a NATO fellowship and by the Centre National de la Recherche Scientifique.

References

1. J. Krupinski, F. Coussen, H. A. Bakalyar, W.-J. Tang, P. G. Feinstein, K. Orth, C. S. Slaughter, R. R. Reed, and A. G. Gilman, *Science* **244,** 1558 (1989).
2. H. A. Bakalyar and R. R. Reed, *Science* **250,** 1403 (1990).
3. P. G. Feinstein, K. A. Schrader, H. A. Bakalyar, W.-J. Tang, J. Krupinski, A. G. Gilman, and R. R. Reed, *Proc. Natl. Acad. Sci. U.S.A.* **88,** 10173 (1991).
4. B. Gao and A. G. Gilman, *Proc. Natl. Acad. Sci. U.S.A.* **88,** 10178 (1991).
5. J. Parma, D. Stengel, M. H. Gannage, M. Poyard, R. Barouki, and J. Hanoune, *Biochem. Biophys. Res. Comm.* **179,** 455 (1991).
6. M. Yoshimura and D. M. F. Cooper, *Proc. Natl. Acad. Sci. U.S.A.* **89,** 6716 (1992).
7. R. T. Premont, J. Chen, H. W. Ma, M. Ponnapalli, and R. Iyengar, *Proc. Natl. Acad. Sci. U.S.A.* **89,** 9808 (1992).
8. Y. Ishikawa, S. Katsushika, L. Chen, N. J. Halnon, J. I. Kawabe, and C. J. Homcy, *J. Biol. Chem.* **267,** 13553 (1992).
9. S. Katsushika, L. Chen, J. I. Kawabe, R. Nilakantan, N. J. Halnon, C. J. Homcy, and Y. Ishikawa, *Proc. Natl. Acad. Sci. U.S.A.* **89,** 8774 (1992).
10. J. Krupinski, T. C. Lehman, C. D. Frankenfield, J. C. Zwaagstra, and P. A. Watson, *J. Biol. Chem.* **267,** 24858 (1992).
11. K. Hellevuo, M. Yoshimura, M. Kao, P. L. Hoffman, D. M. F. Cooper, and B. Tabakoff, *Biochem. Biophys. Res. Comm.* **192,** 311 (1993).
12. C. E. Glatt and S. H. Snyder, *Nature* (*London*) **361,** 536 (1993).
13. J. J. Cali, J. C. Zwaagstra, N. Mons, D. M. F. Cooper, and J. Krupinski, *J. Biol. Chem.* **269,** 12190 (1994).
14. P. A. Watson, J. Krupinski, A. M. Kempinski, and C. D. Frankenfield, *J. Biol. Chem.* **269,** 28893 (1994).
15. N. Defer, O. Marinx, D. Stengel, A. Danisova, V. Iourgenko, I. Matsuoka, D. Caput, and J. Hanoune, *FEBS Lett.* **331,** 109 (1994).
16. D. M. F. Cooper, N. Mons, and K. Fagan, *Cell. Signalling* **6,** 823 (1994).
17. R. Taussig and A. G. Gilman, *J. Biol. Chem.* **270,** 1 (1995).
18. Z. Xia, E. J. Choi, F. Wang, C. Blazynski, and D. R. Storm, *J. Neurochem.* **60,** 305 (1993).
19. Z. Xia, E. J. Choi, F. Wang, and D. R. Storm, *Neurosci. Lett.* **144,** 169 (1992).
20. Z. Xia, C. D. Refsdal, K. M. Merchant, D. M. Dorsa, and D. R. Storm, *Neuron* **6,** 431 (1991).
21. N. Mons, M. Yoshimura, and D. M. F. Cooper, *Synapse* **14,** 51 (1993).
22. T. Furuyama, S. Inagaki, and K. Tagaki, *Brain Res.* **19,** 165 (1993).

23. N. Mons and D. M. F. Cooper, *Mol. Brain Res.* **22,** 236 (1994).
24. N. Mons and D. M. F. Cooper, *J. Neuroendocrinol.* **6,** 665 (1994).
25. I. Matsuoka, G. Giuli, M. Poyard, D. Stengel, J. Parma, G. Guellaen, and J. Hanoune, *J. Neurosci.* **12,** 3350 (1992).
26. T. V. P. Bliss and G. L. Collingridge, *Nature (London)* **361,** 31 (1993).

[3] Site-Directed Mutagenesis of G Protein α Subunits: Analysis of α_z Phosphorylation and Fatty Acid Acylation

Hazem Hallak, Karen M. Lounsbury,
and David R. Manning*

Introduction

The α subunit of G_z [α_z (or α_x (1)] exhibits a number of unusual features, including a markedly diminished capacity to hydrolyze GTP relative to other α subunits, resistance to ADP-ribosylation by bacterial toxins, and constraint to cells primarily of hematopoietic and neuronal origin (1–3). α_z is also unique among α subunits as a substrate for protein kinase C (PKC) (4, 5), and like several other subunits is subject to both N-myristoylation and palmitoylation (6–8).

The occurrence of co- and/or posttranslational modifications for α_z engenders questions regarding functional significance. Such questions can be approached in part through site-directed mutagenesis. The systematic alteration of amino acid residues, for example, can be used to determine both the site of modification and the consequences of its elimination. In this chapter, the use of site-directed mutagenesis to define the sites of phosphorylation and fatty acid acylation in α_z are discussed. The prevailing technique for generating mutations in our laboratory is the polymerase chain reaction (PCR) with mismatched primers (9). Other techniques have also been used and are discussed in the last section.

Mutagenesis

Materials, Polymerase Chain Reaction, and Sequencing

The advantages of PCR using mismatched primers relative to other techniques of mutagenesis are simplicity and speed. Mutations can be introduced

---

* To whom correspondence should be addressed.

Methods in Neurosciences, Volume 29

within 1 day, and the mutated proteins can be subcloned and sequenced within an additional 2 or 3 days. A relative disadvantage of the PCR technique is the need to sequence amplified portions of DNA in their entirety. However, sequencing is recommended for any technique.

Primers for PCR are generally 15–18 nucleotides in length, with a G + C composition not exceeding 50–60%. Those used to introduce a particular mutation contain one or more nucleotides that are not complementary to the template ("mismatches"). A mismatch(es) is placed in the middle or near the 5' end of a primer, and in no instance closer than 3 nucleotides to the 3' end. Mismatches near the 3' end diminish the efficiency of extension by *Taq* polymerase and decrease melting temperature. We have successfully included up to four mismatches within the body of an 18-nucleotide primer, but recommend no more than two. Restriction site sequences can be appended, when desired, to the 5' end of a 15- to 18-nucleotide primer, as discussed below. In this instance, several additional nucleotides are typically added 5' to the restriction sequence to ensure efficiency of digestion by the restriction enzyme (10). Purification of primers by high-performance liquid chromatography (HPLC) or polyacrylamide gel electrophoresis (PAGE) is recommended, but is not essential.

In our work with α_z, we have used bovine retinal α_z cDNA (2) inserted into pDP5 or Bluescript (Stratagene, La Jolla, CA) as the template for PCR. pDP5 is a pUC-based plasmid carrying an ampicillin-resistance gene and a human cytomegalovirus promoter, and has been employed by us and others as an expression vector in human embryonic kidney 293 cells (8, 11, 12). The techniques of PCR and protein expression described here are nevertheless applicable to a variety of double-stranded plasmids. For use as templates, plasmids are purified by cesium chloride banding or as minipreparations (10). About 1 μg of denatured plasmid is required per PCR reaction.

For PCR, we routinely use the GeneAmp PCR reagent kit (Perkin-Elmer Cetus, Norwalk, CT). The reaction mixture contains 25 mM tris(hydroxy-methyl)methylaminopropanesulfonic acid (TAPS, sodium salt), pH 9.3, 50 mM KCl, 2 mM MgCl$_2$, 1 mM 2-mercaptoethanol, 200 μM each dATP, dGTP, dTTP, and dCTP, activated salmon sperm DNA, 0.1–5 μM of each primer, ~1 μg template, and 2.5 units AmpliTaq polymerase in a 50-μl final volume. The concentration of primers depends on purity. A concentration of 0.1–0.5 μM is recommended for HPLC- or PAGE-purified primers, and 10-fold higher concentrations for primers used directly following synthesis. Temperatures employed for denaturation, annealing, and extension will depend on the primers [i.e., T_m (13)] and size of the product. For the work discussed here, a PCR cycle using a BioTherm Cycler II typically consists of denaturation at 94°C for 30 sec, annealing at 55°C for 30 sec, and extension at 72°C for 30–90 sec (30 sec per 500 bp). Because increasing the number of cycles increases the chances of unwanted errors by the *Taq* polymerase, more than 25 cycles is not recommended.

Sequencing of the amplified DNA is essential. Tindall and Kunkel (14) report that *Taq* polymerase incorporates an incorrect nucleotide once every 9000 nucleotides and that a frame shift occurs once every 41,000 nucleotides. Sequencing therefore not only ensures that the desired mutation has been achieved, but that other mutations have not been introduced. For sequencing, the PCR-amplified DNA is first subcloned into a convenient plasmid [e.g., TA cloning vector pCR 1000 (Invitrogen, San Diego, CA) or pDP5], then sequenced by the dideoxy chain termination method using a plasmid miniprep as the template.

A Typical Mutation: Ser → Ala in Analysis of Phosphorylation

α_z is unique among G protein α subunits in its capacity to be phosphorylated by PKC. Phosphorylation in human platelets challenged with the phorbol ester PMA, or catalyzed *in vitro* by PKC, approaches 1 mol phosphate per mole of subunit, and is selective: phosphorylation does not occur to an appreciable extent for subtypes of α_i and α_q (5, 11). Phosphoamino acid analysis and CNBr peptide mapping reveal that the phosphorylation in human platelets occurs at a serine residue(s) within the N-terminal 52 residues of the subunit (5). A phosphorylation-sensitive antiserum, generated against residues 24–33 (antibody 6354; see later), moreover, recognizes normal, but not phosphorylated, subunits, suggesting Ser-25 or Ser-27 as the site of phosphorylation.* Site-directed mutagenesis was therefore used to determine which serine residue was relevant to the phosphorylation event (11). Construction of Ser → Ala mutants illustrates the use of a several-stage PCR method to accomplish mutations within the body of the subunit.

PCR

Construction of a Ser-27 → Ala-27 mutant of α_z is described here. A Ser-25 → Ala-25 mutant, for which data are also presented, is fashioned in a similar manner. For the Ser-27 → Ala-27 mutation, two overlapping PCR products are first generated using α_z cDNA in Bluescript KS as the template—one product with the M13FD (sense) primer and the mismatched antisense primer 3′G GTG GAC GCG AGT CTC CGG GTC GCC GTT GCG5′, and the other with a mismatched sense primer 5′GC TCA GAG

---

* Residues are numbered beginning with the initiating methionine. Nucleotides are numbered from the initiating codon.

GCC CAG CGG CAA CGC CGC GAA AT3′ and the antisense "Gz708a" primer 3′T GCT CCT ATT GGT CTG TTC AGC CTA CC5′ (Fig. 1). The nucleotide sequences here are divided into triplets corresponding to codons, and mismatches are double-underlined; mismatches in Fig. 1 are denoted by asterisks. M13FD is 5′AACAGCTATGACCATG3′, and hybridizes to a vector sequence upstream of α_z. The 3′ end of Gz708a terminates at nucleotide 708 of the α_z cDNA. The two PCR products are each purified by electrophoresis in 1.2% (w/v) agarose (containing ethidium bromide) followed by extraction using a GeneClean II kit (Bio 101, Vista, CA). The two products are denatured, reannealed as a mixture, and subjected to a third round of PCR using the outside primers, M13 forward and Gz708a, to obtain a 750-bp PCR product containing the mutation and encompassing the HindIII and BamHI restriction sites used for eventual subcloning.

Ligation and Subcloning

The PCR product is inserted directly (without purification) into the TA cloning vector using the manufacturer's materials and instructions, i.e., with ligation accomplished using T4 DNA ligase at 12°C overnight at a 1 : 1–1 : 3 molar ratio of vector : insert. The mixture (3 μl) is used to transform DH5α-competent bacterial cells, which are then plated on LB agar plates supplemented with ampicillin (100 μg/ml) and X-Gal. Solid white, ampicillin-resistant colonies are isolated, and plasmid minipreparations are screened for the PCR insert using HindIII/BamHI digestion.

A minipreparation of the vector containing the PCR insert is digested with HindIII and BamHI to yield a fragment beginning with nucleotide −27 (corresponding to the Bluescript vector) and ending with nucleotide 635 of the α_z coding sequence. The 662-bp fragment is resolved by electrophoresis in 0.7% agarose, extracted with a GeneClean II kit, and combined directly with pDP5 that has been similarly digested and purified (the polycloning site of pDP5 contains HindIII and BamHI restriction sites, among others). Ligation is accomplished with 100 ng of the pDP5 vector, 200 ng of the 662-bp fragment, and 1 unit of T4 DNA ligase in ligation buffer containing 50 mM Tris-HCl, pH 7.5, 10 mM MgCl$_2$, 10 mM dithiothreitol (DTT), 1 mM ATP, and 25 μg/ml BSA (10 μl total). The ligation is allowed to proceed overnight at 16°C. The mixture (5 μl) is then used to transform DH5α-competent bacterial cells, which are then plated on LB plates supplemented with 100 μg/ml ampicillin. Ampicillin-resistant colonies are screened for the pDP5 vector containing the 662-bp insert using plasmid minipreparations and HindIII/BamHI digestion.

Final Construction

A minipreparation of pDP5 containing the HindIII/BamHI insert is digested with BamHI to achieve a single cut, then with alkaline phosphatase to prevent

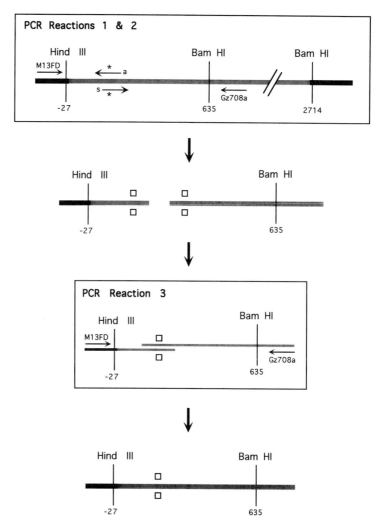

FIG. 1   Creation of the Ser-27 → Ala-27 mutation used in the analysis of α_z phosphorylation. The template for the first two sets of PCR reactions is α_z cDNA (gray) in Bluescript (black). The primers are described in the text, with mismatches (2 base pairs in the case of this mutation) noted by asterisks. The introduced mutations are denoted by squares. The templates for the third PCR reaction are the denatured and reannealed products from the first two sets. Adapted from Ref. 11, with permission.

religation. A BamHI/BamHI α_z cDNA fragment, comprising nucleotides 636–2684 of α_z cDNA plus a small portion of vector DNA, is obtained through a separate digestion of $\alpha_z \cdot$ pDP5. The two fragments are combined and ligated as described above. The final construct is then used to transform DH5α bacteria, and proper orientation of the construct is confirmed by digestion with SmaI (which cleaves α_z cDNA at nucleotides 532 and 1080).

Mutation in Which Restriction Site Is Introduced: Gly-2 → Ala-2 in Analysis of Fatty Acid Acylation

The α subunit of G$_z$, like those of G$_i$ and G$_o$, is subject to N-myristoylation and palmitoylation. A feature common to substrates of myristoyl-CoA : protein N-myristoyltransferase is an N-terminal glycine (i.e., Gly-2) and a serine as residue 6. In order to examine the relationship of N-myristoylation to anchorage and palmitoylation of α_z, we wished to substitute alanine for the N-terminal glycine (8). The construction of this mutant illustrates the insertion of a restriction sequence immediately prior to the initiating codon, eliminating unwanted upstream sequences and the need for double-PCR for mutations within the body of the subunit (Fig. 2).

PCR

Our template is $\alpha_z \cdot$ pDP5, in which the HindIII site in the polycloning sequence is too far upstream to be useful. We therefore use as a forward primer the oligonucleotide 5'CAGAAGCTT **ATG** GCA TGT CGG CAA3', which contains a HindIII restriction site (single underline), the initiating methionine codon (bold), and a codon for alanine instead of glycine (the mutation is double-underlined). The reverse primer is 3'CCG ATG CTG GAC TTT GAG ATG5', which is complementary to nucleotides 688–708 of the α_z coding sequence (codons 230–236) (the Gz708a primer described above could be used as easily). Following PCR, the amplified DNA is resolved on 0.7% agarose gel and extracted.

Ligation, Subcloning, and Final Construction

The amplified DNA is digested with HindIII and BamHI to yield a fragment extending from −5 to +635. The fragment is purified on 0.7% agarose and combined with pDP5 that has been similarly digested and purified. Ligation, subcloning, and final insertion of the BamHI/BamHI α_z cDNA fragment is accomplished essentially as described previously.

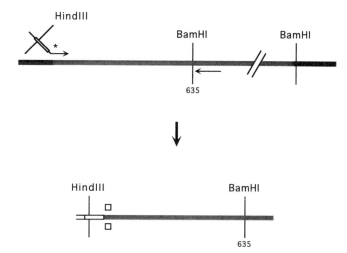

FIG. 2   Creation of the Gly-2 → Ala-2 mutation used in the analysis of α_z fatty acid acylation. The template for the PCR reaction is α_z cDNA (gray) in pDP5 (black). The forward primer contains one mismatch (denoted by an asterisk) that converts the glycine codon (residue 2, adjacent to the initiating methionine) to an alanine codon, and a *Hind*III restriction site (open rectangle) at the 5′ end together with three additional 5′ nucleotides to facilitate eventual digestion for subcloning. The introduced mutation is denoted by squares.

Expression of Normal and Mutant Subunits

Human Embryonic Kidney 293 Cells as Model for Transfection/Expression

The choice of a cell model to be used in the expression of G protein subunits is directed by several considerations: (1) efficiency of transfection and/or capacity of the cell to express the subunit from transfected DNA, (2) endogenous subunit background, and (3) the existence of appropriate signaling components, i.e., receptors, effectors, and regulatory proteins, in the case of functional studies. For our work involving the α subunit of G_z, we have chosen human embryonic kidney 293 cells. These cells were isolated from primary or early passage secondary human embryonic cells transformed with sheared fragments of adenovirus type 5 DNA (15). They replicate rapidly in monolayer culture and take up DNA efficiently. Importantly, they do not contain endogenous α_z, yet express high levels of α_z on transfection. In studies of null mutations, e.g., those inhibiting a covalent modification, the absence of endogenous subunit is particularly advantageous.

Transfection Protocol

Transfection of 293 cells is accomplished by the calcium phosphate precipitation method (16) using freshly fed cells at approximately 70% confluence. Transfection solution (1 ml) containing 1–20 μg of DNA, 125 mM $CaCl_2$, 25 mM N,N-bis[2-hydroxyethyl]-2-aminoethanesulfonic acid, pH 6.95, 140 mM NaCl, and 0.75 mM Na_2HPO_4 is added dropwise to 10 ml of medium covering approximately 2×10^6 cells in monolayer. The cells are incubated overnight at 3% (v/v) CO_2, then fed and incubated with normal medium at 5% (v/v) CO_2 for 24 hr. Expression of protein is maximal at 30–40 hr from the time of addition of DNA, after which levels of protein decrease slowly for the next 24 hr.

The amount of DNA used is dependent on the plasmid vector employed and the sensitivity of techniques for detecting protein. With pDP5, 1 μg of DNA is sufficient for detection of α_z using immunoblotting or immunoprecipitation techniques with approximately 50 μg of membrane protein. As a standard control, other sets of cells can be subjected to the conditions of transfection with no added DNA or with pDP5 containing no insert. In our experiments, protein expression was linear between 0.5 and 10 μg DNA/plate.

Protocols for Analysis of Expression by Antibodies

Two techniques have been employed for detecting expression of α_z in transfected 293 cells: immunoblotting and immunoprecipitation of [^{35}S]methionine-labeled protein. Both rely on rabbit antibodies generated with peptides that correspond to selected sequences in α_z (Table I). Prior to immunoblotting or immunoprecipitation, transfected cells are washed and then harvested by scraping into hypotonic lysis buffer (10 mM Tris-HCl, pH 8.0, 1 mM $MgCl_2$; 1 ml/plate) and lysed by repeated passage through a 27-gauge needle. Membranes are prepared by centrifuging the homogenate at 5000 g for 5 min to remove nuclei and unbroken cells, then centrifuging the supernatant at 100,000 g for 1 hr to isolate membranes as a pellet and cytosolic protein as a supernatant. The membrane pellet is usually resuspended in 1 ml of 10 mM Tris-HCl, pH 8.0, and 1 mM EDTA, centrifuged again and resuspended in 50 μl of the same buffer. All procedures are conducted at 0–4°C.

Immunoblotting

For immunoblotting (17), approximately 50 μg of the desired protein fractions is subjected to SDS-PAGE and transferred to 0.45 μm nitrocellulose membrane. The nitrocellulose membrane is incubated sequentially with 3% gela-

TABLE I   Antibodies Used in Analysis of α_z[a]

Antiserum number[b]	Antigen	Specificity
2919	[KLH]CTGPAESKGEITPELL (111–125)	α_z
2921	[KLH]QNNLKYIGLC (346–355)[c]	α_z
6354	[KLH]CRSESQRNRRE (24–33)	α_z[d]
8645	[KLH]C*F*DVGGQRSERKK (200–211)	α_i, $\alpha_0 > \alpha_z > \alpha_q$, α_{11}, α_{13} (α_{14}, α_{15}, α_{16})
1521	[KLH]CLERIA*Q*SDYI (160–169)	$\alpha_{i2} > \alpha_z$

[a] Antibodies of specificity for α_z (which, in some instances, cross-react with other α subunits) were generated in rabbits using KLH-conjugated peptides corresponding to different regions within the subunit(s). Residues in italics are not present in α_z, as is the case where the peptide was designed from the sequence of another subunit. Number in parentheses refer to positions in α_z. Reactivity of 8645 with α_{14}, α_{15}, and α_{16} is presumed.
[b] The antisera are described in Carlson *et al.* (4) and Lounsbury *et al.* (5).
[c] C terminus.
[d] Reactivity of the antibody to α_z is inhibited by phosphorylation of the subunit.

tin, 1:100 rabbit antiserum, 1:3000 biotinylated goat antirabbit immunoglobulin G (IgG), and 1:3000 avidin–horseradish peroxidase (HRP) (Vector Laboratories), with intervening washes in buffer containing 0.5% Tween 20. Visualization is accomplished with H_2O_2 and 4-chloro-1-naphthol. In our work, colorimetric intensities are analyzed with an LKB Ultroscan XL laser densitometer. This technique is optimal in work with expressed α_z. Greater sensitivity may be afforded with protocols using [^{125}I]Fab as a secondary means of detection or enhanced chemiluminescence (ECL; Amersham).

Immunoprecipitation

Immunoprecipitation of radiolabeled subunits, although more laborious than immunoblotting, is a primary tool in the analysis of expression and covalent modification. Immunoprecipitation allows a direct comparison of subunit expression (using [^{35}S]methionine) to phosphorylation or fatty acid acylation (using [^{32}P]phosphate or ^{3}H-labeled fatty acids, respectively). Immunoprecipitation of radiolabeled subunits is also extremely sensitive, and an almost essential prerequisite to isoelectric focusing as a means of analyzing α subunit subtypes (17).

To examine expression of α_z in 293 cells by immunoprecipitation, cells are incubated with 50 μCi/ml (150 μCi per 10-cm plate) Tran^{35}S-label (Amersham) in methionine-free medium beginning 24 hr following introduction of DNA. After 15–18 hr, the cells are harvested and processed to yield membrane and cytosolic fractions as discussed above. Fractions (50 μl, or multiples thereof) are combined with 17 μl of 2.5% (w/v) sodium dodecyl sulfate

(SDS), 250 mM sodium phosphate, pH 8.0, and 10 mM EDTA, and 17 μl of 25 mM NaF, 50 mM sodium pyrophosphate, 10 mM sodium vanadate, 5% (w/v) aprotinin, and 1 mg/ml leupeptin, then heated at 90°C for 5 min. The denaturation step is not essential, but improves the efficiency of immunoprecipitation and diminishes background. It also helps to alleviate concerns regarding adventitious enzymatic activities. Sample tubes are transferred to ice and diluted by addition of 55 μl of a RIPA-like buffer containing the various inhibitors of enzymatic activity to yield (final concentrations) 50 mM sodium phosphate, pH 7.2, 1% sodium deoxycholate, 1% Triton X-100, 0.5% SDS, 150 mM NaCl, 2 mM EDTA, 5 mM NaF, 2 mM Na$_4$P$_2$O$_7$, 2 mM Na$_3$VO$_4$, 1% aprotinin, and 200 μg/ml leupeptin in a total sample volume of ~140 μl. A "preclear" step is then performed to reduce contamination of subsequent immunoprecipitates with nonspecific IgG-binding proteins. This step entails incubation of the above sample with 2 μl of normal rabbit serum for 1 hr, followed by addition of 150 μl washed, boiled Pansorbin cells (Calbiochem) for 2 hr. The cells, together with IgG and nonspecifically bound protein, are removed by centrifugation in a microfuge at full speed (14,000 g) for 2 min. Depending on titer, 2–15 μl of rabbit antisera is added to the precleared sample (~150 μl), and the mixture is incubated typically for 16 hr at 4°C. The use of more than 15 μl of an antiserum in any one sample can result in distortion of subunit bands by IgG during SDS-PAGE. Antibody–subunit complexes are precipitated by addition of 200 μl protein A-Sepharose (Sigma) prepared as a 20% solution in 150 mM NaCl, 50 mM sodium phosphate, pH 7.2, 2 mM EDTA, 1 mM DTT, 0.5% SDS, 1% sodium deoxycholate, and 1% Triton X-100. Immunoprecipitates are collected 2 hr thereafter and washed three times with 800 μl of 150 mM NaCl, 50 mM sodium phosphate, pH 9.0, 2 mM EDTA, and 0.5% Triton X-100 by sequential resuspension and centrifugation. The final pellet is resuspended in ~30 μl Laemmli sample buffer, boiled for 3 min, and analyzed by 11% SDS-PAGE. Gels are fixed in 50% methanol for 30 min, dried under vacuum, and exposed to film.

End Point Analysis

Phosphorylation

Our primary objective in studies of α_z phosphorylation is to determine the site at which the phosphate is incorporated. CNBr peptide mapping and reactivity of a phosphorylation-sensitive antibody using α_z from human platelets has suggested Ser-25 or Ser-27 as the potential phosphorylation site (5). Each of these two residues is therefore changed to alanine, and the two

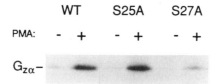

Fig. 3   Phosphorylation of wild-type and mutant α_z in 293 cells. The 293 cells were transfected with pDP5 encoding normal (WT), Ser-25 → Ala-25 (S25A), or Ser-27 → Ala-27 (S27A) forms of α_z. After 24 hr, the cells were incubated with ortho[^{32}P]phosphate (50 μCi/ml) for 3 hr, then with 100 nM PMA or vehicle for 10 min. Extracts were prepared and immunoprecipitated with 2919, followed by washes, SDS-PAGE, and autoradiography. Equivalent expression of wild-type and mutant proteins was confirmed by immunoblotting.

mutant subunits are expressed in 293 cells (11). We first establish, using immunoblots and immunoprecipitation of [^{35}S]methionine-labeled protein, that normal and mutant forms of α_z are expressed at similar levels. We next examine phosphorylation. To do so, the concentration of serum in the medium is dropped from 10 to 1% 24 hr after transfection to diminish "basal" phosphorylation (i.e., that probably promoted by factors stimulating PKC). Cells are incubated 24 hr thereafter in phosphate-free medium containing 50 μCi/ml ortho[^{32}P]phosphate for 2 hr at 37°C. Vehicle (ethanol) or PMA (100 nM) is subsequently added for 10 min. Membranes are prepared, and immunoprecipitation is achieved using the C-terminus-directed antibody 2919. Immunoprecipitates are analyzed by autoradiography following SDS-PAGE.

As shown in Fig. 3, normal α_z was phosphorylated in response to incubation of the cells with 100 nM PMA. The increase (fold) relative to vehicle was comparable to that observed in platelets. Also, in experiments not shown, phosphorylation eliminated reactivity with the phosphorylation-sensitive antibody 6354. As demonstrated for platelets and for purified proteins in solution, the subtypes of α_i present in 293 cells were not phosphorylated. Thus, the pattern of α_z phosphorylation observed in 293 cells parallels that established in platelets and *in vitro*.

Importantly, the Ala-25 mutant was phosphorylated to an extent comparable to the normal subunit. Phosphorylation in the case of the Ala-27 mutant, however, was considerably diminished. As determined with an AMBIS radioanalytical imaging system in multiple experiments, the Ala-27 mutant incorporated 35% of the normal amount of radiolabel. The simplest explanation of these results is that Ser-27 is the preferred site of phosphorylation. Notably, the inhibition of phosphorylation was not complete, suggesting that a small amount of phosphorylation was occurring elsewhere in the Ala-27 mutant, probably at Ser-16 (11).

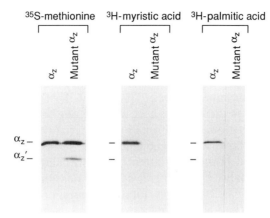

FIG. 4 Expression, myristoylation, and palmitoylation of normal and Ala-2 α_z in 293 cells. The 293 cells were transfected with pDP5 encoding normal α_z or the Gly-2 → Ala-2 mutant. Cells were incubated with [^{35}S]methionine, [^{3}H]myristic acid, and [^{3}H]palmitic acid, as indicated, and the subunits were subsequently immunoprecipitated with antiserum 2919. Immunoprecipitated proteins were resolved by SDS-PAGE and analyzed for incorporated radiolabel by autoradiography (^{35}S) or fluorography (^{3}H). Times of exposure were 4 hr ([^{35}S]methionine), 2 days ([^{3}H]myristate), and 20 days ([^{3}H]palmitic acid). α_z refers to the position of the full-length subunit (M_r 41,000); α_z' refers to the protein probably resulting from secondary initiation of the mutant. Adapted from Ref. 8, with permission.

Fatty Acid Acylation

The importance of N-myristoylation in the anchorage of α subunits has been documented for α_{i1} and α_{o1}, wherein substitution of alanine for the N-terminal glycine prevents not only myristoylation, but attachment of the subunits to membrane (6, 18). We wished to demonstrate a similar role for the myristoylation of α_z, and to determine whether N-myristoylation, as a modification tightly linked to subunit synthesis, had any impact on palmitoylation, a posttranslational modification.

The 293 cells were transfected with pDP5 encoding normal α_z or the Gly-2 → Ala-2 mutant (8). As demonstrated by incorporation of [^{35}S]methionine (Fig. 4, left panel), normal and mutant α_z were expressed to a similar extent. A lower molecular weight protein (referred to here as α_z') was also evident in cells expressing the mutant, and is probably the product of secondary initiation at Met-53 due to manipulation of the 5' untranslated sequence.

The data shown in the middle panel demonstrate that only the normal subunit incorporated [^{3}H]myristic acid. N-Myristoylation was analyzed here by immunoprecipitation, but following incubation of the 293 cells with [^{3}H]myristic acid (NEN, Boston, MA; 0.3 mCi/ml) beginning 18 hr before harvesting. In separate experiments, we demonstrated that >90% of normal α_z sediments with membrane, whereas 70–90% of the Gly-2 → Ala-2 mutant does not. Thus, as is the case with α_{i1} and α_{o1}, the N-terminal glycine is essential for myristoylation, and abrogation of myristoylation prevents subunit anchorage. As shown in the right panel of Fig. 4, the Gly-2 → Ala-2 mutation not only prevented N-myristoylation, but also palmitoylation. Because palmitoylation probably occurs at Cys-3, two possibilities exist for the effect of the Gly-2 → Ala-2 mutation. First, the mutation may simply disrupt recognition of α_z by a palmitoyltransferase. Second, N-myristoylation may be a prerequisite to palmitoylation. N-Myristoylation, for example, may allow presentation of α_z to a membrane-bound palmitoyltransferase by virtue of conferring hydrophobicity to the subunit.

As a cautionary note, we found that α_z', though lacking sites for fatty acid acylation, sedimented with membrane. Of the full-length, unacylated α_z mutant, moreover, 10–30% also sedimented with membrane. We suspect that denaturation can result in nonspecific adsorption. In the case of α_z', which lacks a perhaps stabilizing G-1 GTP-binding/hydrolysis domain, the denaturation would be complete. Overexpression of full-length subunit may normally be associated with a small amount of denaturation.

Other Approaches and Considerations

We have discussed here the analysis of covalent modifications in α_z by site-directed mutagenesis achieved with PCR and mismatched primers. This and other techniques have been used in the development of structure–activity relationships for α subunits in general. An analysis of α subunit sequences relevant to discrimination of receptors by these subunits has been performed by Conklin et al. (19), in which the C-terminal residues of α_q were systematically exchanged for those of α_{i2}, α_o, α_z, and α_s using (primarily) PCR with a 3' primer containing the entire added sequence and an appended restriction site. Previous analyses had included construction of chimeras between α_i and α_s using naturally occurring restriction sites (20), and similarly between normal α_s and the S49 unc variant (21). Regions of α subunits (and of α_s in particular) relevant to interaction of the subunits with effectors have been defined through construction of chimeras between α_s and α_i using naturally occurring restriction sites (22), PCR with mismatched primers (23), and by oligonucleotide-directed mutagenesis using a uracil-containing template (24). Regions of α subunits required for interaction with $\beta\gamma$ have been examined through PCR with mismatched primers (using *in vitro* translation as a means of expression) (25), and by construction of chimeras between α_i and α_s using naturally occurring restriction sites and/or a double-PCR strategy based on complementarity in overlapping primers (22). Residues relevant to $\beta\gamma$ interaction have additionally been examined by mutations induced randomly with PCR (26).

Of considerable utility are mutations that cause "constitutive" activation or inhibition of G protein α subunits. A frequently used activating mutation (e.g., see Ref. 27) involves substitution of leucine for glutamine in the conserved DVGGQ motif [i.e., the G-3 GTP-binding/hydrolysis domain (28)], which limits the ability of an α subunit to hydrolyze GTP and consequently maintains the subunit in an active conformation. The mutation has been used for virtually all α subunits in paradigms of transfection, typically in the context of transformation or analysis of effector stimulation or inhibition. In some instances, an appropriate agonist may be required to achieve the initial exchange of GTP for GDP. In contrast to the activating mutation, substitution of alanine for glycine at the adjacent residue within the DVGGQ motif of α_s prevents agonist activation of adenylylcyclase (29). This glycine is felt to serve as a pivot point in conformational changes induced by GTP, based on X-ray crystallographic data for elongation factor EF-Tu and Ras. The other glycine (DVGGQ) may serve a similar role or otherwise participate in GDP/GTP exchange phenomena. Osawa and Johnson (30), for example, have demonstrated that substitution of threonine for the latter in α_s (Gly-225 $\rightarrow$ Thr-225) serves as a dominant negative mutation, i.e., on overexpression in COS cells, the mutant subunit prevents agonist activation of adenylylcyclase. A Gly-203 $\rightarrow$ Thr-203 mutation in α_{i2} (analogous to the α_s mutation) suppresses activation of phospholipase A_2 (PLA_2) by thrombin and ATP (31).

Another useful modification is epitope tagging, which has been accomplished successfully for α_s, wherein a hemagglutinin sequence has been substituted for part of the insert differentiating α_{s-L} from α_{s-S} (24), and for α_q, in which the sequence has been placed either at the N terminus (32) or at residues 125–130 (33). A hexahistidine sequence has been added to the N terminus of α_s to facilitate purification from Sf9 cells using a nickel-agarose affinity resin (7).

As a final comment, expression of proteins following transfection may be associated with a certain level of improper folding or denaturation. As discussed above, we suspect that the small amount of Gly-2 $\rightarrow$ Ala-2 α_z mutant cosedimenting with membrane is denatured and therefore nonspecifically adsorbed. Where the target enzyme or ion channel of the

α subunit is known, competency of the subunit can be determined simply through assays of second messenger regulation using agonist or activators such as GTPγS. Another useful technique involves GTPγS stabilization of the subunit to proteolysis. GTPγS induces a conformational change in α subunits that protects these subunits from cleavage with trypsin at all points but near the N terminus (34). Thus the ability of a mutant to fold properly and assume an activating conformation can be assessed for expressed proteins regardless of effector interactions by the stability conferred by GTPγS (7, 24, 33).

Acknowledgments

The authors acknowledge the support of NIH Grants GM51196, HL45181, and GM53156 for studies involving mutational analysis and the development of antibodies. We are also grateful to Drs. Lawrence Brass and Mortimer Poncz for their many contributions and advice. HH is supported by a Postdoctoral Training Grant MH14654. DRM is an Established Investigator of the American Heart Association.

References

1. M. Matsuoka, H. Itoh, T. Kozasa, and Y. Kaziro, *Proc. Natl. Acad. Sci. U.S.A.* **85,** 5384 (1988).
2. H. K. W. Fong, K. K. Yoshimoto, P. Eversole-Cire, and M. I. Simon, *Proc. Natl. Acad. Sci. U.S.A.* **85,** 3066 (1988).
3. P. J. Casey, H. K. W. Fong, M. I. Simon, and A. G. Gilman, *J. Biol. Chem.* **265,** 2383 (1990).
4. K. E. Carlson, L. F. Brass, and D. R. Manning, *J. Biol. Chem.* **264,** 13298 (1989).
5. K. M. Lounsbury, P. J. Casey, L. F. Brass, and D. R. Manning, *J. Biol. Chem.* **266,** 22051 (1991).
6. S. M. Mumby, R. O. Heukeroth, J. I. Gordon, and A. G. Gilman, *Proc. Natl. Acad. Sci. U.S.A.* **87,** 728 (1990).
7. M. E. Linder, P. Middleton, J. R. Hepler, R. Taussig, A. G. Gilman, and S. M. Mumby, *Proc. Natl. Acad. Sci. U.S.A.* **90,** 3675 (1993).
8. H. Hallak, L. F. Brass, and D. R. Manning, *J. Biol. Chem.* **269,** 4571 (1994).
9. R. Higuchi, *in* "PCR Protocols" (M. A. Innis, D. H. Gelfand, J. J. Sninsky, and T. J. White, eds.), p. 177. Academic Press, San Diego, 1990.
10. J. Sambrook, E. F. Fritsch, and T. Maniatis, "Molecular Cloning." Cold Spring Harbor Laboratory Press, Cold Spring Harbor, New York, 1989.
11. K. M. Lounsbury, B. Schlegel, M. Poncz, L. F. Brass, and D. R. Manning, *J. Biol. Chem.* **268,** 3494 (1993).
12. F. G. Szele and D. B. Pritchett, *Mol. Pharmacol.* **43,** 915 (1993).
13. F. M. Ausubel, R. Brent, R. E. Kingston, D. D. Moore, J. G. Seidman,

J. A. Smith, and K. Struhl, "Current Protocols in Molecular Biology." Current Protocols, Somerset, New Jersey, 1993.

14. K. R. Tindall and T. A. Kunkel, *Biochemistry* **27**, 6000 (1988).
15. F. L. Graham, J. Smiley, W. C. Russell, and R. Nairn, *J. Gen. Virol.* **36**, 59 (1977).
16. C. Chen and H. Okayama, *Mol. Cell. Biol.* **7**, 2745 (1987).
17. A. G. Williams, M. J. Woolkalis, M. Poncz, D. R. Manning, A. M. Gewirtz, and L. F. Brass, *Blood* **76**, 721 (1990).
18. T. L. Z. Jones, W. F. Simonds, J. J. Merendino, M. R. Brann, and A. M. Spiegel, *Proc. Natl. Acad. Sci. U.S.A.* **87**, 568 (1990).
19. B. R. Conklin, Z. Farfel, K. D. Lustig, D. Julius, and H. R. Bourne, *Nature (London)* **363**, 274 (1993).
20. S. B. Masters, K. A. Sullivan, R. T. Miller, B. Beiderman, N. G. Lopez, J. Ramachandran, and H. R. Bourne, *Science* **241**, 448 (1988).
21. K. A. Sullivan, R. T. Miller, S. B. Masters, B. Beiderman, W. Heideman, and H. R. Bourne, *Nature (London)* **330**, 758 (1987).
22. S. Osawa, N. Dhanasekaran, C. W. Woon, and G. L. Johnson, *Cell* **63**, 697 (1990).
23. H. Itoh and A. G. Gilman, *J. Biol. Chem.* **266**, 16226 (1991).
24. C. H. Berlot and H. R. Bourne, *Cell* **68**, 911 (1992).
25. B. M. Denker, E. J. Neer, and C. J. Schmidt, *J. Biol. Chem.* **267**, 6272 (1992).
26. V. Z. Slepak, T. M. Wilkie, and M. I. Simon, *J. Biol. Chem.* **268**, 1414 (1993).
27. Y. H. Wong, B. R. Conklin, and H. R. Bourne, *Science* **255**, 339 (1992).
28. H. R. Bourne, D. A. Sanders, and F. McCormick, *Nature (London)* **349**, 117 (1991).
29. R. T. Miller, S. B. Masters, K. A. Sullivan, B. Beiderman, and H. R. Bourne, *Nature (London)* **334**, 712 (1988).
30. S. Osawa and G. L. Johnson, *J. Biol. Chem.* **266**, 4673 (1991).
31. S. Winitz, S. K. Gupta, N.-X. Qian, L. E. Heasley, R. A. Nemenoff, and G. L. Johnson, *J. Biol. Chem.* **269**, 1889 (1994).
32. N.-X. Qian, S. Winitz, and G. L. Johnson, *Proc. Natl. Acad. Sci. U.S.A.* **90**, 4077 (1993).
33. P. B. Wedegaertner, D. H. Chu, P. T. Wilson, M. J. Levis, and H. R. Bourne, *J. Biol. Chem.* **268**, 25001 (1993).
34. B. K.-K. Fung and C. R. Nash, *J. Biol. Chem.* **258**, 10503 (1983).

[4]   G Protein Activity in Solubilized Membrane Preparations

Irene Litosch* and Hong Xie

Introduction

Phosphoinositide-specific phospholipase C (PLC) plays a key role in the transmembrane signaling system that regulates intracellular Ca^{2+} levels (1). Activation of PLC results in the rapid hydrolysis of phosphatidylinositol 4,5-bisphosphate (PIP_2) and generation of inositol 1,4,5-trisphosphate and diacylglycerol. Inositol 1,4,5-trisphosphate (IP_3) triggers the release of Ca^{2+} from intracellular stores. Diacylglycerol activates protein kinase C. Three major PLC isozymes have been purified and cloned from mammalian tissue. These have been designated as PLC-β, PLC-γ, and PLC-δ based on a comparison of their deduced amino acid sequences and immunological cross-reactivity (2). Receptor regulation of PLC-β and PLC-γ has been demonstrated *in vivo* and *in vitro*. PLC-γ is stimulated by growth factor receptors via tyrosine phosphorylation (3, 4). PLC-β is regulated by heterotrimeric regulatory GTP-binding (G) protein-linked receptors. The $G_{q/11}$ family of pertussis toxin-insensitive G proteins functions in this signaling mechanism. Both $G_{q\alpha}$ and $\beta\gamma$ subunits contribute to the regulation of PLC-β (5–8).

Many other aspects of PLC regulation by G proteins remain to be elucidated. These include the mechanism by which G proteins mediate inhibition of PLC activity (9–12), regulation of PLC-γ by G-protein-linked receptors such as the angiotensin II receptor (13), and the role of G_i in regulating PLC-γ activity (14). Furthermore, the regulation of PLC activity by kinases and the potential involvement of G proteins in this mechanism is not well understood. Cyclic AMP-dependent protein kinase A phosphorylates PLC-γ *in vivo* and *in vitro* (15). Increases in cyclic AMP levels inhibit PLC activity in some cell types, but whether cyclic AMP-dependent protein kinase A affects PLC-γ activity directly is not known (15). Similarly, PLC-β is phosphorylated by protein kinase C (16). Protein kinase C attenuates receptor stimulation of PLC activity *in vivo*. Direct effects of protein kinase C on PLC-β activity have not been demonstrated (16).

Elucidation of these regulatory mechanisms requires a cell-free approach

---

* To whom correspondence should be addressed.

Methods in Neurosciences, Volume 29

that minimizes the potential contribution of indirect cellular compensatory mechanisms. A solubilized membrane system has been used in this laboratory to study G protein regulation of PLC activity (11, 17). A nondenaturing detergent, such as sodium cholate, is used to extract membrane proteins into solution. Peripheral membrane proteins as well as trapped cytosolic proteins are also released into solution. If successful, the detergent-extracted membrane proteins retain functional activity and can be used to study the interaction between PLC and G proteins in a detergent solution. Solubilization is often the first step in a purification protocol. The biochemistry of the system prior to final purification can provide insight as to whether the properties of the system have been affected by the purification scheme as might occur through the loss of a regulatory component.

Using protocols similar to those originally developed for the solubilization of G proteins linked to adenylylcyclase (18–22), studies in this laboratory have shown that similar approaches result in the solubilization of a G-protein-regulated PLC (11, 17). Thus, specific guanine nucleotide-dependent regulation of PLC activity is retained in the solubilized state. The following method for solubilization has been slightly modified from the original protocols (11, 17) to decrease the time needed to perform the entire procedure.

Preparation of Membranes

Membranes are isolated from calf bovine brain because bovine brain is a good source of G proteins as well as PLC. For the most part, either fresh or frozen bovine calf brains can be used. In either case, the cerebellum and major blood vessels are removed prior to homogenization. All procedures are carried out at 5°C. Fresh brains are preferable because of the greater ease of handling and less protease activation. The fresh brains are packed in ice and processed as soon as possible. The brains are chopped into small pieces and transferred to homogenization buffer consisting of 280 mM sucrose, 1 mM ethylenediaminetetraacetic acid (EDTA), 10 mM Tris-HCl (pH 7.4), 1 mM dithiothreitol (DTT), and protease inhibitor cocktail (200 μM phenylmethylsulfonyl fluoride, 2 μg/ml soybean trypsin inhibitor, 2 μg/ml leupeptin, and 2 μg/ml aprotinin). If fresh material cannot be obtained, then frozen brains such as those sold by Pel Freeze can be used instead. In this case, the stripped brains are a better choice because most of the surface blood vessels have been removed prior to freezing, and this greatly facilitates subsequent tissue preparation. The frozen brains are split into sections with a blow of a heavy hammer and then chopped into smaller pieces after partial thawing. The tissue is transferred into cold homogenization buffer containing protease inhibitors.

After this step, the procedure is the same for both fresh and frozen brains. The minced tissue from one bovine brain is subjected to four washes with 1 liter of the ice-cold homogenization buffer. This wash step removes red cells and debris from the brain tissue. The wash step is best accomplished by filtering the tissue through two to four layers of cheesecloth. The buffer is discarded and the tissue is resuspended into fresh homogenization buffer. The tissue is subsequently homogenized with a Brinkmann Polytron (setting 1–5 for 5 min) (Brinkmann Instruments, Westbury, NY) in a cold room in 4 liters of ice-cold homogenization buffer. The homogenate is centrifuged for 30 min at 17,000 g. The supernatant is discarded. The membranes (from one brain) are resuspended in 1 liter of homogenization buffer with a Dounce homogenizer using one to two passes with pestle B. This is followed by another centrifugation for 30 min at 17,000 g. The membranes are rehomogenized and subjected to an additional centrifugation. This procedure is repeated three times and results in the removal of a substantial amount of cytosolic proteins. The membranes are finally resuspended to a protein concentration of 10–20 mg/ml in homogenization buffer containing 25% (v/v) glycerol and frozen in a dry ice–ethyl alcohol bath. The frozen membranes are stored at −80°C until use. In initial studies, the membranes were further purified by centrifugation on a sucrose gradient as described (23). However, this step is now omitted. Omitting the sucrose gradient step improves the yield of membranes for solubilization as well as decreases the time needed for membrane preparation.

Membrane Solubilization

Brain membranes contain an enormous amount of PLC activity. The hydrolytic capacity of PLC in crude membranes is so great that it may rapidly hydrolyze all the substrate in an unregulated manner. Because the regulatory capacity of G proteins *in vitro* may be low compared to the catalytic activity of PLC, a reduction in the amount of catalytic PLC should facilitate the study of regulation. This is accomplished by extracting the membranes with low salt and detergent buffer before solubilization to remove adsorbed cytosolic PLC activity.

Membranes are thawed and resuspended by vortexing in four volumes of buffer consisting of 0.1% (w/v) sodium cholate, 50 mM NaCl, 1 mM EDTA, 10 mM Tris-HCl (pH 8.0), 1 mM DTT, and protease inhibitor cocktail. In general, 4 ml of membranes at a protein concentration of 10 mg/ml is added to 16 ml of buffer. The resuspended membranes are maintained on ice for 15 min followed by centrifugation at 32,000 g for 30 min. The supernatant is discarded. The membranes are resuspended in the same buffer and this

TABLE I   Content of PLC-β_1 in Membranes and Cholate Extract[a]

Sample	Protein (μg)	PLC-β_1	PLC-γ_1
Membranes	25	++++	—
First wash	25	++++	—
Second wash	15	++++	—
Washed membranes	25	None detected	—
Solubilized extract	88	None detected	—
	150	+	None detected

[a] Three ml of bovine brain membranes (10 mg/ml) were preextracted twice with 3 ml of preextraction buffer containing 50 mM NaCl and 0.1% cholate as described (11). The respective washes were concentrated. The indicated amount of protein was separated on 7.5% SDS-PAGE followed by immunoblotting with pooled monoclonal antibodies to PLC-β_1 or polyclonal antibodies to PLC-γ_1. The intensity of the alkaline phosphatase stain in the starting membrane preparation is indicated by ++++. The amount of PLC-β_1 in the other fractions is expressed relative to that of the membranes. The amount of protein loaded per lane is indicated. Total protein amounts in the samples are as follows: membranes (30 mg), combined wash (12 mg), and washed membranes (18 mg). Results are representative of two experiments.

step is repeated twice. As shown in Table I, this protocol results in the removal of approximately 40% of the membrane protein as well as a major fraction of membrane-associated PLC-β activity. Some PLC-β activity is nonetheless retained on the membranes and it now appears that this activity corresponds to the regulated activity. PLC-γ is not detected in the preparation.

After preextraction, the membranes are brought to a final protein concentration of 10 mg/ml in preextraction buffer and sufficient sodium cholate is added to stirred membranes to bring the final cholate concentration to 1%. Ethylene glycol is also added to a final concentration of 0.1%. Ethylene glycol has been reported to stabilize G proteins (24). Membranes are stirred vigorously for 1.5 to 2 hr at 5°C. At the end of the solubilization, the ethylene glycol concentration is increased to 10%. The mixture is centrifuged at 100,000 g for 90 min. The supernatant containing solubilized PLC and G proteins is collected. In some cases, the resultant supernatant may be turbid due to the presence of lipid and lipid–detergent micelles. The supernatant may be clarified by chromatography on a 1-ml DEAE anion-exchange column equilibrated in column buffer consisting of 1% cholate, 1 mM EDTA, 10 mM Tris-HCl (pH 8.0), 10% ethylene glycol, 1 mM DTT, and protease inhibitor cocktail. The resin is poured into a siliconized glass Pasteur pipette with a glass wool plug or into a small plastic column. Prior to application, the solubilized extract is diluted threefold into column buffer. This dilution reduces the salt concentration in the extract, which might interfere with the binding of protein to the DEAE resin. The diluted sample is next applied to a 1-ml DEAE column equilibrated in column buffer. The unbound material that elutes from the column may be turbid and is discarded. The resin is washed with 2 ml of column buffer to remove unbound protein. The bound activity, containing PLC as well as G proteins, is eluted with 2 ml of column buffer containing 250 mM NaCl. This step accomplishes two goals. The extract is clarified and the sample is concentrated. The solubilized preparation is stored frozen in 100-μl aliquots at $-80°$C. The entire procedure is done in 1 day.

Measurement of Phospholipase C Activity

The solubilized preparation is usually assayed immediately for catalytic and regulatory activities. Stimulatory G protein regulation of PLC activity is unstable and decays with time. Loss in G protein stimulatory regulation of PLC on solubilization has also been noted by others (25). PLC is also regulated by a G-protein-mediated inhibitory mechanism. The inhibitory regulation is slightly more stable, although it also deteriorates with storage.

PLC catalyzes the hydrolysis of phosphatidylinositol 4,5-bisphosphate to diacylglycerol and inositol 1,4,5-trisphosphate. Radioactive PIP$_2$ ([^{3}H]PIP$_2$) is used to monitor the enzymatic activity. Activity is determined by measuring the production of labeled water-soluble inositol phosphates from the hydrolysis of PIP$_2$. The enzymatic rate is increased by Ca^{2+} and thus a Ca^{2+}–EGTA buffer system is used to set the Ca^{2+} concentration. In addition, receptors and G proteins regulate the activity of PLC-β. Both stimulatory and inhibitory G protein regulation of PLC activity has been described. Hydrolysis-resistant guanine nucleotides such as guanosine 5-O-(3-thiotriphosphate) (GTPγS) and guanylyl imidodiphosphate [Gpp(NH)p] are used to activate G proteins. Mg^{2+} is required for G protein regulation and is included in the assay. LiCl is added to inhibit phosphatases that degrade inositol phosphates to inositol.

The PIP$_2$ substrate is prepared by drying both the labeled and unlabeled PIP$_2$ under a stream of N$_2$. Final concentration of PIP$_2$ in the assay tube is 3.5–10 μM. Deoxycholate is added to a final concentration of 0.02% (w/v). The mixture is sonicated on ice in 200 μl of buffer containing 50 mM N-(2-hydroxyethyl)piperazine-N'-(2-ethanesulfonic acid) (HEPES; pH 6.75) and 25 mM LiCl. The sonicated mixture is adjusted to final volume with the same buffer. An aliquot (5–10 μl) of the substrate is removed and the radioactivity of the substrate is checked by liquid scintillation counting.

Phospholipase C activity is generally assayed at 350 nM Ca^{2+} in the absence or presence of 100 μM Gpp(NH)p or GTPγS. Generally three to four time points are used to establish the assay conditions that generate a linear enzymatic rate, i.e., 5, 10, 15, and 30 min. G protein regulation is evidenced by a guanine nucleotide-dependent change in PLC activity. A serial dilution of the solubilized preparation is done in buffer containing 25 mM HEPES (pH 6.75), 1 mM EDTA, 10% ethylene glycol, and 1 mM DTT. A concentration of solubilized extract is selected that demonstrates a linear enzymatic activity under basal and G-protein-stimulated conditions. From this information, an appropriate incubation time is selected for the proposed experiments. Whenever possible, this initial characterization is immediately followed by experiments to investigate the regulation of PLC. G protein regulation of PLC may diminish the next day. Loss in guanine nucleotide-dependent regulation is not accompanied by any detectable change in PLC activity, suggesting that the loss in regulation is due to an effect on the G protein rather than on the PLC. Some preparations may demonstrate G protein regulation of PLC activity for at least 1 week. The regulatory capacity of the solubilized extract is also affected by the age of the membranes (11).

G protein regulation of PLC activity is measured as follows. A volume of 5 μl of appropriately diluted solubilized preparation is added to buffer containing (final concentrations) 25 mM HEPES (pH 6.75), Mg^{2+} as indicated, Ca^{2+} as indicated, 12 mM LiCl, and 3.5–10 μM PIP_2 (~100–150 cpm/pmol) in a final volume of 50 μl. Total radioactivity per tube is approximately 30,000 cpm. ATP and adenylyl imidodiphosphate [App(NH)p; final concentration of 0.1 mM] are included to prevent degradation of guanine nucleotides by nonspecific nucleotidases. Except where indicated, the free Ca^{2+} concentration is maintained at 350 nM and set by a Ca^{2+}–EGTA buffer using 3 mM ethylene glycol bis(β-aminoethyl ether)N,N,N',N'-tetraacetic acid (EGTA). The free Ca^{2+} concentration is determined from the total concentration by solving equilibrium binding equations (26) using published stability constants for EDTA, EGTA, and adenosine 5-triphosphate (ATP) (27). Incubation is conducted in plastic 10 × 75-mm tubes that have been coated with bovine serum albumin (fraction V) to minimize nonspecific adsorption of PLC or substrate to the plastic tubes. Plastic tubes are rinsed successively with a solution of 0.01% (w/v) bovine serum albumin once, 1 M NaCl once, and water several times, then air dried. Incubation is initiated by the rapid transfer of tubes containing experimental components from an ice bath to an incubator. Incubation is conducted at 24° or 30°C. The incubation is terminated by the transfer of tubes to an ice bath followed by the addition of acidified 1.25 ml methanol : chloroform (2 : 1, v : v), 0.5 ml chloroform, and 0.5 ml H_2O. Inositol phosphates are isolated by chromatography on Dowex formate resin (17). Radioactivity is determined by liquid scintillation counting. The incubation can also be terminated by the addition of 150 μl of ice-cold trichloroacetic acid and 75 μl of 10 mg/ml bovine serum albumin. The tubes are centrifuged and an aliquot of the supernatant is removed for analysis by liquid scintillation counting (5). Each experiment contains a set of background tubes that contain all experimental components but have not been incubated. This background value indicates the amount of water-soluble radioactivity present in the substrate. This value is subtracted from all experimental points. Background values are generally less than 0.2% of the total radioactivity present in the tube. The data are analyzed for statistical significance using the Student's *t*-test for paired analysis.

One difficulty with this assay is the establishment and maintenance of linear kinetics. This is due, in part, to the lack of a substrate-regenerating system. Thus the levels of substrate are constantly decreasing. A decrease in substrate levels to values below the K_m of the enzyme for substrate will result in a decrease in enzymatic activity. This can lead to nonlinear kinetics. A departure from linear kinetics can occur when 3% of the substrate is hydrolyzed.

The ability of both $G_{q\alpha}$ and $\beta\gamma$ subunits to stimulate PLC-β activity indicates that either or both subunits may be responsible for the stimulatory regulation observed. The components involved in mediating G protein inhibitory regulation of PLC-β activity have not been identified at this time. Attempts to identify the relevant components by immunoprecipitation have not been successful. This is likely due to the relatively poor affinity of antibodies in detergent solutions.

In summary, this solubilized preparation provides the opportunity to study the interaction of PLC-β with regulatory G proteins in a detergent solution prior to purification. Furthermore, this system has the potential to be used in reconstitution approaches for the study of interactions between receptors, G proteins, and PLC.

Acknowledgments

Research supported by Grant DK37007 from the National Institutes of Health and a grant-in-aid from the American Heart Association.

References

1. R. H. Michell, *Biochim. Biophys. Acta.* **415,** 81 (1975).
2. S. G. Rhee and K. D. Choi, *J. Biol. Chem.* **267,** 12393 (1992).
3. S. Nishibe, M. I. Wahl, S. Hernandez-Sotomayor, N. K. Tonks, S. G. Rhee, and G. Carpenter, *Science* **250,** 1253 (1990).
4. M. I. Wahl, G. A. Jones, S. Nishibe, S. G. Rhee, and G. Carpenter, *J. Biol. Chem.* **267,** 10447 (1992).

5. S. J. Taylor, J. A. Smith, and J. H. Exton, *J. Biol. Chem.* **265**, 17150 (1990).

6. A. V. Smrcka, J. R. Helper, K. O. Brown, and P. C. Sternweis, *Science* **251**, 804 (1991).

7. D. Park, D. Y. Jhon, C. W. Lee, K. H. Lee, and S. G. Rhee, *J. Biol. Chem.* **268**, 4573 (1993).

8. A. V. Smrcka and P. C. Sternweis, *J. Biol. Chem.* **268**, 9667 (1993).

9. I. Litosch, *Biochem. J.* **261**, 245 (1989).

10. C. Bizzarri, M. Di Girolamo, M. C. D'Orazio, and D. Corda, *Proc. Natl. Acad. Sci. U.S.A.* **87**, 4889 (1992).

11. I. Litosch, I. Sulkholutskaya, and C. Weng, *J. Biol. Chem.* **268**, 8692 (1993).

12. Y. Cherifi, C. Pigeon, M. L. Romancer, A. Bado, F. Reyl-Desmars, and M. J. M. Lewin, *J. Biol. Chem.* **267**, 25315 (1990).

13. M. B. Marrero, W. G. Paxton, J. L. Duff, B. C. Berk, and K. E. Bernstein, *J. Biol. Chem.* **269**, 10935 (1994).

14. L. J. Yang, S. G. Rhee, and J. R. Williamson, *J. Biol. Chem.* **269**, 7156 (1994).

15. U.-H. Kim, J. W. Kim, and S. G. Rhee, *J. Biol. Chem.* **264**, 20167 (1989).

16. S.-H. Ryu, U.-H. Kim, M. I. Wahl, A. B. Brown, G. Carpenter, K.-P. Huang, and S. G. Rhee, *J. Biol. Chem.* **265**, 17941 (1990).

17. I. Litosch, *J. Biol. Chem.* **266**, 4764 (1991).

18. T. Pfeuffer, *J. Biol. Chem.* **252**, 7224 (1977).

19. E. M. Ross, A. C. Howlett, K. M. Ferguson, and A. G. Gilman, *J. Biol. Chem.* **253**, 6401 (1978).

20. J. K. Northup, P. C. Sternweis, M. D. Smigel, L. S. Schleifer, E. M. Ross, and A. G. Gilman, *Proc. Natl. Acad. Sci. U.S.A.* **77**, 6516 (1980).

21. P. C. Sternweis, J. K. Northup, M. D. Smigel, and A. G. Gilman, *J. Biol. Chem.* **256**, 11517 (1981).

22. E. Perez-Reyes and D. M. F. Cooper, *J. Neurochem.* **46**, 1508 (1986).

23. C. W. Cotman, *in* "Methods in Enzymology," (S. Fleischer and L. Packer, eds.), Vol. XXXI, p. 445. Academic Press, New York, 1974.

24. J. Codina, J. D. Hildebrandt, R. D. Sekura, M. Birnbaumer, J. Bryan, C. R. Manclark, R. Iyengar, and L. Birnbaumer, *J. Biol. Chem.* **259**, 5871 (1984).

25. G. Berstein, J. L. Blank, A. V. Smrcka, T. Higashijima, P. C. Sternweis, J. H. Exton, and E. M. Ross, *J. Biol. Chem.* **267**, 8081 (1992).

26. D. D. Perrin and I. G. Sayce, *Talanta* **14**, 833 (1967).

27. R. M. Smith and A. E. Martell, *in* "Critical Stability Constants," (R. M. Smith and A. E. Martell, eds.), Vol. 1, p. 1. Plenum, New York, 1974.

[5]   Cross-Linking of Synaptoneurosome G Proteins

Sherry Coulter, Saleem Jahangeer, and Martin Rodbell

Introduction

Heterotrimeric GTP-binding proteins (G proteins) are essential transducing elements that convey information between surface receptors and a variety of signal-producing elements (adenylylcyclase, ion transporters, phospholipases, phosphodiesterases, etc). Composed of GTP-binding subunits (α proteins) and a tightly linked couplet ($\beta\gamma$), G proteins are coupled to receptors on the inner face of the plasma membrane. Target size (irradiation) analysis (1) coupled with cross-linking studies (2) suggests they may be multimeric proteins, similar in structure to cytoskeletal elements such as actin and tubulin (3).

One means of probing the organizational structure of G proteins in their native membrane environment is to examine the structures obtained after treatment of the membranes with cross-linking agents. The greater the stringency of the cross-linking reactions, the more credible the interpretations of the cross-linked structures. For this purpose a sulfhydryl-based monofunctional cross-linker, p-phenylenedimaleimide (p-PDM) is appropriate both because of the short phenyl group as spacer between the maleimide or sulfhydryl reactive groups and because it is known that G-proteins contain multiple sulfhydryl residues for potential cross-linking. Synaptoneurosome membranes are employed because they contain essentially all of the known species of G-proteins and are a relatively specific type of membrane from the brain cortex.

Procedures

Materials

Synaptoneurosome preparation buffer: 118.5 mM NaCl, 4.7 mM KCl, 1.18 mM $MgSO_4$, 1.0 mM $CaCl_2$, 10 mM glucose, 20 mM HEPES, 9 mM Tris, pH 7.4.

Phosphate-buffered saline (PBS): 8.1 mM sodium phosphate, 1.5 mM potassium phosphate, 137 mM NaCl, 3 mM KCl, pH 7.1.

Methods in Neurosciences, Volume 29

p-Phenylenedimaleimide (p-PDM, Aldrich, Milwaukee, WI): 2.5 mM
stock solution in dimethylformamide (DMF).

Column elution buffer: 10 mM Tris-HCl, pH 7.4, 150 mM NaCl, 0.1%
(w/v) sodium dodecyl sulfate (SDS)

Digitonin: Stock solution (5%, w/v) is prepared by boiling followed by
cooling to room temperature for 2–3 days and filtration through a
0.45-μm filter to remove insoluble material; store solution at room
temperature. Dilute solutions just before use.

Antisera: Sources of antisera that react selectively with $G_{\alpha s}$, $G_{\alpha i}$, $G_{\alpha o}$,
and the β subunits (β_1, β_2) of G proteins have been described (4).

Preparation of Rat Synaptoneurosome Membranes

Rats are decapitated (5), their brains removed, and cortical material isolated
and placed in synaptoneurosome preparation buffer. The tissue is blotted,
divided into 1-g portions, and cut into small pieces. The tissue is then homogenized in 7 ml of cold buffer using five strokes of a Duall (Kontes, Vineland,
NJ) homogenizer. The homogenate is diluted with an additional 30 ml of
buffer and filtered through three layers of 160-μm nylon mesh (Tetko, Inc.,
Glinsford, NY), then divided into two portions and filtered through a 10-μm
Mitex filter (Millipore, Bedford, MA, type LC). The filtrate is centrifuged
for 15 min at 4°C at 1000 g. The supernatant is discarded and the pellet
resuspended gently in a small amount of buffer with a Teflon homogenizer.
The homogenate is again diluted with 30 ml of cold buffer and centrifuged
for 15 min at 4°C at 1000 g. The supernatant is discarded and the pellet
resuspended in PBS in a volume (in ml) twice the original tissue weight (in
grams). Protein is measured by the method of Lowry and membranes are
aliquoted and stored in liquid nitrogen.

Cross-Linking Conditions

It is essential to employ cross-linking reagents that do not cause extensive,
generalized cross-linking of membrane proteins. p-PDM is a homobifunc-
tional sulfhydryl linking agent with a relatively short (12–14 Å) linker or
spacer region between the two maleimide reactive groups. With its high
specificity for SH groups and short spacer, p-PDM does not produce significant cross-linking of the major proteins (other than actin and tubulin) detectable by Coomassie blue staining of synaptoneurosome proteins. Optimal
conditions for cross-linking synaptoneurosome G proteins are determined
by treating the membranes with p-PDM, and then extracting with Laemmli
buffer after stopping cross-linking by the addition of 2-mercaptoethanol (final

concentration, 7 mM). After boiling and electrophoresis on 10% SDS-PAGE gels, the extents to which the immunoreactive bands shift to higher apparent molecular weights on SDS-PAGE gels relative to untreated membranes provide estimates of cross-linking efficacy. Concentrations of p-PDM in the range of 50 to 100 μM (in 10% (v/v) DMF) give the most efficacious and selective cross-linking when incubated with 1 mg/ml of membrane protein in PBS for 30 min on ice; controls are treated with 10% DMF.

The time and temperature of incubation with p-PDM necessary for optimal cross-linking differ among the different α and β proteins. For example, α_i shows the greatest susceptibility; i.e., significant decreases in immunochemical labeling of its 41-kDa immunoreactive band occurs within 30 min when treated at 5°C or within 5 min when treated at room temperature. α_o shows cross-linking characteristics similar to α_i. α_s is less sensitive, with only minor cross-linking evident even at 60 min on ice; extensive cross-linking occurs at room temperature. The β subunits do not cross-link noticeably at 5°C and require incubations at room temperature of at least 30 min.

Separation and Sizing of Cross-Linked Material

Although SDS-PAGE electrophoretic changes in mobility are useful for determining optimal cross-linking conditions, this procedure does not provide an estimate of the sizes of the cross-linked material. Under optimal conditions, all of the G protein subunits cross-link to form structures that fail to enter beyond the stacking gel. Separation requires the use of sizing gels. Acceptable separation of a broad spectrum of sizes is obtained by stacking in tandem a Bio-Gel (Bio-Rad, Richmond, CA) A-50m column for the very large cross-linked structures with a Bio-Gel A-5m column that separates smaller structures. Both columns (1.5 × 40 cm each) are kept at room temperature during the entire chromatographic procedure to maintain solubility of SDS in the elution buffer.

In a typical experiment, 5 mg of rat synaptoneurosome membrane is diluted to 1 mg/ml in PBS, and p-PDM is added to give 75 μM. After incubation at room temperature for 60 min, 2-mercaptoethanol is added to 7 mM and the membranes are pelleted at 8000 rpm for 15 min in a Sorvall RC-5B, SM-24 rotor. The supernatant is discarded and the pellet resuspended in a small volume (usually 300 μl) of PBS. An equal volume of 10% SDS is added and the membranes are allowed to solubilize for 30 min at room temperature. Bromphenol blue is added as a tracking dye, and the material is injected onto the tandem columns. Elution is at room temperature with elution buffer at 0.1 ml/min. Protein elution is monitored at 275 nm with an OD_{max} of 0.05. Fractions (1 ml) are collected. Complete elution requires approximately 36 hr.

Aliquots (100 μl) of each fraction are spotted onto nitrocellulose using a Minifold microfiltration apparatus (Schleicher and Schuell, Keene, NH). Blots are immunostained using antibody specific to each protein and are-quantitated with a Chromoscan 3 (Joyce-Loebl Ltd, Gatshead, England) densitometer in reflectance mode or with a Microtek 600Z scanner (Microtek Lab, Inc., Redondo Beach, CA) and NIH Image software for more precise measurements.

Whereas non-cross-linked material separates into two major protein peaks centered at around fractions 130 and 180 on these columns, cross-linked materials also produce a peak centered around fraction 50. All of the G protein subunits examined appear predominantly in the peak at 130, with the α_o and α_i subunits exhibiting similar cross-linking patterns (Fig. 1). The α subunits shift to a large extent to the fraction-50 peak and an additional peak centered at around fraction 80. The β subunits also shift to the larger peak, but do not exhibit the intermediate peaks. Pooling fractions for subsequent separation by SDS-PAGE followed by electroblotting and immunostain also shows clear evidence of a variety of intermediate cross-linked species.

Because precise sizes of cross-linked material cannot be obtained from the sizing procedures, it is useful to compare the apparent sizes with known multimeric proteins such as tubulin and actin. In synaptoneurosome preparations, cross-linking of these proteins produces structures with comparable elution profiles seen for cross-linked G proteins as determined either with specific antibodies or by Coomassie blue staining (2).

Cross-Linking G Proteins in Detergent Extracts

Detergents exert varying effects on the organizational structures of G proteins (6). For example, extraction of membranes with Lubrol yields heterotrimeric structures (monomers) of G proteins whereas extraction with digitonin yields large, presumably multimeric structures. Cross-linking of Lubrol and digitonin extracts demonstrates this point.

Synaptoneurosomes (1 mg) suspended in 250 μl of PBS medium are extracted with either 50 μl of 5% (w/v) digitonin or 25 μl of 10% (v/v) Lubrol for 1 hr on ice. Layer 200 μl of total reaction mixture over a prechilled 5–20% (w/w) sucrose gradient in 20 mM HEPES/NaOH, pH 7.4 with 150 mM NaCl, 1 mM EDTA, 1 mM DTT, and either 1% Lubrol or 0.5% digitonin. After centrifugation at 50,000 rpm for 15 hr at 4°C in a Beckman SW60 rotor, about 22 fractions of 200 μl each are collected (fraction 22 contains pelleted material).

Cross-linking is carried out by the addition of 6 μl of 2.5 mM p-PDM (final, 75 μM) to each fraction. Incubations are 1 hr at room temperature with

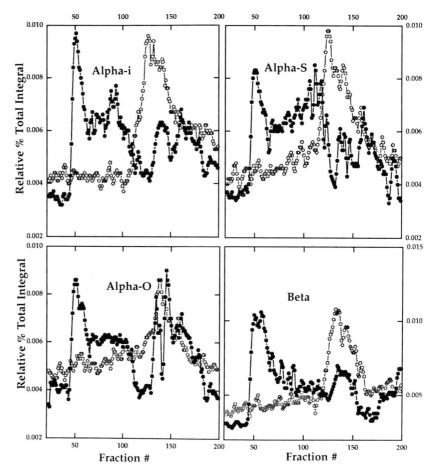

FIG. 1   Cross-linking of α and β subunits of G_s, G_i, and G_o in synaptoneurosomes. ○, Control; ●, cross-linked.

gentle shaking. Appropriate control fractions are given 6 μl of DMF alone and incubated in parallel.

After 1 hr the cross-linking reaction is terminated by the addition of 100 μl of a 3× Laemmli sample loading buffer to each fraction. These samples are boiled for 5 min and then analyzed for α and β subunits of G proteins by SDS-PAGE and Western blotting using various antisera as indicated above. As markers for relative S values ($s_{20,w}^0$), cytochrome c (2.1 S, 12.5 kDa), bovine serum albumin (BSA, 4.4 S, 68 kDa), aldolase (8 S, 158 kDa), and catalase (11.2 S, 240 kDa) are added in parallel to the same gradients used with the detergent extracts of synaptoneurosomes. Both cytochrome c and catalase also function as indicator proteins due to their colors and can be visualized directly.

In Lubrol extracts of synaptoneurosomes essentially no cross-linking is observed, i.e., the electrophoretic patterns displayed by both the various α and the β subunits are indistinguishable from the controls. On the other hand, all of the G proteins extracted with digitonin and treated with p-PDM fail to traverse the stacking gel and are not immunodetectable. Large molecular weight complexes of cross-linked proteins can, however, be visualized after Coomassie blue staining of whole gels containing sample wells and stacking gel.

References

1. W. Schlegel, E. S. Kempner, and M. Rodbell, *J. Biol. Chem.* **254,** 5168 (1979).
2. S. Coulter and M. Rodbell, *Proc. Natl. Acad. Sci. U.S.A.* **89,** 5842 (1992).
3. M. Rodbell, *Curr. Top. Cell. Regul.* **32,** 1 (1992).
4. S. Nakamura and M. Rodbell, *Proc. Natl. Acad. Sci. U.S.A.* **87,** 6413 (1990).
5. E. B. Hollingsworth, E. T. MacNeal, J. L. Burton, R. J. Williams, J. W. Daly, and C. R. Creveling, *J. Neurosci.* **5,** 2240 (1985).
6. S. Jahangeer and M. Rodbell, *Proc. Natl. Acad. Sci. U.S.A.* **90,** 8782 (1993).

[6] Quantitative Reconstitution and ADP-Ribosylation Assays for G Proteins

Joel Abramowitz

Introduction

Many transmembrane signaling systems are multicomponent in nature and are composed of receptors, effectors, and signal-transducing G proteins (1). Due to the component nature of these signal-transducing systems, responsiveness to a hormone or neurotransmitter is the result of the interactions among these components. Thus, being able to assess changes in the function of receptors, G proteins, and effectors is desirable in defining the responsiveness of a given tissue to a specific agonist. Several volumes in this series have dealt with assessing receptor (2, 3) and effector (4, 5) function. The reader is also referred to several reviews that address the structure, function, and molecular diversity of G proteins (1, 6, 7). This chapter describes methods for the functional assessment and quantitation of the heterotrimeric G proteins that regulate adenylylcyclase activity, G_s and G_i. Two types of assays will be described; one involves the reconstitution of G_s-stimulated adenylylcyclase activity in the cyc^- variant of the S49 mouse lymphoma cell line and the other is the ability of the α subunits of these G proteins to be ADP-ribosylated by cholera and pertussis toxins. We have used these assays to quantitate changes in G_s and G_i in corpora lutea after treatment of pseudopregnant rabbits with human chorionic gonadotropin (8) and epinephrine (9).

Reconstitution Assays for G_s

G proteins are responsible for conferring guanine nucleotide and Mg^{2+} sensitivity to receptors and effectors and for coupling agonist-occupied receptors to their effectors (1). The development of functional reconstitution assays for G_s by Ross and Gilman (10) allowed for the characterization of the properties of G_s and was instrumental in the purification of G_s (11, 12). We have modified these assays so that they can be used to quantitate G_s functionally (8).

Methods in Neurosciences, Volume 29

Materials

Reagents are purchased from the indicated suppliers: Inorganic ^{32}P and [^{3}H]cAMP (10–20 Ci/mmol), International Chemical and Nuclear Corp. (Irvine, CA); ATP (Catalog No. A-2383), GTP, EDTA, Tris, creatine phosphate, creatine phosphokinase, myokinase, HEPES, Norit A, (−)-isoproterenol, dithiothreitol (DTT), and cholic acid, Sigma Chemical Co. (St. Louis, MO); NaF, Fisher Scientific Co. (Fair Lawn, NJ); Sephadex G-25 PD-10 columns, Pharmacia (Piscataway, NJ).

The components of the nucleoside triphosphate-regenerating system used in the adenylylcyclase assay are subjected to purification steps to decrease contamination with "guanine nucleotide-like" compounds as follows: myokinase (1 mg/ml of ammonium sulfate precipitate, 2.5 ml) and creatine phosphokinase (10 mg/ml, 2.5 ml) are individually passed over a Sephadex G-25 PD-10 column, the void volume is collected, and the protein content of the eluate is determined by the method of Lowry et al. (13). Creatine phosphate (5 ml, 500 mM) is mixed with activated charcoal (25 mg of Norit A) for 10 min at 0–4°C. The charcoal is removed by centrifugation at 10,000 g for 30 min at 4°C, followed by filtration through Whatman No. 1 filter paper (Whatman, Clifton, NJ).

[α-^{32}P]ATP (specific activity, >50 Ci/mmol) is synthesized from inorganic ^{32}P according to the procedure of Walseth and Johnson (14) or can be purchased commercially.

Cell Cultures and Preparation of Membranes

The cyc^- variant of the S49 mouse lymphoma cell line (obtained from the University of California at San Francisco Cell Culture Center after obtaining consent from Dr. Henry R. Bourne, Department of Pharmacology, UCSF School of Medicine) are grown in Dulbecco's modified Eagle's basal medium supplemented with 10% heat-inactivated horse serum, 100 U/ml penicillin, and 75 U/ml streptomycin in an atmosphere of 5% CO_2–95% air at 37°C in suspension culture, as described by Coffino et al. (15), and cyc^- membranes are prepared according to Ross et al. (16). Briefly, cells are harvested by centrifugation at 3000 g for 15 min at 4°C and prepared for lysing by washing three times with Puck's G saline (without divalent cations). The resuspended cells are then lysed in homogenization buffer (20 mM Na-HEPES, 125 mM NaCl, 1.0 mM dithiothreitol, and 1.0 mM EDTA, pH 8.0) by nitrogen cavitation and decompression using a Parr bomb after equilibration with nitrogen at 400 psi for 20 min. The lysate is centrifuged at 1500 g for 5 min at 4°C. The resulting supernatant is subjected to centrifugation at 40,000 g for 20 min at 4°C. The supernatant is discarded and the pellet is washed once with 20 mM Na-HEPES, 1.0 mM dithiothreitol, and 1.0 mM EDTA, pH 8.0. The resulting cyc^- membranes are resuspended in the wash buffer at an approximate protein concentration of 10–15 mg/ml, aliquoted, and stored at −70°C in 100-μl fractions until used.

Cholate Extraction of Membranes

Cholic acid is dissolved in hot 100% ethanol and slowly cooled, allowing the cholic acid to recrystallize. The resulting crystals are filtered over Whatman No. 1 filter paper, washed with ice-cold double distilled water, and dried. This procedure is repeated once. Twice recrystallized cholic acid is dissolved in 10 mM Tris-HCl at a final concentration of 5% (w/v), and the final pH is adjusted to 7.5 with Tris base.

Test membranes containing the G$_s$ to be analyzed are suspended at a protein concentration of 10–15 mg/ml in 10 mM Tris-HCl, pH 8.0. To this membrane suspension enough recrystallized Tris–cholate is added to give a final concentration of 1% Tris–cholate. The mixture is left on ice for 30 min with frequent vortexing. This suspension is centrifuged at 100,000 g for 60 min at 4°C. The supernatants (cholate extracts) are transferred to clean tubes. The resulting supernatants have a protein concentration of approximately 3 mg/ml and are stored at −70°C until assayed.

Assay of G$_s$-Stimulated Adenylylcyclase Activity

The cholate extracts are incubated at 32.5°C for 20 min to inactivate endogenous adenylylcyclase activity. The heated cholate extracts are diluted fivefold in an ice-cold solution containing 0.5 mM ATP, 1.5 mM MgCl$_2$, 0.1 mM GTP, and 10 mM Tris-HCl, pH 8.0 (Solution 1). Two additional serial dilutions are made with 0.2% Tris–cholate in Solution 1, giving rise to a total of three solutions with protein concentrations ranging from approximately 150 to 600 μg/ml. It is important that these dilutions be made with 0.2% Tris–cholate and that the final detergent concentration is the same in all assay tubes, because the detergent will have an inhibitory effect on adenylylcyclase activity. The diluted cholate extracts are incubated on ice for 30 min in the absence and presence of S49 cyc^- membranes (4 mg/ml), which are suspended in an equal volume of Solution 1. Samples (10 μl) containing cyc^- membranes alone, dilutions of cholate extract alone, and cyc^- membranes plus the dilutions of cholate extract are assayed for adenylylcyclase activity in triplicate. These incubations are carried out at 32.5°C for 20 min. The final assay conditions are 0.1 mM ATP with 2.5×10^7 cpm [α-^{32}P]ATP, 20 mM MgCl$_2$, 1 mM EDTA, 1 mM cAMP with 10,000 cpm [^{3}H]cAMP, 0.02 mg/ml myokinase, 0.2 mg/ml creatine kinase, 20 mM creatine phosphate, 0.1 mM GTP, 0.02% Tris–cholate, 25 mM Na-HEPES (pH 8.0), and, when present, 20 μg cyc^- membrane protein, and 0.75–3 μg cholate extract protein. To assess NaF stimulation of G$_s$, 10 mM NaF is added to the assay, and to assess the ability of G$_s$ to interact with the β-adrenergic receptor present in the cyc^- membranes, 100 μM isoproterenol is added to the assay. The assays are stopped by the addition of 100 μl of "stopping solution" consisting of 10 mM cAMP, 40 mM ATP, and 1% SDS. The [^{32}P]cAMP formed and the [^{3}H]cAMP added to monitor recovery are isolated using the double chromatographic isolation method of Salomon et al. (17) over Dowex and neutral alumina. The final eluates containing the [^{32}P]cAMP and [^{3}H]cAMP are collected in scintillation vials and the ^{32}P and ^{3}H present in the samples determine using a liquid scintillation spectrometer.

Data Analysis

Any residual adenylylcyclase activity present in the heated cholate extracts alone and the cyc^- membranes alone is subtracted from the adenylylcyclase activity obtained in the presence of the heated cholate extracts plus the cyc^- membranes. The resulting reconstituted adenylylcyclase activity is plotted versus cholate extract protein as shown in Fig. 1. The resulting curves should yield straight lines, indicating a linear relationship between added cholate extract and reconstituted adenylylcyclase activity. It is important that all three points fall on the same straight line, especially the data point from the cholate sample containing the highest protein concentration. If there is an indication that at the highest protein concentrations the curves deviate from linearity, it may be an indication that the system is becoming saturated and that lower cholate extract protein concentrations should be used. The slopes of these straight lines are calculated to determine reconstituted adenylylcyclase activity, which is expressed as picomoles cAMP formed per milligram per 20 min.

If many samples are to be analyzed over an extended period of time, it is recommended that a "standard" cholate extract containing G$_s$ activity be made. This extract is made as above except that a large quantity is made from untreated tissue. The resulting cholate extract should be aliquoted such that a standard curve could be constructed from the aliquoted extract. The "standard" cholate extract is stored at $-70°$C until used. This "standard" will allow for monitoring interassay variability, which should not be greater than 10%. It is also important to note that reconstituting ability of different

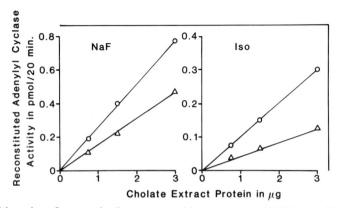

FIG. 1 Linearity of reconstitution assays with respect to cAMP formed by the S49 *cyc⁻* adenylylcyclase system reconstituted with varying amounts of Gₛ-containing cholate extract protein. Pseudopregnant New Zealand White rabbits were injected intravenously with either 100 IU human chorionic gonadotropin (hCG, △) or 0.9% saline (control, ○). The rabbits were killed 24 hr after treatment, the ovaries were removed and placed in ice-cold Krebs–Ringer bicarbonate (pH 7.4) until dissection of the corpora lutea (CL). The dissected corpora lutea were homogenized in ice-cold 27% w/w sucrose in 10 mM Tris-HCl and 1 mM EDTA (pH 7.5) and membrane particles were prepared as described by Birnbaumer *et al.* (32) and stored at −70°C until used. Gₛ function from corpora lutea from control and hCG-treated animals was assessed by reconstituting NaF-stimulated (left) and isoproterenol-stimulated (Iso, right) adenylylcyclase activity in the *cyc⁻* variant of the S49 mouse lymphoma cell line, as described in the text. Note that linearity exists over the range of cholate extract protein used (1–3 μg) whether the source of the extract is corpora lutea from control or human chorionic gonadotropin-treated rabbits and whether the reconstituted system is stimulated with NaF or isoproterenol. Values represent the means of triplicate determinations.

batches of *cyc⁻* membranes can vary significantly (up to threefold). The reason for these variations is not known, but may be due in part to differences in growth rate and cell density at the time of cell harvesting as well as differences in the heat-inactivated horse serum used in the culture media. Therefore, it is recommended that a single batch of *cyc⁻* membranes be used to analyze samples from a single experiment.

ADP-Ribosylation Assays

Cholera (18) and pertussis (19) toxins have been useful tools in the identification and anaylsis of the structural and functional properties of Gₛ and Gᵢ. Furthermore, pertussis toxin was instrumental in the purification of Gᵢ (12,

20). Using NAD as cosubstrate, cholera and pertussis toxins act by ADP-ribosylating the α subunits of G_s and G_i, respectively. ADP-ribosylation of $G_{\alpha s}$ by cholera toxin results in the inhibition of the GTPase activity associated with G_s and potentiation of the actions of GTP on adenylylcyclase (21). In contrast, ADP-ribosylation of $G_{\alpha i}$ by pertussis toxin inhibits G_i function such that receptors are uncoupled from there effectors (22), but does not result in the inhibition of the GTPase activity associated with G_i (23). When used with [^{32}P]NAD cholera toxin, [^{32}P]ADP ribosylates G_s α subunits that are between 42,000 and 55,000 Da (24), whereas pertussis toxin [^{32}P]ADP ribosylates G_i α subunits that are between 40,000 and 41,000 Da (19). We have taken advantage of the abilities of cholera and pertussis toxins to utilize [^{32}P]NAD and to [^{32}P]ADP-ribosylate the α subunits of G_s and G_i, respectively, to develop quantitative ADP-ribosylation assays for these G protein α subunits (8). Although these assays were originally developed to quantitate G_s and G_i α subunits, the ability of pertussis toxin to ADP-ribosylate the α subunits of transducin (25) and G_o (26) should allow the use of these assays to quantitate these latter G protein α subunits as well.

Materials

Sources for some reagents are given above. Additional required reagents are purchased from the indicated suppliers: DEAE-Sephadex A-25, NAD, arginine, thymidine, ADP-ribose, NAD glycohydrolase, and cholera toxin, Sigma Chemical Co. (St. Louis, MO); sodium dodecyl sulfate, Gallard-Schlesinger (Carle Place, NY); acrylamide, Eastman Kodak Co. (Rochester, NY); N,N'-methylenebisacrylamide, ammonium persulfate, N,N,N',N'-tetramethylethylenediamine (TEMED), 2-mercaptoethanol, bromphenol blue, Coomassie Brilliant Blue R-250, lysozyme, soybean trypsin inhibitor, carbonic anhydrase, ovalbumin, bovine serum albumin, and phosphorylase b (all used as molecular weight standards), Bio-Rad Laboratories (Richmond, CA); pertussis toxin, List Biological Laboratories (Campbell, CA). [^{32}P]NAD is synthesized from [α-^{32}P]ATP according to the procedure of Cassel and Pfeuffer (27) or is purchased commercially. [^{32}P]ADP-ribose is synthesized by incubating 2 mCi [^{32}P]NAD, 1 U/ml NAD glycohydrolase, 15 mM MgCl$_2$, 50 mM phosphate buffer, pH 7.0, at 32.5°C for 30 min. The [^{32}P]ADP-ribose formed is purified on a DEAE-Sephadex A-25 (HCO$_3^-$ form) column (0.6 × 23.0 cm, 7-ml bed volume) with a linear 0–1.0 M ammonium bicarbonate gradient with a total volume of 600 ml (Fig. 2). Fractions containing [^{32}P]ADP-ribose are pooled, lyophilized, resuspended in distilled water, and stored frozen until used.

The DEAE-Sephadex A-25 is converted to the HCO$_3^-$ form as follows: DEAE-Sephadex A-25 (20 g) is added to 200 ml of 1 M NH$_4$HCO$_3$ and

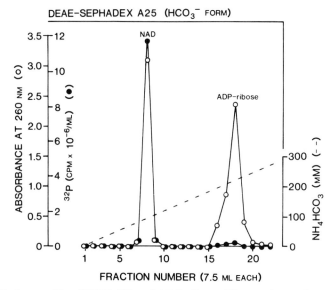

FIG. 2   Elution profile of NAD (●) and ADP-ribose (○) on column chromatography over DEAE-Sephadex A-25. A solution of 2 mM ADP-ribose, 2 mM NAD, and 10^8 cpm of [^{32}P]NAD in a volume of 1 ml was applied to a DEAE-Sephadex A-25 column prepared as described in the text. The column was washed with 4 ml of water followed by a linear 0–1.0 M ammonium bicarbonate gradient with a total volume of 600 ml.

incubated overnight at 4°C. This slurry is transferred to a 1-liter graduated cylinder containing 800 ml of water, mixed, and the DEAE-Sephadex A-25 is allowed to settle for 30 min. The supernatant is decanted; the DEAE-Sephadex A-25 is resuspended in 1 liter of water and allowed to settle for another 30 min. The supernatant is decanted, the DEAE-Sephadex A-25 is resuspended in 500 ml of water, transferred to a suction filter flask, and kept under vacuum overnight. Transfer the slurry to a graduated cylinder, let settle for 1 hr, and adjust the volume to twice the volume of the DEAE-Sephadex A-25. The DEAE-Sephadex A-25 now in the HCO$_3^-$ form and is stored refrigerated until used.

Activation of Cholera and Pertussis Toxins

Cholera (24) and pertussis (28) toxins are multisubunit bacterial toxins consisting of A and B protomers. When used in cell-free systems these toxins require treatment to dissociate the A and B protomers. The ADP-ribosyltransferase activity associated with the A protomer of the toxins becomes activated on dissociation. In order to activate cholera and pertussis toxins they are dissolved in water to give concentrations of 8 mg/ml and 200 μg/ml, respectively. The toxin solutions are made 20 mM in dithiothreitol and incubated for 10 min at 32.5°C. A white precipitate may form, especially with the cholera toxin. If this occurs, cool the samples to 4°C, subject the suspension to centrifugation at 20,000 g for 30 min at 4°C, and save the supernatant. To remove the excess dithiothreitol, the toxin-containing supernatant fluid is passed over a Sephadex G-25 PD-10 column that has been equilibrated with 10 mM Tris-HCl and 1 mM EDTA, pH 7.5. Fractions (0.4 ml each) are collected and the toxin peak is located by protein determination using the procedure of Lowry *et al.* (13). Fractions with toxin are stored at −20°C until used.

[^{32}P]ADP-Ribosylation

Three concentrations of membranes (approximately 40, 80, and 160 μg) are incubated in a total volume of 100 μl containing 1 mM ATP, 0.5 mM GTP, 15 mM thymidine, 5 mM ADP-ribose, 20 mM arginine, 5 mM dithiothreitol, 25 mM Tris-HCl, (pH 7.5), and 15 μM [^{32}P]NAD (specific activity, 1 Ci/mmol). Incubations are carried out for 30 min at 32.5°C. When present, cholera toxin is at a concentration of 40 μg/ml and pertussis toxin is at a concentration of 2.5 μg/ml. Samples are diluted fivefold with ice-cold 10 mM Tris-HCl (pH 7.5) and 1 mM EDTA, and the membranes are precipitated by centrifugation in a microfuge for 15 min at 4°C. The supernatant is removed, and the membranes are resuspended in 50 μl of Laemmli's (29) sample buffer with 5% 2-mercaptoethanol, boiled for 1 min, and subjected to SDS-polyacrylamide gel electrophoresis. The quantity of protein loaded on the gel is also determined by the method of Lowry *et al.* (13).

The assay conditions described are designed to reduce endogenous ADP-ribosylation, which can be considerable (30). In order to determine whether the [^{32}P]ADP-ribosylation obtained is due to the ADP-ribosyltransferase activity associated with the toxins, the above incubations are conducted substituting [^{32}P]ADP-ribose for [^{32}P]NAD, and the results using the two incubation conditions compared. If the covalent modification is due to ADP-ribosyltransferase activity, there should be no labeling from incubations using [^{32}P]ADP-ribose compared to those using [^{32}P]NAD.

SDS-Polyacrylamide Gel Electrophoresis

Electrophoresis is in slabs (1.5 mm × 15 cm × 16 cm) using Laemmli's (29) running buffer. The gels contain a stacking gel of approximately 1 cm and a separating gel of approximately 14 cm. The stacking gel is 5.0% acrylamide,

0.13% bisacrylamide, 62.5 mM Tris-HCl (pH 6.8), 0.1% SDS, 3.3 mM TEMED, and is polymerized at 0.1% ammonium persulfate. The separating gel is 10.0% acrylamide, 0.26% bisacrylamide, 187.5 mM Tris-HCl (pH 8.8), 0.1% SDS, 2.2 mM TEMED, and is polymerized at 0.05% ammonium persulfate. The gels are prerun for 4 hr at a constant voltage of 100 V prior to loading the samples. After the samples are loaded, they are run at 50 V until the dye front enters the separating gel, after which the gels are run at a constant voltage of 100 V for 6–7 hr. The gels are fixed and stained in glacial acetic acid : methaol : water (1 : 5 : 5, v : v : v) with 0.1% Coomassie Brilliant Blue overnight at room temperature. The gels are then destained for 2 hr in same solution without Coomassie Brilliant Blue followed by incubation for 2 hr in 7.5% glacial acetic acid in 5% ethanol. Finally the gels are incubated for 2 hr in 0.1% glycerol prior to drying. After drying, the gels are exposed to Kodak-Omat AR (XAR-5) film in the presence of Dupont Cronex lightning plus intensifying screens at −70°C for 4 days.

In addition to the electrophoretic separation procedure outlined above, Codina *et al.* (31) have described a urea gradient/SDS-polyacrylamide gel electrophoresis procedure that has the capability of resolving αi1, αi2, and αi3. Using the standard SDS-polyacrylamide gel electrophoresis procedure described above, αi1, αi2, and αi3 migrate as a single band. Therefore, if the tissue to be analyzed contains more than one αi subunit and changes in a given αi subunit are to be determined, the urea gradient/SDS-polyacrylamide gel electrophoresis procedure should be used.

Data Analysis

The bands corresponding to the α subunits of G_s and G_i on the autoradiogram are scanned with a laser densitometer in order to quantitate the extent of labeling. The area under the curves corresponding to a given G protein α subunit is plotted versus the membrane protein concentration loaded onto the gel as illustrated in Fig. 3. This should result in a straight line, indicating a linear relationship between membrane protein loaded on the gel and the extent of [^{32}P]ADP-ribosylation by cholera or pertussis toxins. The slopes of these straight lines are calculated to determine the extent of [^{32}P]ADP-ribosylation expressed as relative units/milligram membrane protein. Note that these lines will go through zero only if background labeling has been subtracted from the peaks. As with the reconstitution assays, it is important that all three points fall on the same straight line. This is especially true for the data point from the membrane sample containing the highest protein concentration. If the data point from the highest membrane protein concentration deviates from linearity, it may indicate one of two problems. The

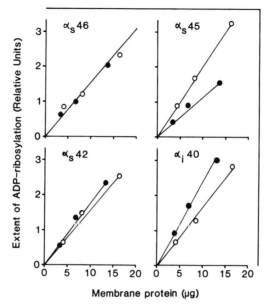

FIG. 3   Linearity of cholera and pertussis toxin-mediated ADP-ribosylation of rabbit luteal G protein α subunits. Rabbits were treated and luteal membranes were prepared as described in Fig. 1, except that the treatment time was 6 hr. The extent to which the various G protein α subunits present in corpora lutea (CL) from control ($\bigcirc$) and human chorionic gonadotropin (hCG)-treated ($\bullet$) rabbits were ADP-ribosylated by cholera and pertussis toxin was determined as described in the text. Note that linearity exists over the range of membrane protein used whether the source of the membranes was from corpora lutea from control or human chorionic gonadotropin-treated rabbits and whether the membranes were treated with cholera or pertussis toxin or whether a given α subunit isomorph was analyzed. Membrane protein values refer to the amount of protein loaded onto the gel.

first is that the autoradiogram is overexposed so that the sensitivity of the film has been exceeded. This can be simply rectified by reducing the exposure time of the autoradiogram. If this does not rectify the problem, then it may indicate that there is either too much membrane protein in the incubation or not enough toxin.

If the samples to be analyzed are from treated animals or treated cells, it is important that the appropriate control samples are loaded on the same gel as the treated samples. Then the resulting data for a given autoradiogram are expressed as a percentage of the control value. This will correct for variations in runs from gel to gel, differences in exposure time to the film, and differences in development of the autoradiograms.

References

1. L. Birnbaumer, J. Abramowitz, and A. M. Brown, *Biochim. Biophys. Acta* **1031,** 163 (1990).
2. This Series, Vol. XI.
3. This Series, Vol. XII.
4. This Series, Vol. XVIII.
5. This Series, Vol. XIX.
6. M. I. Simon, M. P. Strathmann, and N. Gautam, *Science* **252,** 802 (1991).
7. L. Birnbaumer, Cell **71,** 1069 (1992).
8. B. P. Jena and J. Abramowitz, *Endocrinology (Baltimore)* **124,** 1932 (1989).
9. B. P. Jena and J. Abramowitz, *Endocrinology (Baltimore)* **124,** 1942 (1989).
10. E. M. Ross and A. G. Gilman, *Proc. Natl. Acad. Sci. U.S.A.* **74,** 3715 (1977).
11. J. K. Northup, P. C. Sternweis, M. D. Smigel, L. S. Schleifer, E. M. Ross, and A. G. Gilman, *Proc. Natl. Acad. Sci. U.S.A.* **77,** 6516 (1980).
12. J. Codina, J. D. Hildebrandt, R. D. Sekura, M. Birnbaumer, J. Bryan, C. R. Manclark, R. Iyengar, and L. Birnbaumer, *J. Biol. Chem.* **259,** 5871 (1984).
13. O. H. Lowry, N. J. Rosebrough, A. L. Farr, and R. J. Randall, *J. Biol. Chem.* **193,** 265 (1951).
14. T. F. Walseth and R. A. Johnson, *Biochim. Biophys. Acta* **562,** 11 (1979).
15. P. Coffino, H. R. Bourne, and G. M. Tomkins, *J. Cell. Physiol.* **85,** 603 (1975).
16. E. M. Ross, M. E. Maguire, T. W. Sturgill, R. L. Biltonen, and A. G. Gilman, *J. Biol. Chem.* **252,** 5761 (1977).
17. Y. Salomon, C. Londos, and M. Rodbell, *Anal. Biochem.* **58,** 541 (1974).
18. D. M. Gill and R. Meren, *Proc. Natl. Acad. Sci. U.S.A.* **75,** 3050 (1978).
19. T. Katada and M. Ui, *Proc. Natl. Acad. Sci. U.S.A.* **79,** 3129 (1982).
20. G. M. Bockoch, T. Katada, J. K. Northup, M. Ui, and A. G. Gilman, *J. Biol. Chem.* **259,** 3560 (1984).
21. D. Cassel and Z. Selinger, *Proc. Natl. Acad. Sci. U.S.A.* **74,** 3307 (1977).
22. H. Kurose, T. Katada, T. Amano, and M. Ui, *J. Biol. Chem.* **258,** 4870 (1983).
23. T. Sunyer, B. Monastirsky, J. Codina, and L. Birnbaumer, *Mol. Endocrinol.* **3,** 1115 (1989).
24. D. M. Gill, *in* "ADP-Ribosylation Reactions, Biology and Medicine" (O. Hayaishi and K. Ueda, eds.), p. 593. Academic Press, New York, 1982.
25. C. Van Dop, G. Yamanaka, F. Steinberg, R. D. Manclark, L. Stryer, and H. R. Bourne, *J. Biol. Chem.* **259,** 23 (1984).
26. P. C. Sternweis and J. D. Robishaw, *J. Biol. Chem.* **259,** 13806 (1984).
27. D. Cassel and T. Pfeuffer, *Proc. Natl. Acad. Sci. U.S.A.* **75,** 2669 (1978).
28. M. Tamura, K. Nigimori, S. Murai, M. Yajima, K. Ito, T. Katada, M. Ui, and S. Isahi, *Biochemistry* **21,** 5516 (1982).
29. U. K. Laemmli, *Nature (London)* **227,** 680 (1970).
30. J. Abramowitz and B. P. Jena, *Int. J. Biochem.* **23,** 549 (1991).
31. J. Codina, D. Grenet, K. J. Chang, and L. Birnbaumer, *J. Receptor Res.* **11,** 587 (1991).
32. L. Birnbaumer, P. C. Yang, M. Hunzicker-Dunn, J. Bockaert, and J. M. Duran, *Endocrinology* **99,** 163 (1976).

[7] Functional Assays to Study Coupling of Dopamine D2 Receptors to G Proteins

Ja-Hyun Baik, Madhav Bhatia, Roberto Picetti,
Graziella Thiriet, and Emiliana Borrelli*

Introduction

The seven transmembrane (7TM) domain G-protein-coupled receptors constitute a large family of membrane receptors (1). These proteins possess the conserved feature of presenting seven domains of hydrophobic amino acids of similar length, which have been proposed to be the regions of the protein crossing the plasma membrane. These membrane receptors have been shown to be activated by the interaction with a large variety of ligands, from the catecholamines to peptide hormones.

In the presence of the appropriate ligand, 7TM receptors activate signal transduction mechanisms, which, in turn, modulate the physiological responses of the cell. Activated 7TM receptors transduce the signal by stimulating the heterotrimeric GTP-binding proteins, or G proteins (2). The activation of the G proteins induces the generation of intracellular second messengers by the action of cellular effectors such as adenylylcyclase (AC) and phospholipase C (PLC). The generation of second messengers is the starting point of fundamental secondary events that include the phosphorylation of cellular factors and the modulation of gene expression. The following inactivation of the receptor brings the cell to the unstimulated state.

The 7TM receptors display high levels of homology at the amino acid level in segments of the protein corresponding to the transmembrane regions. This property has led to the recognition of these proteins as a family of related genes (3). Presently, more than 100 members constitute this family. In addition, it is thought that there are more than 1000 members in the large subfamily of olfactory receptors (4). The G proteins that interact with these receptors also represent a family of proteins made up of approximately 20 members sharing a high degree of homology. The molecular cloning of the genes corresponding to the various members of the G-protein-coupled receptor family has evidenced the presence of more than one receptor for a specific ligand. These receptors displayed functional characteristics that appeared

* To whom correspondence should be addressed.

Methods in Neurosciences, Volume 29

to be in agreement with former biochemical and pharmacological studies. The sequence similarity between members of the same subfamily could be as high as 70% at the amino acid level in the transmembrane regions. Interestingly, the signal transduction pathways to which they are linked could be the same. In addition, it has to be mentioned that the same receptor has the potential to activate more than one transduction pathway. For example, some 7TM receptors whose activation results in a down-regulation of AC can also activate the PLC pathway or ion channels. These observations suggest that there is a large choice of possible interactions of receptors with different G proteins, with the consecutive stimulation of alternative pathways, and underline the complexity of the 7TM receptor-mediated signal transduction. It is therefore important to discern the features underlying receptor–G protein interaction. An important advance in this sense has been made by the first analysis at the molecular level of the cloned receptors. In particular, exchange of intracellular domains from one receptor to another has shown a crucial role for these regions in the coupling to the G protein and especially the importance of the third intracellular loop (5). In this review we present methods aimed to the study of the coupling characteristics of the dopamine D2 receptors. Particular attention will be given to the function of the third intracellular loop of these receptors in the interaction with the G proteins.

Dopamine Receptors

Dopamine is a neuromediator that controls many physiological functions. In the central nervous system (CNS) dopamine is involved in the coordination of movements, visual perception, cognition, and emotion, and in the neuroendocrine control of the pituitary gland. The activation by dopamine of membrane receptors generates the formation of intracellular second messengers.

Dopamine receptors belong to the family of 7TM domain G-protein-coupled receptors (6). Pharmacological and biochemical studies describe the existence of two classes of dopamine receptors, D1 and D2. D1 is classified as activator of AC, in contrast to D2, which is described as an inhibitor of the same cellular effector. These two types of receptors differ in their pharmacological affinity for classical dopamine receptor antagonists such as SCH23390 for D1, and spiperone for D2. The molecular cloning of the cDNA for the D2 receptor (7) has allowed low-stringency screenings of cDNA and genomic libraries, with the isolation of four additional dopamine receptors. Two of them, D3 (8) and D4 (9), display phamacological characteristics similar to D2 receptors, namely, picomolar to nanomolar affinity for the antagonist spiperone. The other two, D1 (10) and D5 (11), display typical pharmacological features of the classically described D1 receptors. The cloned receptor cDNAs have been expressed in cell lines and their pharmacological profiles have been defined as well as their properties to transduce the signal inside the cell. As expected, D1 and D5 receptors are stimulators of AC; D5 has a higher affinity for dopamine than D1. The picture is less clear for the D2-like receptors group. At present, the mechanism by which D3 and D4 receptors transduce the signal remains unknown, although they present a typical D2-like pharmacological profile. A role for D3 as an autoreceptor has been proposed on the basis of a higher affinity for dopamine as compared to D2 (6). D4 has a higher affinity (about 10-fold) than D2 and D3 for clozapine, an antipsychotic, at concentrations that are active *in vivo,* and it has been proposed to be the target of this drug. The absence of the identification of a transduction pathway for these receptors could be due to the absence of specific G proteins in the cell lines tested.

The whole subfamily of dopamine receptors is widely expressed in the CNS, in particular in the caudate putamen, nucleus accumbens, hypothalamus, and olfactory tubercule. The common pharmacological characteristics between the D1 and D2 groups of dopamine receptors are also reflected at the amino acid level, where a homology of more than 40% is observed on the overall sequence between members of each group.

D2 Receptors

The first D2 receptor cDNA isolated encoded for a putative protein of 415 amino acids and it was isolated from a brain cDNA library (7). The pharmacological profile obtained from transfected cells clearly identified it as a dopamine D2 receptor. Shortly after, another cDNA was isolated from different sources, which was 100% homologous to the first one, excepted for an in-frame insertion of 29 amino acids in the putative third intracellular domain of the receptor. This cDNA therefore encoded a protein of 444 amino acids, which was found in rat, mouse, bovine, and human specimens (the human cDNA has 445 amino acids) (6, 12). The analysis of genomic DNA demonstrated that both cDNAs were products of the same gene, generated by an alternative splicing mechanism that adds a small exon of 87 bp in the longer mRNA. These receptors have been named D2S, for the short form of the receptor, and D2L, for the longer one.

The structure of the genomic fragment containing the D2 gene has been elucidated. Seven coding exons have been found in all species analyzed, with an astonishing feature represented by the presence of an extremely large intron ($\geq$30 kb) located upstream of the first coding exon. The presence of such a large intronic sequence might underlie a mechanism of control over the expression of the D2 gene. The cloning of the promoter region for the D2 gene has been reported; although no particular feature has been described yet, it seems to bear the characteristics of a housekeeping promoter. Interestingly, both isoforms of D2 receptors are coexpressed in all the tissues tested, with a ratio that normally favors D2L (12). At the pharmacological level, these two isoforms have similar profiles and bind the antagonist spiperone with similar affinity. The activation of these D2 receptors principally results in the inhibition of AC. However, it appears that they are also able to activate other signal transduction pathways (6). These observations, together with the fact that the two receptors differ only for the alternative insertion in the third intracellular loop, suggested that the two receptors might differ in their ability to couple to one or different G proteins. We concentrated our efforts in studying the effect of D2 receptor activation on the inhibition of cAMP levels. In particular, by analyzing a large number of cell types, we found one, the JEG3 cell line (a human choriocarcinoma cell line), in which the two receptor isoforms behaved differently. We showed that these cells did not contain detectable level of the α subunit of G_{i2}. These findings indicated that one functional difference between the two isoforms resided in the specificity of coupling to the G proteins, due to the different third intracytoplasmic domain.

Characterization of D2 Receptor in Vitro

To elucidate the function of the two isoforms of the D2 receptor, we studied their binding properties as well as their ability to inhibit endogenous cAMP levels. These experiments were conducted by comparing the function of the two receptors in mammalian cells transiently transfected with expression vectors containing either D2L or D2S.

In order to perform these experiments the cell lines used were previously characterized. As first requirement the cells should not endogenously express the receptor, or subtypes of it, that could be activated by the same agonist, in our case dopamine. Accordingly, binding tests and cAMP assays were performed on untransfected cells. Second, to avoid transfection artifacts, which could have misled the interpretation of the results, care was taken to always have positive and negative controls.

Cells in a 100-mm dish are usually transfected with 8–10 μg of total DNA, which is composed by 2 μg of each analyzed expression vector, the remainder being the carrier DNA. For transfections in smaller or larger cell dishes the quantity of DNA transfected was decreased or increased with respect to the dish surface available, as compared to the 100-mm dish.

In transfection experiments we always keep constant the molar ratio of promoter-containing expression vectors per point analyzed. For example, consider the case of the transfection experiments in which we compare the cAMP inhibition of D2L in the absence or presence of G_{i2}. In the absence of G_{i2}, the expression vector without insert is added at the same concentration. This is done in order to avoid possible differences in the transcription efficiency of transfected DNA due to the utilization of larger quantities of promoter-containing expression vectors that might be noted in the titration of the endogenous and available transcription factors.

Construction of Expression Vectors

The cloned cDNAs for D2S and D2L, as well as all the other cDNAs used in these studies, are subcloned in the vector pSG5 (13). pSG5 contains the simian virus 40 (SV40) early promoter upstream of a multiple cloning site. The polyadenylylation site of SV40 is located downstream of the site of insertion of the cDNAs. This vector contains a powerful promoter that is functional in most cell types. Moreover, when transfected in COS 7 cells, which express the SV40 large T antigen, this vector is able to replicate, giving rise to a very high number of transcripts.

Transfection of Cells by $CaPO_4$ Technique

The expression vectors are prepared in a large-scale plasmid preparation, using the alkaline lysis method as described in Sambrook *et al.* (14). We purify the plasmids on two consecutive CsCl gradients.

JEG3 cells are purchased from American Type Culture Collection (ATCC, Rockville, MD), and transfected with the $CaPO_4$ precipitation method. This method is effective to transfect many stable cell lines.

The solutions used are $CaCl_2$, $2M$; $2\times$ HBS (NaCl, 280 mM; HEPES, 50 mM; Na_2HPO_4, 1.5 mM; pH 7.05); these solutions were sterilized through passage in Millipore (Bedford, MA) filters, 0.22 μm.

For ligand binding assays on whole cells and cAMP measurements, transfections are performed in six-well Costar plates. Seed 1×10^5 cells/well in a six-well Costar plate (35 mm $\times$ 6) with 1.5 ml/well of the appropriate cell culture medium supplemented with serum, 12 hr before transfection. A total of 1.6 μg of DNA per well is transfected; this is made up by 400 ng of specific DNA and 1.2 μg of carrier DNA (we use pBluescript SK, Stratagene). To make ligand binding assays or cAMP measurements in triplicate, a minimum of four six-well plates per tested receptor is needed. We prepare a tube of DNA precipitate for each six-well plate; indeed, we have found that preparation of larger quantities of precipitate, for example, for the total of four plates in the same tube, does not give good results. For a six-well plate we add 9.6 μg of total DNA to 600 μl of 0.24 M $CaCl_2$ in sterile H_2O in a small

Falcon tube. To this solution 600 μl of 2× HBS is added dropwise, while vortexing the tube. The final solution is left at room temperature for 30 min. Then 200 μl of the $CaPO_4$–DNA precipitate is added to each well and the plates are transferred back in the incubator. At 8 to 10 hr after transfection the medium is removed, the cells are washed once with cell culture medium, then fresh medium is added and the cells are grown for an additional 16 to 18 hr.

To perform binding experiments on isolated membranes of transfected cells, a larger amount of cells must be transfected. In order to have enough material to perform Scatchard and competition analysis, a minimum of 10 100-mm plates of JEG3 cells is transfected with a plasmid expressing each isoform of dopamine D2 receptor: 1×10^6 cells are plated per 100-mm dish and are transfected with a total of 8 μg DNA/plate. The total amount of DNA to be transfected can be entirely specific DNA or part specific DNA and part carrier plasmid. We find that, when comparing the pharmacology of different receptors and receptor mutants, the amount of transfected specific DNA may need to be varied in order to obtain expression of a similar number of sites for all of them. These conditions are established experimentally, starting by transfecting the same amount of specific DNA for each receptor. Cells are plated in 7 ml of medium per plate. The DNA precipitate in each plate is formed by mixing 8 μg of the total DNA mixture in 500 μl of 0.24 M $CaCl_2$; to this solution 500 μl of 2× HBS is added as described before. After 30 min the precipitate is added to the cells. The cells are washed and harvested, following the same time schedule as the one already described for the six-well plates.

Ligand Binding Assays on Membranes or Whole Cells

Materials

Bovine serum albumin (BSA), polyethyleneimine (PEI), sodium ascorbate (Sigma, St. Louis, MO), (+)-butaclamol and R(−)-propylnorapomorphine (NPA) (Research Biochemical Inc., Natick, MA), [^{3}H]spiperone (benzene ring-^{3}H; 921.3 GBq/mmol) (New England Nuclear; England), scintillation fluid (Ready Safe, Beckmann Instrument, Palo Alto, CA).

Stock Solutions

Chlorides (NaCl, 3 M; KCl, 125 mM; $CaCl_2$, 50 mM; $MgCl_2$, 25 mM); 1% (w/v) sodium ascorbate; 1 M Tris-HCl, pH 7.7; (+)-butaclamol, 10^{-3} M in ethanol (w/v); NPA, 10^{-2} M in ethanol (v/v). Tritiated ligand stock solutions are diluted in 0.1% sodium ascorbate.

For ligand binding on whole cells, the binding buffer is composed of 40 mM Tris-HCl, pH 7.7; 96 mM NaCl; 4 mM KCl; 1.6 mM $CaCl_2$; 0.8 mM

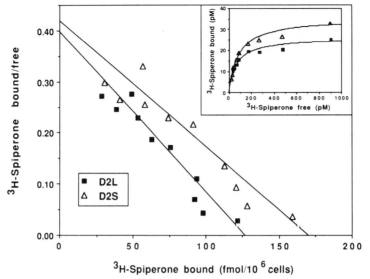

FIG. 1   Scatchard plots of [³H]spiperone saturation experiments on pSVD2L- and pSVD2S-transfected JEG3 cells. [³H]Spiperone binding experiments on whole cells were performed as described in the text. Concentrations of [³H]spiperone ranging from 0.02 to 1 nM were used, and nonspecific binding was determined in the presence of 1 μM (+)-butaclamol. Results of one experiment are representative of three independent ones, each performed in duplicate. As estimated from computer-assisted analysis, in this experiment, the dissociation constants (K_d) were 66 and 88 pM, and the maximum binding capacities were 127 and 170 fmol/10^6 cells for D2L and D2S receptors, respectively. *Inset:* Saturation binding isotherms of this experiment.

MgCl$_2$; 0.1% sodium ascorbate; 1% BSA; 0.001% PEI. All operations are performed at 4°C.

Transfected cells are washed twice with ice-cold binding buffer. A saturation curve is determined by the incubation of each sample with [³H]spiperone in increasing amounts, ranging from 0.02 to 1 nM, in 1 ml/well of binding buffer for 90 min at room temperature. Nonspecific binding is determined in the presence of 1 μM (+)-butaclamol; the stereospecific binding is calculated by subtracting the nonspecific binding observed in the presence of (+)-butaclamol from the total binding observed in its absence (Fig. 1). For competition experiments, cells are incubated with 100 pM [³H]spiperone and increasing concentrations of NPA, from 10^{-11} to 10^{-4} M in binding buffer. NPA is added at the same time as [³H]spiperone. Cells are then washed three times with 3 ml of ice-cold Tris-HCl, 50 mM, pH 7.7, and 0.1% BSA. SDS (1%, 1.5 ml) is then added to each well to recover the cells. Membrane-bound [^{3}H]spiperone is measured in the presence of 5 ml of Ready Safe scintillation fluid with a counting efficiency of 48%.

For ligand binding on membranes of transfected cells, membranes are prepared by washing the plates twice with ice-cold phosphate-buffered saline (PBS), harvested in 5 ml of PBS per plate, and centrifuged at 500 g for few minutes. The pooled cell pellets from 10 plates are resuspended in 5 ml of 10 mM Tris-HCl, pH 7.5, and 5 mM EDTA and homogenized in a Dounce homogenizer with 10–12 strokes of a glass pestle, type B; the homogenate is centrifuged at 1000 g for 10 min. The supernatant is recovered while the pellet is washed once again in the same volume of Tris-EDTA and centrifuged at the same speed. The two supernatants are pooled and centrifuged at 45,000 g for 40 min. The pellet obtained is resuspended in 10 ml of the same buffer and centrifuged again at the same speed for the same time. The final pellet is resuspended in approximately 5 ml of 50 mM Tris-HCl, pH 7.7, and stored in aliquots at $-80°C$ till further use.

Bindings are performed in a 1.25-ml/tube, containing the following components (final concentrations): 40 mM Tris-HCl, pH 7.7; 96 mM NaCl; 4 mM KCl; 1.6 mM CaCl$_2$; 0.8 mM MgCl$_2$; 0.1% sodium ascorbate; [^{3}H]spiperone in 0.1% sodium ascorbate. Care must be taken in determining the amount of ascorbate to be added, by subtracting the amount of it already added with the ligand solution. The concentrations of ligands used are the same as the one used for binding on whole cells. The binding reaction is started by the addition of membranes, 15–30 μg/tube. Each tube is incubated for a total of 30 min at 37°C in a water bath, therefore membranes are added to each tube with a defined interval of time, which makes the following manipulations easier and the incubation time constant for each point. The reaction is stopped by filtration through Whatman GF/B filters on a filtrating apparatus followed by three washes with 2 ml ice-cold 50 mM Tris-HCl, pH 7.7. The radioactivity bound to the filter is counted in 5 ml of Ready Safe scintillation fluid after an overnight elution at 4°C.

All the binding data are analyzed with the EBDA-Ligand program (Elsevier-BIOSOFT), using a one-site fitting model for saturation experiments and either a one- or two-site fitting model for competition experiments.

Effect of D2 Dopamine Receptor Activity on Inhibition of cAMP

To measure the activity of dopamine D2 receptors we proceeded to analyze the intracellular cAMP in different cell lines. In order to observe an inhibition of the intracellular cAMP levels, the cAMP pathway has to be activated first. This cannot be achieved in transient transfection assays by using activators of the cAMP pathway, for example, forskolin. Forskolin is a diterpene compound that constitutively activates AC. The percentage of transfected cells in a transient assay varies from one type to another, but is never greater than 50%. Therefore, agents such as forskolin are not advised, because they activate AC in all the cells, thus masking the amplitude of the inhibitory effect that is obtained only in transfected cells. Consequently, cells are cotransfected with the β_2-adrenergic receptor together with either isoform of the D2 receptors. The β_2-adrenergic receptor is a 7TM domain G-protein-coupled receptor that stimulates AC on addition to the isoproterenol medium, a β_2 agonist.

Measurement of Intracellular cAMP Levels

Cells are transfected in six-well Costar plates as described previously. Each point (in triplicate) is composed of 0.4 μg of pSVD2L or pSVD2S, 0.4 μg of pKSVTF (the expression vector for the β_2-adrenergic receptor), and 0.8 μg of carrier DNA, to maintain a total of 1.6 μg per 35-mm plate. At 45 min prior to the induction, cells are incubated in serum-free medium containing 25 mM HEPES, pH 7.5. At 20 min, isobutylmethylxanthine (IBMX), an inhibitor of cAMP phosphodiesterase, is added to each plate, to a final concentration of 0.5 mM. Cells are then induced with 10 μM (−)-isoproterenol and the appropriate concentration of dopamine (from 10^{-10} to 10^{-5} M). After an incubation for 45 min at 37°C, the medium is aspirated and 1 ml of 60% (v/v) ethanol (in H_2O) is added to the cells for 1 hr at room temperature. The extracts are lyophilized and resuspended in a buffer supplied with the cAMP RIA kit (New England Nuclear).

cAMP in each sample is quantified according to the manufacturer's instructions. The values obtained are normalized for the estimated number of cells per dish.

Effect of Dopamine D2 Receptor Activity at Transcriptional Level

We established a system to study the transcriptional effect due to the activation of the dopamine receptors on a reporter gene (15). Three vectors were coexpressed in the same cells: pSVD2L or pSVD2S, pKSVTF, and pSomCAT. The first three vectors are the expression vectors for D2L, D2S, and β_2-adrenergic receptors, respectively and have been already described. pSomCAT is the reporter plasmid that contains, upstream of the herpes virus thymidine kinase promoter, the cyclic AMP-responsive element (CRE) sequence of the rat somatostatin promoter (16). This fusion promoter drives the expression of the bacterial chloramphenicol acetyltransferase gene (17). The CRE sequence binds transcription factors that are stimulated by intracellular modifications of the cAMP levels. Therefore the simultaneous expression of the β_2-adrenergic receptor and pSomCAT, followed by the addition

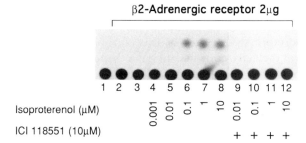

FIG. 2 Stimulation of CAT activity by the β_2-adrenergic receptor. The JEG3 cells were transfected with 2 μg of the β_2 expression vector, pKSVTF, and pSomCAT. Lane 1, pSomCAT alone; lanes 2 and 3, pSomCAT and pKSVTF in the absence of ligand; lanes 4–8, pSomCAT and pKSVTF in the presence of increasing concentrations of (−)-isoproterenol as indicated; lanes 9–12, same as lanes 5–8 plus 10 μM ICI118551, a β_2-specific antagonist.

into the media of isoproterenol, results in the stimulation of AC by the activation of the G protein G_s, with production of cAMP. As a consequence, the cAMP-dependent protein kinase A is stimulated. One of the functions of this kinase is the phosphorylation of cAMP-responsive transcription factors, which bind CRE sequences present in the reporter gene promoter, stimulating its transcription. Consequently, an accumulation of the CAT protein takes place inside the cell, which can be measured by performing a CAT assay (Fig. 2). The magnitude of the stimulation, compared to nonstimulated cells, can be then calculated. In successive points this stimulation is challenged by the cotransfection of one of the D2 receptors, inhibitors of the AC activity, in the presence of dopamine and isoproterenol (Fig. 3). The corresponding reduction of CAT activity gives the value of the inhibition obtained by the activation of the D2 receptor in the presence of determinate concentrations of dopamine. This system works at physiological ligand concentrations, is reproducible, and has the advantage to amplify the response of at least 10% when compared to direct measurements of cAMP levels. The inhibition by D2 receptors can be blocked by butaclamol and by pretreatments of the cells with pertussis toxin, which inactivates G_i and G_o members of the G protein family. Moreover, the reporter CAT gene can be exchanged with any other suitable and convenient reporter (i.e., luciferase).

Differential Inhibition of Intracellular Level of cAMP in JEG3 Cells by D2L and D2S

Expression Vectors

pSVD2L, pSVD2S, pKSVTF, pSomCAT, and pCH110 (Pharmacia, Piscataway, NJ). The vector pCH110 contains the β-gal gene under the control of

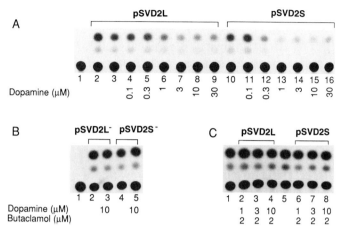

FIG. 3 Specific inhibition of CAT-stimulated activity by D2 receptor isoforms. (A) JEG3 cells were cotransfected with 2 μg of pSomCAT, pKSVTF, and either pSVD2L or pSVD2S. Lane 1, pSomCAT alone; lane 2, pSomCAT and pKSVTF in the presence of 10 μM (−)-isoproterenol; lanes 3–9 and 10–16, as in lane 2 plus pSVD2L (lanes 3–9) or pSVD2S (lanes 10–16), with increasing concentrations of dopamine as indicated. (B) Antisense plasmid pSVD2L or pSVD2S cotransfected with 2 μg of pSom-CAT and pKSVTF in the presence of 10 μM (−)-isoproterenol. Lane 1, pSomCat alone; lanes 2 and 4, pSomCAT and pKSVTF plus 2 μg of either one the dopamine receptor antisense vectors as indicated; lanes 3 and 5, as in lanes 2 and 4 plus 10 μM dopamine. No inhibition is observed as expected, showing the specificity of the system. (C) (+)-Butaclamol (2 μM) completely blocks D2-induced inhibition of CAT activity. Same points as in A, (lanes 2, 6, 7, 8, 10, 13, 14, and 15), but with the addition of (+)-butaclamol.

the SV40 promoter and is used to normalize the transfection efficiency. pBluescript SK or pSG5 is used as carrier DNA.

Method

JEG3 cells are seeded at 30–40% of confluency in 100 mm (1–2 × 10⁶/plate) in 7 ml of the appropriate medium containing 10% (v/v) fetal calf serum (FCS) 12 hr before transfection. The following plasmids are used: 2 μg of pSVD2L or pSVD2S, 2 μg of pKSVTF, 1 μg of pCH110, and pBluescript SK up to 8 μg/plate. The transfection is performed as described previously and 1 ml of precipitate is added per plate. At 8 hr after washing the precipitate, the cells are induced with 10 μM isoproterenol and dopamine from 10^{-7} to 10^{-4} M; 8 hr after the induction the cells are harvested in 1 ml of TNE (Tris-HCl, 40 mM, pH 7.4; EDTA, 1 mM, pH 8; NaCl, 150 mM), pelleted at low-speed centrifugation, then resuspended in 50 μl of 0.25 M Tris-HCl, pH 8. To prepare the homogenate, three cycles of freezing–thawing of the suspensions are performed. To eliminate cellular debris the tubes are spun for 1 min in a bench centrifuge at 12,000 rpm.

β-Gal assays (18) are performed on 5 μl of the extracts (about 5% of the total extract); the values obtained are then adjusted on the sample containing the lowest activity, by diluting the more concentrated ones with 0.25 M Tris-HCl, pH 8. This assay is used to control the transfection efficiency in each plate, and avoids artifact variations.

CAT assays are performed with about 40–50% of the cell extract per reaction. This amount may be lowered if a high CAT activity is obtained. The reaction is carried out in 150 μl, which contains 40 μl of extract, 70 μl of 1 M Tris-HCl, pH 7.8, 19 μl of H_2O, 1 μl of [^{14}C]chloramphenicol (45 mCi/mmol), 20 μl of 4 mM acetyl-CoA. After mixing, the samples are left to incubate for 2 hr at 37°C in a water bath. The acetylated forms of chloramphenicol are then extracted with 500 μl of ethyl acetate by vortexing each sample for 30 sec; these manipulations are performed under a fume hood. The tubes are then centrifuged in a bench microfuge for 1 min and the top phase is recovered (the interface is discarded). The ethyl acetate is subsequently lyophilized in a speed vacuum apparatus. The dried pellet is resuspended in 20 μl of ethyl acetate and spotted on a thin-layer chromatographic plate (Merck). Preparation of the TLC plate includes drawing a line at 2 cm from the edge with a pencil along the axis of the 10 $\times$ 20-cm silica plate. The locations, spaced every 1.5 cm, where the samples will be spotted, are labeled along this line. To avoid spreading of the samples, they are deposed on the plate dropwise; drops are allowed to dry before applying additional drops. After all the samples are spotted, the plate is allowed to dry while a solution containing chloroform–methanol (95 : 5) is prepared and placed in a chromatography chamber in which two sheets of filter paper of the size of the TLC plate have been placed along the sides. The chloroform–methanol solution should fill the chamber to a depth of 1 cm. To saturate the atmosphere and avoid dilution of the eluent, the chamber should be well sealed; the chloroform–methanol solution is routinely placed into the chamber 10–20 min before the chromatography is run. The plate is placed in the solution, sample side in the eluent, to allow the solution to ascend the plate. When the front reaches the top the chromatography is stopped and the plate is dried. This takes about 1 hr. The plate is exposed to X-ray Kodak film. Other methods to perform CAT assays have also been described (19, 20).

This type of analysis performed in JEG3 cells evidences that D2L inhibits AC activity less efficiently than D2S, even though the pharmacological profile for both receptors is similar (Figs. 1 and 3). These data suggest that the difference observed is probably due to the coupling of the receptors to the

TABLE I   Dopamine EC_{50} Values Obtained in JEG3 and NCB20 Cells Transfected
by Either Isoform of D2 Receptors and by $G_{i\alpha2}$ Expression Vectors[a]

JEG3			NCB20		
Receptor	G protein	EC_{50} (nM)	Receptor	G protein	EC_{50} (nM)
D2L	—	80 ± 7	D2L	—	20 ± 2
D2S	—	28 ± 3	D2S	—	31 ± 3
D2L	$G_{i\alpha2}$	10 ± 1	D2L	$ASG_{i\alpha2}$[b]	260 ± 10
D2S	$G_{i\alpha2}$	32 ± 2	D2S	$ASG_{i\alpha2}$	16 ± 2

[a] EC_{50} values were determined after transient expression of the appropriate receptors in JEG3 and NCB20
cells. EC_{50} values represent the dopamine concentration required to observe half-maximal CAT activity.
[b] $ASG_{i\alpha2}$ is the antisense construction of $G_{i\alpha2}$.

G proteins. The analysis of the mRNAs of these cells for their content in
inhibitory G proteins reveals the absence of the $G_{i\alpha2}$ subunit. Consequently,
the expression vector containing $G_{i\alpha2}$ is cotransfected together with D2L in
JEG3 cells, to investigate whether it is able to restore the same inhibitory
activity observed with D2S. These experiments are performed and give the
expected result (Table I). In this experiment, 2 μg of the expression vector
containing $G_{i\alpha2}$, pSVGi2, is added to the DNA mixture containing 2 μg of
pSVD2L or pSVD2S, 2 μg of pSomCAT, 2 μg of pKSVTF, 1 μg of pCH110,
and 1 μg of the carrier DNA, pBluescript SK. In plates in which the activity
of D2L or D2S is tested on pSomCAT, in the absence of pSVGi2, 2 μg of
the vector pSG5 without inserts is used, to keep the amount of promoter-
containing vectors equivalent for each point.

These experiments, performed in JEG 3 cells, show that D2L needs $G_{i\alpha2}$
to inhibit adenylylcyclase to a greater extent. JEG3 cells are therefore an
ideal model system to study such interaction.

The specificity of interaction between receptors and G proteins can also
be addressed by the use of the antisense expression vectors. These vectors
contain the G protein of interest inserted in the opposite orientation as
related to the promoter, consequently an antisense RNA is transcribed,
which hybridizes to the endogenous mRNA of the corresponding G protein.

To confirm the results found in JEG3 cells, we perform experiments with
the antisense G_{i2} in NCB20 cells (21). To perform these experiments, the
same combinations of expression vectors used in JEG3 cells to study the
effect of G_{i2} are utilized, except that the sense G_{i2} vector is substituted with
2 μg of the antisense vector. NCB20 cells constitutively express G_{i2} and, in
fact, in this cell line no difference in the efficiency of inhibition between D2
receptors is observed (Table I). In addition, if the antisense G_{i2} construct

(pSVASGi2) is cotransfected with either D2L or D2S, the activity of D2L is impaired but not that of D2S (Table I). This method offers the advantage that the receptor/G protein coupling can be studied in any cell type.

Antisense oligonucleotides can be used for the same purpose and they have been shown to give good results, as in the case of the study of the somatostatin receptor (21). The use of this technique requires the availability of microinjection equipment and the skills to perform it. Oligonucleotides have also been transfected using different methods; however, it is a very expensive technique, considering that micro- to millimolar amounts have to be used.

Conclusions

We have described here the methods used to study the coupling of the D2 dopamine receptors to the G proteins. These studies allowed us to show that D2L couples preferentially to the α subunit of G_{i2} in the inhibition of adenylatecyclase. The results obtained with this method are reproducible and afford the advantage of studying receptor/G protein coupling in different types of cells. Considering that each cell type contains its own combination of specific G proteins, many cell types should be tested in order to get a complete picture of the signal transduction pathway utilized by each receptor. Performing these studies via the establishment of stable cell lines, although perhaps useful for some cases, is a long and time-consuming approach. In addition, the comparison of subtypes of the same receptor requires the establishment of cells expressing the same amounts of the receptors. The advantage of transient assays resides in the possibility to manipulate the system at any level. For example, by varying the concentration of transfected receptors it is possible to obtain different numbers of receptor sites in binding assays. In addition, the interaction of different proteins can be tested by cotransfecting the appropriate expression vectors altogether. Finally, the transcriptional assay has the advantage of amplifying the response to the activation of membrane receptors and enabling study of their effect at the transcriptional level. This method is rapid and easy to perform. The combined use of the techniques described perhaps will help to characterize the interactions of membrane receptors with members of the G protein family and to obtain insights into the complexity of signal transduction mechanisms mediated by the 7TM domain G-protein-coupled receptors.

Acknowledgments

We acknowledge Jean-Pierre Montmayeur for setting up the transcriptional assay and Janique Guiramand for the binding on whole cells. We thank Paolo Sassone Corsi for discussions and critical reading of the manuscript. JHB and RP were recipients of fellowships from Fondation Fyssen and AIRC, respectively. This work was supported by grants from Association pour la Recherche contre le Cancer, CNRS, INSERM, and Rhone Poulenc Rorer.

References

1. H. K. Dohlman, J. Thorner, M. G. Caron, and R. J. Lefkowitz, *Annu. Rev. Biochem.* **60,** 653 (1991).
2. M. L. Simon, M. P. Strathmann, and N. Gautam, *Science* **252,** 802 (1991).
3. W. C. Probst, L. A. Snyder, D. I. Schuster, J. Brosius, and S. C. Sealfon, *DNA Cell Biol.* **11,** 1 (1992).
4. L. Buck and R. Axel, *Cell* **65,** 175 (1991).
5. M. G. Caron and R. J. Lefkowitz, *Recent Prog. Hormone Res.* **48,** 277 (1993).
6. J. A. Gingrich and M. G. Caron, *Annu. Rev. Neurosci.* **16,** 299 (1993).
7. J. R. Bunzow, H. H. M. Van Tol, D. Grandy, P. Albert, J. Salon, M. D. Christie, C. A. Machida, K. A. Neve, and O. Civelli, *Nature* (*London*) **336,** 783 (1988).
8. P. Sokoloff, B. Giros, M. P. Martres, M. L. Bouthenet, and J. C. Schwartz, *Nature* (*London*) **347,** 146 (1990).
9. H. H. M. Van Tol, J. R. Bunzow, H. C. Guan, R. K. Sunahara, P. Seeman, H. B. Niznik, and O. Civelli, *Nature* (*London*) **350,** 610 (1991).
10. A. Dearry, J. A. Gingrich, P. Falardeau, R. T. Fremeau, Jr., M. D. Bates, and M. G. Caron, *Nature* (*London*) **347,** 72 (1990).
11. D. K. Grandy, Y. Zhang, C. Bouvier, Q.-Y. Zhou, R. A. Johnson, L. Allen, K. Buck, J. R. Bunzow, J. Salon, and O. Civelli, *Proc. Natl. Acad. Sci. U.S.A.* **88,** 9175 (1991).
12. J.-P. Montmayeur, P. Bausero, N. Amlaiky, L. Maroteaux, R. Hen, and E. Borrelli, *FEBS Lett.* **278,** 239 (1991).
13. S. Green, I. Issemann, and E. Sheer, *Nucl. Acids Res.* **16,** 369 (1988).
14. J. Sambrook, E. F. Fritsch, and T. Maniatis, "Molecular Cloning: A Laboratory Manual." Cold Spring Harbor Laboratory, Cold Spring Harbor, New York, 1989.
15. J.-P. Montmayeur and E. Borrelli, *Proc. Natl. Acad. Sci. U.S.A.* **88,** 3135 (1991).
16. E. Borrelli, J.-P. Montmayeur, N. S. Foulkes, and P. Sassone Corsi, *CRC Rev. Oncogenesis* **3**(4), 321 (1992).
17. C. M. Gorman, L. F. Moffat, and B. H. Howard, *Mol. Cell. Biol.* **2,** 1044 (1982).
18. P. Herbomel, B. Bourachot, and M. Yaniv, *Cell* **39,** 653 (1984).
19. M. J. Sleigh, *Anal. Biochem.* **156,** 251 (1986).
20. D. E. Hruby, and E. M. Wilson, *in* Methods in Enzymology," Vol. 216, p. 369. Academic Press, San Diego, 1992.
21. J.-P. Montmayeur, J. Guiramand, and E. Borrelli, *Mol. Endocrinol.* **7,** 161 (1993).
22. C. Kleuss, J. Herscheler, C. Hewel, W. Rosenthal, G. Schultz, and B. Wittig, *Nature* (*London*) **353,** 43 (1991).

[8] Assays for G Protein $\beta\gamma$ Subunit Activity

Joyce B. Higgins and Patrick J. Casey*

Introduction

Heterotrimeric guanine nucleotide-binding regulatory proteins (G proteins) couple transmembrane receptors to intracellular effectors, resulting in the modulation of second messenger concentrations (1). The G protein subunits are designated G_α, G_β, and G_γ, in order of decreasing mass, with the G_α subunit containing the guanine nucleotide-binding site. On activation, the G_β and G_γ subunits dissociate from the GTP-bound G_α subunit as a complex. The G_β and G_γ subunits do not separate under nondenaturing conditions and therefore the $G_{\beta\gamma}$ complex acts functionally as a monomer. The $G_{\beta\gamma}$ complex is required for efficient association of the G_α subunit with the plasma membrane and for activation of the G_α subunit by the receptor. Additionally, a number of signaling functions have been directly ascribed to the $G_{\beta\gamma}$ complex released on G protein activation (2). Among the activities ascribed to the $G_{\beta\gamma}$ complex, the association of $G_{\beta\gamma}$ with G_α is the most well-defined protein–protein interaction. One consequence of delineating the properties associated with $G_\alpha \cdot G_{\beta\gamma}$ association is that assays based on these interactions can be used to quantitate accurately the amount of "active" $G_{\beta\gamma}$ in a preparation. In addition, molecules that associate with either G_α or $G_{\beta\gamma}$ to disrupt $G_\alpha \cdot G_{\beta\gamma}$ interactions can be identified using these types of assays.

The $G_{\beta\gamma}$ complex modulates the activity of the G_α subunit by supporting ADP-ribosylation of G_α by pertussis toxin (PTx) and by decreasing the ability of G_α to exchange guanine nucleotides, thereby inhibiting the turnover of GTP to GDP in the GTPase reaction (3). PTx catalyzes the covalent modification of certain G_α subunits on a cysteine residue near the carboxyl terminus (4), and it is the intact heterotrimer rather than the free G_α polypeptide that serves as the substrate. Modification of this particular residue of G_α hinders the coupling of the heterotrimer to its receptor (5). A select group of G_α isotypes (i.e., G_i, G_o, G_t) serve as PTx substrates. One substrate, $G_{o\alpha}$, is particularly abundant in brain, constituting about 1–2% of the membrane

* To whom correspondence should be addressed.

Methods in Neurosciences, Volume 29

protein in this tissue. This chapter presents protocols for measuring both the pertussis toxin-catalyzed ADP-ribosylation of the G_α subunit of G_o ($G_{o\alpha}$) and the intrinsic GTPase activity of $G_{o\alpha}$. These two protocols are used to determine the quantity of active $G_{\beta\gamma}$ in samples containing this complex. The sensitivity of the GTPase assay for $G_{\beta\gamma}$ is in the low picomole range, whereas the ADP-ribosylation assay is more sensitive and can be used to quantitate femtomole amounts of $G_{\beta\gamma}$. Included are adaptations of the standard protocols that are useful in competition studies designed to detect and/or quantitate molecules that interact with either $G_{o\alpha}$ or $G_{\beta\gamma}$. The advantage of these assays is that they measure $G_{\beta\gamma}$ accurately, reproducibly, and rapidly, and the purification of both $G_{o\alpha}$ (6) and $G_{\beta\gamma}$ (7) (the G protein subunits required for the assays) from bovine brain is readily achievable.

Protocol 1: Pertussis Toxin-Catalyzed ADP-Ribosylation of $G_{o\alpha}$

Reagents

Stock Solutions
Prepare in distilled H_2O unless otherwise indicated.

   1 *M* Tris-HCl, pH 7.7; store at 4°C.
   0.1 *M* Na-EDTA, pH 8.0; store at 4°C.
   100 μM NAD; store at −20°C.
   1 *M* Dithiothreitol (DTT); store at −20°C.
   1 *M* MgCl$_2$; store at 4°C.
   10 m*M* GDP; store at −20°C.
   100 μg/ml Pertussis toxin (PTx, List Biologicals, Campbell, CA); add
      0.5 ml of 2 *M* urea in 100 m*M* potassium phosphate, pH 7.0, to one
      vial (50 μg) PTx; store at 4°C, stable for at least 2 months.
   30 m*M* Dimyristoylphosphatidylcholine (DMPC, Sigma, St. Louis, MO);
      suspend 10 mg DMPC in 0.5 ml of 20 m*M* HEPES, pH 8.0, 2 m*M*
      MgCl$_2$, 1 m*M* Na-EDTA; store at 4°C and sonicate 5 min at room
      temperature before each use to dispense the lipid.
   [^{32}P]NAD (NEN, Boston, MA, 800 Ci/mmol); store at −20°C.
   10% Lubrol (ICN, Costa Mesa, CA); resuspend 100 g in 1 liter distilled
      H_2O and deionize by stirring overnight with 5 g mixed-bed AG 501
      ion-exchange resin (Bio-Rad, Richmond, CA); filter to remove resin
      and store at 4°C.
   2% Sodium dodecyl sulfate (SDS); store at room temperature.
   30% and 6% trichloroacetic acid (TCA); store at room temperature.

Materials

BA85 filters (Schleicher and Schuell, Keene, NH)
12 × 75-mm polypropylene tubes

Working Solutions

Buffer A: 50 mM Tris-HCl, pH 7.7, 1 mM Na-EDTA, 1 mM DTT, 0.025% Lubrol.

PTx mix, 400 μl (for 25-tube assay); prepare fresh each day.
20 μl 1 M Tris-HCl, pH 7.7.
26.7 μl 100 μM NAD.
2.2 μl 1 M DTT.
2 μl 1 M MgCl$_2$.
10 μl 10 mM GDP.
53.3 μl 100 μg/ml PTx.
19 μl 30 mM DMPC.
2 × 10^7 cpm [^{32}P]NAD.
Distilled water to 400 μl final volume.

G$_{o\alpha}$: Dilute stock G$_{o\alpha}$ to 3.75 μM (0.5 μM for modified protocol) in Buffer A; keep on ice during use.

G$_{\beta\gamma}$: Dilute stock G$_{\beta\gamma}$ solution to the desired concentration in Buffer A; keep on ice during use.

Standard Procedure

The assay is conducted by mixing G$_{o\alpha}$ (4 μl, 15 pmol) and the solution containing G$_{\beta\gamma}$ (standards and samples) with Buffer A to a final volume of 25 μl in 12 × 75-mm polypropylene tubes on ice. The typical range for the standard curve is 0.05 to 2.0 pmol G$_{\beta\gamma}$. At timed intervals, the reactions are initiated by addition of 15 μl of the PTx mix (final volume of the assay is 40 μl) and the tubes immediately transferred to a 30°C water bath. After 20 min, the reactions are stopped by addition of 0.5 ml 2% SDS and are removed from the water bath. The proteins are precipitated by addition of 0.5 ml 30% TCA. Each tube is vortexed and then incubated at room temperature for at least 15 min. The reactions are diluted with 2 ml of 6% TCA and filtered through BA85 nitrocellulose filters. The tubes are rinsed with an additional 2 ml of 6% TCA, this rinse is filtered, and the filters are washed with an additional 3 × 2 ml of 6% TCA. The filters are dried (use of a heat lamp will decrease the drying time), transferred to a scintillation vial, suspended in 4 ml of scintillation fluid, and subjected to liquid scintillation spectrometry. To quantitate the specific activity of the [^{32}P]NAD, 5 μl of the PTx mix is

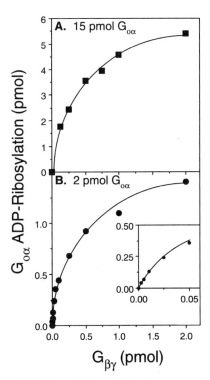

FIG. 1 $G_{\beta\gamma}$-supported ADP-ribosylation of $G_{o\alpha}$ by pertussis toxin. The indicated amounts of $G_{\beta\gamma}$ from bovine brain were assayed for their ability to support ADP-ribosylation of either 15 pmol (A) or 2 pmol (B) $G_{o\alpha}$. *Inset* (in B) The initial region of the dose–response curve at 2 pmol $G_{o\alpha}$ in expanded form for clarity. Symbols represent the average of at least four determinations.

spotted on a dry BA85 filter and counted with the samples (5 μl = 33.3 pmol NAD).

Results with and Modifications to ADP-Ribosylation Assay

A typical standard curve for the assay is depicted in Fig. 1A. This assay is most accurate for samples containing between 0.05 and 0.5 pmol $G_{\beta\gamma}$, because the reaction is linearly dependent on $G_{\beta\gamma}$ in this range. Samples with concentrations below this range can be more accurately detected by utilizing a revised protocol that reduces the quantity of $G_{o\alpha}$ present in the assay mixture to 2 pmol. The results obtained with the procedure using the lower amount of $G_{\beta\gamma}$ are shown in Fig. 1B. This type of assay is most accurate with $G_{\beta\gamma}$ samples containing between 2 and 25 fmol of the protein. Because $G_{\beta\gamma}$ acts catalytically to support ADP-ribosylation of $G_{o\alpha}$ (8), it is essential that the reaction incubations are accurately timed.

The standard assay described above has proved extremely useful in determining the concentration of active $G_{\beta\gamma}$ in homogeneous preparations as well as in less purified samples. One advantage of using $G_{o\alpha}$ in this assay is that the $G_{o\alpha}$ does not appear to discriminate between $G_{\beta\gamma}$ isolated from different tissue sources (i.e., retinal and brain $G_{\beta\gamma}$) (8). At a constant concentration of $G_{\beta\gamma}$, the ADP-ribosylation of $G_{o\alpha}$ reaches a maximum level at relatively low concentrations of $G_{o\alpha}$ in the 20-min incubation period (8). Therefore, samples that contain significant amounts of PTx-substrate G_α proteins can be assayed for their $G_{\beta\gamma}$ content by conducting the assay at levels of $G_{o\alpha}$ at or above 10 pmol. Using 15 pmol of $G_{o\alpha}$ (i.e., the standard protocol) will minimize the interference from such proteins in the sample; this is especially the case when the sample contains low amounts of $G_{\beta\gamma}$ (below 0.2 pmol $G_{\beta\gamma}$). Moreover, if the $G_{o\alpha}$ supply is not limiting, increasing the amount of $G_{o\alpha}$ in the assay to 25 pmol results in a $G_{\beta\gamma}$ response that is essentially G_α independent up to 1.0 pmol $G_{\beta\gamma}$ (9). It is important to note here, however, the G_α subunits that do not serve as PTx substrates can act as competitors in the assay by associating with the $G_{\beta\gamma}$ in the sample.

Molecules that interact with the G_α or $G_{\beta\gamma}$ proteins can be identified by their abilities to reduce the ADP-ribosylation of $G_{o\alpha}$ if the binding of the molecule to either G protein subunit blocks its capacity to interact with its cognate subunit. We typically use the standard protocol described above for these competition assays. A standard curve is generated, followed by an appropriate number of assay tubes containing 0.25 pmol of $G_{\beta\gamma}$. This concentration of $G_{\beta\gamma}$ is in the middle of the linear region of the $G_{\beta\gamma}$ response and therefore will allow either an inhibition or stimulation of ADP-ribosylation to be detected. Usually, the putative competitor and the subunit with which it interacts are preincubated at room temperature or 30°C for 5 min, followed by addition of the remaining G protein subunit and initiation of the reaction by addition of PTx mix. The preincubation allows any association to occur before the second G protein subunit is added and maximizes the probability of observing competition. Alternatively, when the competitor supply is limiting, the modified protocol (2 pmol $G_{o\alpha}$, 10 fmol $G_{\beta\gamma}$) can be employed so that less competitor is required to see the effect.

We have used this type of assay to detect a direct interaction of the β-adrenergic receptor kinase (βARK) and $G_{\beta\gamma}$ (10). βARK phosphorylates agonist-stimulated β_2-adrenergic receptors, which augments receptor desensitization. On receptor stimulation, βARK is rapidly translocated to the plasma membrane; this translocation is thought to be mediated through the binding of βARK to $G_{\beta\gamma}$ released on G protein activation by the liganded

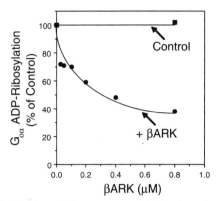

FIG. 2   Competitive inhibition of $G_{\beta\gamma}$-supported ADP-ribosylation of $G_{o\alpha}$ by βARK. The indicated amounts of βARK either native (●) or heat inactivated (control, ■), were incubated with 15 pmol $G_{o\alpha}$ and 0.25 pmol $G_{\beta\gamma}$ and assayed for ADP-ribosylation of $G_{o\alpha}$ by pertussis toxin. The "100% control" value is the level of ADP-ribosylation observed without added competitor. Symbols represent the average of two to four determinations. Adapted with permission from Ref. (10), J. A. Pitcher, J. Inglese, J. B. Higgins, J. L. Arriza, P. J. Casey, C. Kim, J. L. Behovic, M. M. Kwatra, M. G. Caron, and R. J. Lefkowitz. Role of $\beta\gamma$ subunits of G proteins in targeting the β-adrenergic receptor kinase to membrane bound receptors. *Science* **257**, 1264–1267 (1992), © American Association for the Advancement of Science.

receptor. The competition curve for βARK in the ADP-ribosylation assay is shown in Fig. 2. Concentration-dependent inhibition of ADP-ribosylation is observed, with an EC_{50} for the response occurring at a βARK:$G_{\beta\gamma}$ ratio of approximately 1. Because βARK preferentially associates with brain $G_{\beta\gamma}$ over retinal $G_{\beta\gamma}$ (10), an observation that holds for several other proteins that interact with $G_{\beta\gamma}$ (e.g., adenylylcyclase and phospholipase C), the use of $G_{\beta\gamma}$ isolated from brain is recommended for these studies.

The support of ADP-ribosylation assay was designed primarily to determine concentrations of functional $G_{\beta\gamma}$ in purified preparations of this protein, and thus the effect of agents such as detergents on the assay have not been extensively studied. However, the sensitivity of the assay has eliminated most such problems because most samples can be diluted quite far in Buffer A prior to analysis. However, the analysis of samples via competition studies may cause some interference if the samples are not diluted in Buffer A. Due to this concern, it is important to control for the sample buffer by generating the standard curve and testing the samples under equivalent buffer conditions. If there is inhibition due to a detergent, this can be alleviated by adding additional amounts of the lipid DMPC to the PTx mix. A fourfold increase in the DMPC concentration in the assay has no apparent effect on the standard curve. An increased amount of phospholipid will serve to sequester the detergent and thereby reduce its effect on the association of the G protein subunits. In addition to the buffer components, the test molecule may have hydrophobic characteristics that disturb the phospholipid environment of the assay. To diminish this type of problem, the competition assay can be performed either in the presence of excess (fourfold) DMPC or in the absence of DMPC. The latter condition only slightly lowers the level of ADP-ribosylation observed compared to the standard protocol (8) and allows an analysis of the competitor's affect on the subunit association without the complication of the component–lipid interactions.

Protocol II: GTPase Assay

Reagents

Stock Solutions

1 M Tris-HCl, pH 7.7; 1 M MgCl$_2$; 0.1 M Na-EDTA, pH 8.0; 1 M DTT; 10% Lubrol; see reagents for ADP-ribosylation assay.
100 μM GTP; store at $-20°C$.
[γ-^{32}P]GTP (NEN, 6000 Ci/mmol); store at $-20°C$.
0.5 M NaH$_2$PO$_4$; store at 4°C.

Materials

12 × 75-mm polypropylene tubes.

Working Solutions

GTPase mix, 1 ml (for 40-tube assay), prepare fresh: 50 μl 1 M Tris-HCl, pH 7.7; 1 μl 1 M MgCl$_2$; G proteins have differing responses to free magnesium concentration, therefore this concentration will vary for different G$_\alpha$ subunits (3); 10 μl 0.1 M Na-EDTA, pH 8.0; 2 μl 1 M DTT; 10 μl 100 μM GTP; 1 × 10^7 cpm [γ-^{32}P]GTP (for a specific activity of ~10,000 cpm/pmol); distilled water to 1 ml final volume.
Buffer B: 50 mM Tris-HCl, pH 7.7; 1 mM Na-EDTA; 1 mM DTT; 0.1% Lubrol.
Charcoal slurry: 25 g Norit A (Aldrich, Milwaukee, WI) suspended in 500 ml of 50 mM NaH$_2$PO$_4$; stir constantly during use; store at 4°C.
G$_{o\alpha}$: dilute stock G$_{o\alpha}$ to 0.1 μM in Buffer B; the protocol requires 5 μl per assay tube; keep on ice during use.
G$_{\beta\gamma}$: dilute stock G$_{\beta\gamma}$ solution to 0.1 μM in Buffer B; keep on ice during use.

Standard Protocol

The assay is conducted by adding $G_{o\alpha}$ (5 μl, 0.5 pmol) to the solution containing $G_{\beta\gamma}$ (standards and samples) in 12 × 75-mm polypropylene tubes on ice. Buffer B is added to a final volume of 25 μl. The $G_{\beta\gamma}$ standard curve ranges from 0.1 to 1.0 pmol. Contamination of the radioactive GTP by inorganic phosphate, and also the nonenzymatic breakdown of GTP during the reaction, can be corrected for by including a control sample with no added G protein. This background is usually quite low; however, the purity of the water source is crucial because minor contaminants have been known to cause drastic increases in background levels of phosphate release. The reactions are initiated at timed intervals by the addition of 25 μl of the GTPase mix (final volume of the assay is 50 μl) to each tube; the tubes are then incubated at 24°C for a selected time (this is generally 10 min for $G_{o\alpha}$). Typically, a time course generated by stopping samples at 2, 5, 10, and 20 min is conducted to determine the linearity of the assay for the particular preparation of $G_{o\alpha}$ (or other G_{α}) used. Each reaction is stopped by the addition of 750 μl of charcoal slurry (constantly stirred during use); the tube is then immediately vortexed and transferred to ice. After all the tubes have been at least 15 min on ice, the tubes are centrifuged at 1500 g at 4°C for 15 min. A 0.4-ml (~50%) aliquot of the supernatant is carefully transferred from each tube to a scintillation vial. The samples are then mixed with 4 ml of scintillation fluid and subjected to liquid scintillation spectrometry. The specific activity of the [γ-^{32}P]GTP is determined by mixing 10 μl of the GTPase mix with scintillation fluid and counting this with the samples (10 μl = 10 pmol GTP). The amount of P_i released is determined by averaging the radioactivity determined in duplicate samples, dividing by the specific activity of the GTP used, and multiplying by 2 (to correct for analysis of 50% of the final reaction mix).

Results with GTPase Assay

Figure 3 shows a typical standard curve generated for this assay. The maximal inhibition of the steady-state GTPase of 50–60% is typical for most G_{α} subunits; this maximal inhibition occurs at slightly greater than a 1:1 stoichiometry of G_{α}:$G_{\beta\gamma}$. Although the GTPase assay has traditionally been utilized to assess the activity of purified G_{α} proteins, the ability to inhibit the activity in a dose-dependent fashion with $G_{\beta\gamma}$ makes it a useful tool for assessing $G_{\beta\gamma}$ activity and the activities of agents that interact with G_{α} or $G_{\beta\gamma}$. Thus, any molecule that disrupts or reduces the $G_{\alpha} \cdot G_{\beta\gamma}$ interaction (i.e., binds to either G_{α} or $G_{\beta\gamma}$ to block their association) would result in a higher rate of

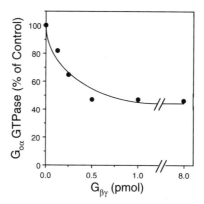

FIG. 3 $G_{\beta\gamma}$ inhibition of GTP hydrolysis by $G_{o\alpha}$. $G_{o\alpha}$ (0.5 pmol) was incubated for 10 min at 24°C in the absence or presence of the indicated amounts of $G_{\beta\gamma}$ and GTP hydrolysis was quantitated. The "100% control" value is the amount of GTP hydrolyzed by $G_{o\alpha}$ in the absence of $G_{\beta\gamma}$. Symbols represent the average of at least six determinations.

GTP hydrolysis than would occur in its absence. Another property of this reaction is that molecules that alter (either inhibit or stimulate) the intrinsic GTPase activity of G_{α} proteins can be detected in this system. As with the ADP-ribosylation assay described above, detergents or other components in the sample solution may interfere with $G_{\alpha} \cdot G_{\beta\gamma}$ interactions and/or alter the GTPase activity of $G_{o\alpha}$, necessitating that standard curves be conducted in the presence of components identical to those used during sample testing.

Conclusions and Applications of Both Methods

The ADP-ribosylation assay is a highly accurate and extremely sensitive assay for measuring $G_{\beta\gamma}$ concentrations. This assay can be utilized to identify molecules that interact with either G_{α}, $G_{\beta\gamma}$, or the heterotrimer. The GTPase assay is useful for further analysis of these potential competitors to determine their ability to disrupt $G_{\alpha} \cdot G_{\beta\gamma}$ interactions. A stable association of G_{α}–GDP to $G_{\beta\gamma}$ is required to see the reduction in GTP hydrolysis in the GTPase assay, whereas a catalytic $G_{\alpha} \cdot G_{\beta\gamma}$ interaction supports ADP-ribosylation by PTx. And, unlike the PTx-catalyzed ADP-ribosylation assay, which involves three molecules and a poorly understood mechanism, the GTPase assay requires only two molecules and is well characterized. Thus, the GTPase assay eliminates the potential problems associated with the PTx assay in which the third molecule (PTx) involved can be affected by the putative competitors. Although each of these two assays has benefits and drawbacks, when both are used, the complementary results provide higher confidence in descriptions of the interactions involved.

Several groups have applied these two types of assays to assess functional activities and levels of $G_{\beta\gamma}$ in their preparations. In addition, the ADP-ribosylation assay has proved extremely valuable in studies that addressed posttranslational modification of the G_γ subunit in $G_{\beta\gamma}$ complexes. The G_γ subunits are prenylated, subjected to carboxyl-terminal proteolysis, and carboxyl methylated (11). One study demonstrated that the ability of $G_{\beta\gamma}$ to support ADP-ribosylation was increased twofold when the $G_{\beta\gamma}$ was methylated (12). This method was employed because analysis of the rate of ADP-ribosylation is the most sensitive method for detecting the interaction of $G_\alpha \cdot G_{\beta\gamma}$ interaction in solution. Other studies have shown that the $G_\alpha \cdot G_{\beta\gamma}$ interaction required prenylation of the G_γ subunit (13, 14). In one of these studies that utilized *in vitro* assembly of $G_{\beta\gamma}$ (14), only femtomole quantities of $G_{\beta\gamma}$ could be produced and the utilization of the ADP-ribosylation assay was essential to its success. These examples, and the βARK study discussed earlier demonstrate the utility and the sensitivity of the ADP-ribosylation assay in particular.

Acknowledgments

This work was supported in part by American Cancer Society Grant BE-117. PJC is an Established Investigator of the American Heart Association.

References

1. J. R. Hepler and A. G. Gilman, *TIBS* **17**, 383 (1992).
2. D. E. Clapham and E. J. Neer, *Nature (London)* **365**, 403 (1993).
3. T. Katada, M. Oinuma, and M. Ui, *J. Biol. Chem.* **261**, 8182 (1986).
4. R. E. West, Jr., J. Moss, M. Vaughan, T. Liu, and T.-Y. Liu, *J. Biol. Chem.* **260**, 14428 (1985).
5. M. Ui, *Trends Pharmacol. Sci.* **5**, 277 (1984).
6. P. C. Sternweis and J. D. Robishaw, *J. Biol. Chem.* **259**, 13806 (1984).
7. D. J. Roof, M. L. Applebury, and P. C. Sternweis, *J. Biol. Chem.* **260**, 16242 (1985).
8. P. J. Casey, M. P. Graziano, and A. G. Gilman, *Biochemistry* **28**, 611 (1989).
9. P. J. Casey, I.-H. Pang, and A. G. Gilman, *in* "Methods in Enzymology" (R. A. Johnson and J. D. Corbin, eds.), Vol. 195, p. 315. Academic Press, San Diego, 1991.

10. J. A. Pitcher, J. Inglese, J. B. Higgins, J. L. Arriza, P. J. Casey, C. Kim, J. L. Benovic, M. M. Kwatra, M. G. Caron, and R. J. Lefkowitz, *Science* **257,** 1264 (1992).
11. P. J. Casey, *Curr. Opin. Cell Biol.* **6,** 219 (1994).
12. Y. Fukada, T. Matsuda, K. Kokame, T. Takao, Y. Shimomshi, T. Akino, and T. Yoshizawa, *J. Biol. Chem.* **269,** 5163 (1994).
13. J. A. Iñiguez-Lluhi, M. I. Simon, J. D. Robishaw, and A. G. Gilman, *J. Biol. Chem.* **267,** 23409 (1992).
14. J. B. Higgins and P. J. Casey, *J. Biol. Chem.* **269,** 9067 (1994).

[9] α_2-Adrenergic Receptor Coupling to G Proteins

Ann E. Remmers

Introduction

Membrane receptors for a large number of neurotransmitters and hormones activate guanine nucleotide-binding proteins (G proteins), causing an alteration in cellular metabolism (for review, see Ref. 1). A requirement for GTP (2) and an observed increase in GTPase activity (3) due to hormonal stimulation of intracellular cAMP accumulation indicates that G proteins are an integral component of receptor-mediated signal transduction. Initially, heterotrimeric G proteins (composed of α, β, and γ subunits) were classified based on their ability to stimulate (G_s) or inhibit (G_i) adenylylcyclase activity. Subtypes of these G proteins, in addition to other G proteins, have been identified (for review, see Ref. 4). In addition to modulation of adenylyl-cyclase activity, G proteins have been shown to alter the function of many other effector enzymes (i.e., phospholipases C and A_2, cGMP phosphodiesterase, sodium/proton exchanger, and ion channels) (5).

G protein α subunits function as molecular switches. The GTP-bound form is "on" interacting with effector enzyme and the GDP-bound form is inactive. Agonist-liganded receptor catalyzes the exchange of GDP for GTP bound to G protein α subunit. The activated G protein α and $\beta\gamma$ subunits can interact with different effectors. The activation signal is terminated by α subunit GTPase activity, forming inactive α-GDP, which reassociates with $\beta\gamma$ subunits. The α_2-adrenergic receptor (α_2AR) has been used as a model system to study G_i protein–receptor interactions and generation of second messengers (6–9). This chapter addresses the α_2AR coupling to inhibitory G proteins, although the methods described are applicable to many other G-protein-coupled receptors.

The affinity of many brain membrane receptor agonists but not antagonists is diminished by nonhydrolyzable GTP analogs. For example, equilibrium binding of [^{3}H]UK 14,304 (5-bromo-6-[2-imidazoline-2-ylamino]quinoxaline; brimonidine), an α_2AR receptor agonist in purified platelet membranes, indicated that high-level agonist affinity for receptor was eliminated by a nonhydrolyzable guanine nucleotide analog, guanosine 5′-(β,γ-imido)triphosphate [Gpp(NH)p; (10)]. In contrast, the binding characteristics of [^{3}H]yohimbine were unaltered by guanine nucleotides. Because agonist stimulates binding of GTP, thermodynamics predict that GTP will lower the binding affinity of agonist.

The inhibitory G proteins (G_{i1}, G_{i2}, G_{i3}), transducin (the retinal G protein), and G_o, the major G protein in brain, serve as substrates for *Bordetella pertussis* toxin. The toxin catalyzes the ADP-ribosylation of a cysteine near the C terminus of G proteins, rendering them unable to exhibit receptor-catalyzed GTP exchange. Thus pertussis toxin sensitivity of a receptor-stimulated effector is an important means to implicate inhibitory G proteins in a cellular signal transduction pathway. For example, neuropeptide Y inhibited the calcium current in rat dorsal root ganglion neurons *in vitro*, an effect that was abolished by pertussis toxin pretreatment (11). Readdition of $G_{o\alpha}$ and GTP restored neuropeptide Y inhibition of calcium current.

Reconstitution of purified proteins is a standard method for identifying and characterizing the mechanisms of protein–protein interaction. Reconstitution of receptor and G protein into phospholipid vesicles followed by assessment of G-protein-dependent high-affinity agonist binding and agonist-mediated receptor activation of GTPase indicates a functional coupling of specific receptors and G proteins. For example, Cerione *et al.* (9) reconstituted α_2AR with purified G_o, red blood cell G_i, and transducin and compared the ability of agonist to stimulate GTPase activity. They found that α_2AR was capable of maximally activating G_o and G_i to the same extent, but was less effective in promoting acitvation of transducin or G_s. These results indicate that one receptor is capable of activating more than one type of G protein. However, one drawback of reconstitution approaches is the physiological significance, i.e., do the receptor–G protein interactions occur in intact cells? Methods for reconstitution of receptors and G proteins in phospholipid vesicles have been described previously (12).

Additional approaches have been developed to investigate recpetor coupling to G proteins. Antibodies that recognize G protein α or β subunits have been utilized to block specific receptor–G protein interactions. Because antibodies exist that are specific for G protein subtypes, specific interactions in cell membranes can be identified (see Refs. 13 and 14 for review).

The observation that a specific receptor and G protein interact in cells can be taken a step further by determining which portion of the receptor is essential for G protein interaction. Peptides containing receptor sequence have been used to modulate receptor–G protein interaction for numerous receptors, including the βAR (15), muscarinic (16), dopamine D_2 (17), and α_2AR (7, 18) receptors and rhodopsin (19), thus elucidating proposed protein–protein contact regions.

To identify specific receptor G protein coupling on the cellular level, antisense oligonucleotides can be injected into cell nuclei. Kluess *et al.* (20, 21) designed oligodeoxynucleotides in an antisense orientation to the respective messenger RNAs for specific G protein α or β subunits. The DNA microinjected into the nuclei of rat pituitary GH3 cells selectively suppressed protein expression of the targeted G protein subunit. They found that G_o protein subtypes G_{o1} and G_{o2} mediate inhibition of voltage-dependent calcium channels through the muscarinic cholinergic and somatostatin receptors, respectively. Although this is a powerful approach to determine specific receptor–G protein–effector interactions, presently it is limited to effector systems that can be monitored at the level of single cells. This chapter will focus on methods utilizing G protein subunit antibodies and receptor sequence peptides to probe α_2AR–G protein interactions.

Use of G Protein Antisera to Detect Receptor–G Protein Coupling

Activation of the α_2AR inhibits adenylylcyclase via the activation of a pertussis toxin-sensitive guanine nucleotide-binding protein G_i (22). Because all three G_i subtypes, as well as G_o, are inactivated by pertussis toxin, a method to discriminate among the G_i subtypes was desirable in order to determine which G protein(s) are essential for α_2AR inhibition of adenylylcyclase. Use of antisera directed against the C-terminal region of G_{i1}/G_{i2} or G_{i3} proteins in a high-affinity α_2AR agonist binding assay and adenylylcyclase assay demonstrated that both G_{i2} and G_{i3} transduce α_2AR inhibition of adenylylcyclase (8). The following approaches are applicable to examining pertussis toxin-sensitive receptor signal transduction.

Preparation of Membranes

Stable clones of Chinese hamster ovary (CHO-K1) cells transfected with human platelet α_{2A}AR are utilized in these experiments (8, 23). Because the CHO-K1 cells do not contain endogenous α_2ARs, the experiments performed assess the ability of one specific receptor subtype to transduce via distinct G proteins. Mu opioid receptor–G protein coupling has been assessed using membrane preparations from rat brain, neuroblastoma glioma hybrid NG108-15 cells, and differentiated human neuroblastoma SH-SY5Y cells (24, 25). Detailed methods for preparation of membranes from tissue culture and primary tissue have been previously described (26). For these experiments, cells are grown in monolayer culture in Dubecco's modified Eagle's medium containing 10% (v/v) fetal calf serum, in a water-saturated 5% (v/v) CO_2 atmosphere, at 37°C. Monolayer cultures of cells are washed with ice-cold phosphate-buffered saline (PBS), lysed hypotonically with 1 mM Tris-HCl, pH 7.4, at 4°C for 15 min, scraped, and pelleted (30 min centrifugation at

TABLE I   G Protein Antibodies[a]

Designation	epitope	Ref.
AS/7	C terminus of $G_{i\alpha1}/G_{i\alpha2}$	44, 45
EC/2	C terminus of $G_{i\alpha3}/G_{o\alpha}$	35
GC/2	N terminus of $G_{o\alpha}$	13

[a] Available from Du Pont/NEN (Boston, MA).

100,000 g). Cell membranes are resuspended in TME buffer and frozen in liquid nitrogen and stored at $-70°C$. Membrane protein is determined by Lowry protein assay with bovine serum albumin (BSA) as standard (27).

Western Blotting

The first step in implicating a specific G protein in a receptor–effector system is to determine which G proteins are present in the membranes. Methods for quantitative immunoblotting of G protein subunits have been described previously (28). G protein antisera are available commercially (see Table I)* (see also Chapter 14, this volume). The experiments described here utilize AS/7 and EC/2 antibodies, which are specific against the α subunits of G_{i1}/G_{i2} and G_{i3}, respectively.

Cell membranes (75 μg membrane protein) from CHO-K1 cells expressing 1–2 pmol of α_2AR/mg membrane protein are mixed with Laemmli sample buffer and run on an 11% (w/v) sodium dodecyl sulfate-polyacrylamide gel (29). Bovine brain G protein, consisting primarily of a G_{oA} (39 kDa) and G_{i1} (41 kDa) mixture (100 ng), is also run on the same gel. Proteins are transferred to Immobilon-P (Millipore, Bedford, MA), and a Western blot is performed with G protein antisera at 1:1000 dilution. After transfer, the Immobilon is washed 3 × 20 min with blocking buffer (1% BSA in 0.1% Tween 20, 20 mM Tris-Cl, 500 mM NaCl, pH 7.4) followed by a 3 × 5 min wash with wash buffer (0.1% (w/v) Tween 20, 20 mM Tris-Cl, 500 mM NaCl, pH 7.4). Antibody (5 μl) is added to 5 ml antibody dilution buffer (1% (w/v) BSA in 0.1% Tween 20, 20 mM Tris-Cl, 500 mM NaCl, 0.02% sodium azide, pH 7.4) and incubated with the Immobilon on a rocker at room temperature. After an overnight incubation, the Immobilon is washed 3 × 10 min with

---

* Biodesign International, Kennebunkport, ME, and Upstate Biotechnology Incorporated, Lake Placid, NY, are additional sources of anti-G protein antibodies.

blocking buffer and 3×5 min with wash buffer. To visualize the bound primary antibody, the biotin–ExtrAvidin–alkaline phosphatase system is used. The Immobilon is incubated with antirabbit IgG (whole molecule) biotin conjugate (Sigma, St. Louis, MO, 1:2000 dilution) in antibody diluton buffer for 60–90 min, followed by 3×10 min wash with blocking buffer and 3×5 min wash with wash buffer. The blot is then incubated 60–90 min with ExtrAvidin–alkaline phosphatase (Sigma, 1:2000 dilution) in antibody dilution buffer, followed by 2×10 min wash with wash buffer and 2×10 min with 20 mM Tris-Cl, 500 mM NaCl, pH 7.4. The blot is developed using development solution. [To make the development solution, first dissolve 6 mg nitro blue tetrazolium (Sigma) and 3 mg 5-bromo-4-chloro-3-indoyl phosphate (Sigma) in 0.2 ml dimethylformamide followed by 0.2 ml of 100 mM Tris-Cl, 1 mM MgCl$_2$, pH 9.5, and then 18.4 ml of the same buffer.] Development is stopped by washing in distilled water. (The diluted antibodies can be stored at 4°C and reused two more times.)

Western blot analysis of CHO-K1 cell membranes expressing α_{2A}AR indicates that G_{i2} and G_{i3} are present in these membranes but not G_o or G_{i1} (8). Thus, the attributes of the G_{i1}/G_{i2} antisera in these membranes are due to its interaction with G_{i2}.

Ligand Binding

Graeser and Neubig (26) provide protocols for preparing plasma membranes as well as methods and experimental considerations for radioligand binding to detect receptor–G protein interactions. Tritiated UK 14,304 and [^{125}I]PIC are full and partial agonists, respectively, for both α_2ARs. Both radiolabeled agonists and antagonist [^{3}H]yohimbine are obtained from DuPont-NEN. Unlabeled PIC is purchased from Research Biochemicals (Natick, MA). Ligand binding studies are performed in TME buffer (50 mM Tris-Cl, 10 mM MgCl$_2$, 1 mM EGTA, pH 7.6) at room temperature. Membranes (0.04 mg/ml for PIC binding and 0.16 mg/ml for [^{3}H]UK 14,304) are incubated for 45 min at room tepmerature with 1:100 dilutions of serum. G protein coupling to other receptors is blocked by a similar concentration of these antisera (25, 30). A control incubation with nonimmune rabbit serum (Sigma) is also included. The binding assay is initiated by addition of 1 nM [^{125}I]PIC or 1 nM [^{3}H]UK 14,304. Following a 60-min incubation, samples are diluted with 5 ml ice-cold TM (50 mM Tris-Cl, 10 mM MgCl$_2$, pH 7.6) and immediately filtered using a Brandel cell harvester and Whatman (Clifton, NJ) GF/C filters. Filters are washed with 3×4 ml TM, dried, and counted in 4 ml ScintiVerse (Fisher Scientific, Fair Lawn, NJ) liquid scintillation cocktail at 35% efficiency. Filters with [^{125}I]PIC are counted in a Beckman LS-5500 gamma counter at

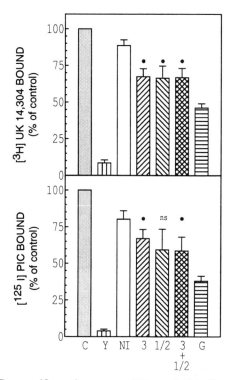

FIG. 1   Effects of $G_{i\alpha}$-specific antisera on α_2AR agonist binding. CHO-K1 cell membranes containing expressed α_{2A}AR were preincubated for 45 min at room temperature with buffer (C), 10 μM yohimbine (Y), 1:100 nonimmune serum (NI), 1:100 G_{i3} antiserum (3), 1:100 G_{i1}/G_{i2} antiserum (1/2), 1:100 G_{i1}/G_{i2} antiserum, and 1:100 G_{i3} antiserum (3 + 1/2), or 10 μM Gpp(NH)p (G). The binding assay was initiated by the addition of 1 nM [^{3}H]UK 14,304 (top) or 1 nM [^{125}I]PIC (bottom). Data are the mean ± SE of three or more experiments performed in triplicate. The asterisks above the columns denote the statistically significant difference from the nonimmune control ($p < 0.05$); ns, no statistically significant difference ($p < 0.05$). Reprinted with permission from Ref. 8, M. A. Gerhardt and R. R. Neubig, *Mol. Pharmacol.* **40**, 707–711.

78% efficiency. Nonspecific binding is defined using 10 μM yohimbine. The G-protein-dependent agonist binding is determined using 10 μM Gpp(NH)p (Sigma). In addition, α_2AR binding conditions for membranes prepared from rat cerebral cortex have been described previously (31, 32). Significance of the antisera effects is determined by calculating p values from paired, one-tailed, t-tests of the results with specific serum compared with a nonimmune control.

Figure 1A shows the effects of $G_{i\alpha}$-specific antisera on α_2AR agonist binding. Compared to the nonimmune control, specific [^{3}H]UK 14,304 binding was decreased 25% by G_{i1}/G_{i2} antiserum, G_{i3} antiserum, or both antisera together. Guanine nucleotide analog reduced agonist binding by 49%. In theory, only the guanine nucleotide-sensitive agonist binding would be decreased by the antisera; thus antisera decrease guanine nucleotide-sensitive agonist binding by 50%. Binding experiments with partial agonist [^{125}I]PIC produced similar results (Fig. 1B). If nonimmune serum decreases agonist binding at these dilutions, saturation binding of anti-G protein antibodies over a range of dilutions can be performed to determine their titers and the appropriate amount of antibody to use (25). The antigenic peptides are available commercially (Du Pont/NEN). To rule out nonspecific effects of the antibody, the peptides can be used to block antibody binding to the G protein.

Adenylylcyclase

To determine if the ability of G_i to couple to the α_2AR in a binding assay has any functional significance, the ability of agonist to inhibit forskolin-stimulated adenylylcyclase has been analyzed. Measurement of adenylylcyclase activity is done in buffer containing (final concentrations) 25 mM Tris-Cl, pH 7.6, 100 mM NaCl, 2.5 mM MgCl$_2$, 0.2 mM ATP (0.5 μCi [α-^{32}P]ATP/tube), 10 μM GTP, 0.1 mM EGTA, 1 mM dithiothreitol, 5 mM phosphocreatine, 0.2 mg creatine kinase/ml, 1 mM cAMP (Tris salt), 0.1 mM isobutylmethylxanthine, and 10 μM propanolol. Membranes are preincubated for 45 min with vehicle, nonimmune, and G protein antisera at room temperature. The room temperature incubation is absolutely required to see the effects of antibody; the antibodies have no effect following incubation on ice. The room temperature incubation, however, decreases the maximal cyclase inhibition observed by α_2AR agonist (from 60 to 35%). The assay is initiated by the addition of reaction cocktail, 10 μM forskolin, and 1 μM UK 14,304 or PIC. Assays are conducted for 15 min at 30°C. The [^{32}P]cAMP formed is measured by the method of Salomon et al. (33). The 5-ml aqueous fraction is counted in 15 ml Ultima Gold liquid scintillation cocktail. Alternatively, cAMP can be determined by radioligand binding assay [Diagnostic Products Corp., Los Angeles, CA (34)].

Reduction of the α_{2A}AR-mediated inhibition of adenylylcyclase by the G_{i1}/G_{i2} antiserum and the G_{i3} antiserum is shown in Fig. 2. Forskolin (10 μM) stimulates cAMP formation approximately 10-fold over basal, and the antisera have little if any effect on forskolin-stimulated adenylylcyclase (data not shown). The full agonist UK 14,304 inhibits forskolin-stimulated cAMP formation slightly more than does partial agonist PIC (34 ± 3% versus 27 ±

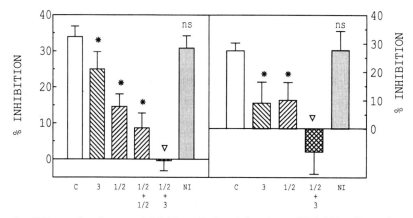

FIG. 2   Effects of antisera on inhibition of adenylylcyclase. CHO-K1 cell membranes containing expressed α_{2A} AR were preincubated for 45 min at room temperature with buffer (C), 1:100 G_{i3} antiserum (3), 1:100 G_{i1}/G_{i2} antiserum (1/2), 1:50 G_{i1}/G_{i2} antiserum (1/2 + 1/2), 1:100 G_{i1}/G_{i2} antiserum, and 1:100 G_{i3} antiserum (1/2 + 3), and 1:100 nonimmune serum (NI). The assay was initiated by the addition of reaction cocktail, forskolin, and 1 μM UK 14,304 (left) or 1 μM PIC (right). Data are presented as mean ± SE from three to five experiments. Individual experiments were performed in duplicate or triplicate. The asterisks above the columns denote the statistically significant difference from the nonimmune control ($p < 0.05$); ns, no statistically significant difference ($p < 0.05$). Nonimmune treatment (NI) was compared with control (C). ∇, Significant difference, compared with the result with G_{i1}/G_{i2} antiserum alone. Reprinted with permission from Ref. 8, M. A. Gerhardt and R. R. Neubig, *Mol. Pharmacol.* **40,** 707–711.

3%). Both G_{i1}/G_{i2} and G_{i3} antisera block α_{2A}AR-mediated inhibition. Statistically, both G_{i3} antiserum ($p = 0.04$) and G_{i2}/G_{i3} antiserum ($p = 0.0006$) treatments decrease UK 14,304 inhibition of cAMP accumulation and are different than the nonimmune control. The effects of the two antisera are additive. Similar results are observed for PIC. These results demonstrate that the α_{2A}AR can couple to both G_{i2} and G_{i3} proteins to inhibit adenylylcyclase in this transfected CHO cell system. Results using platelet membranes indicate that only G_{i2} is involved in adenylylcyclase inhibition but additivity of G_{i2} and G_{i3} antisera has not been tested (35).

It is unlikely that cross-reactivity of the G_{i3} antibody with G_{i2} accounts for these results, although cross-reactivity between the anti-G_{i3} (EC/2) antiserum and G_{i2} has been reported (35). The functional data presented here are inconsistent with an effect of the EC antibody solely on G_{i2}. The G_{i2} antiserum was able to block only partially the platelet and CHO cell adenylylcyclase (35) (Fig. 2). However, in seven different experiments, the combination of

G_{i1}/G_{i2} and G_{i3} antisera resulted in a complete blockade of adenylylcyclase inhibition by α_2AR agonist. In addition, the ability of G_{i1}/G_{i2} and G_{i3} antisera to block adenylylcyclase inhibition was greater than simply doubling the concentration of G_{i1}/G_{i2} antibody (Fig. 2). Thus, G_{i3}, as well as G_{i2}, couples to the α_{2A}AR and participates in the inhibition of adenylylcyclase.

Recent results (20, 36, 37) indicate that β and γ subunits lend specificity to receptor–G protein interaction. These types of antibody experiments have focused on G_α receptor coupling, but the methods are amenable to address G protein β and γ receptor interaction with antibodies to the different β and γ subunits.

Use of Receptor Peptides to Detect Receptor–G Protein Coupling

The G-protein-coupled receptors, including the α_2AR, have a similar topographical arrangement in the membrane. These receptors contain seven membrane-spanning domains connected by three intracellular and three extracellular loops, with the C terminus intracellular (see Fig. 3). Mutagenesis studies as well as studies with chimeric adrenergic receptor and muscarinic receptors implicate the third intracellular loop as a critical, but not sole, determinant of specific G protein–receptor interaction. Although the following methods examine α_2AR–G protein coupling, these approaches are applicable to other G-protein-coupled receptors whereby the receptor sequence and radiolabeled agonist and antagonist are available.

α_2AR Peptide Design

Sequences of 12–13 amino acids from portions of the intracellular loops of the porcine brain α_2AR are synthesized (Fig. 3). Sequences are chosen to include regions near the putative membrane-spanning region because G proteins are membrane bound. Initial screening for effective peptides requires only 2–10 mg of peptide. Peptide synthesis and purification is available commercially.* Peptides used here are synthesized by the University of Michigan Protein Structure Facility or by Cambridge Research Biochemicals (Wilmington, DE) using fmoc chemistry. Peptides containing a sequence of the extracellular portion of the receptor yet with the same overall charge

---

* For example, Cambridge Research Biochemicals Incorporated, Wilmington, DE; see "Peptide Synthesis Services" for a list of suppliers in *American Laboratory* (buyers guide), February, 1995.

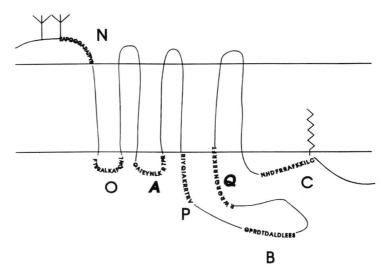

FIG. 3   Locations and sequences of synthetic peptides from the α_{2A}AR. The proposed structure of the α_2AR consists of seven membrane-spanning hydrophobic regions forming three cytosolic and three intracellular loops. Peptides were synthesized from the indicated regions based on the published sequence (43): N, residues 17–29; O, 59–70; A, 137–148; P, 218–229; B, 284–296; Q, 361–373; C, 430–442. Except for peptide C, which has a cysteine in the corresponding region of the α_{2A}AR, all peptides were synthesized with an additional cysteine attached to the N terminus (B, N. O, P) or the C terminus (A, Q). Peptide Q sequence: Arg-Trp-Arg-Gly-Arg-Gln-Asn-Arg-Glu-Lys-Arg-Phe-Thr-Cys. Peptide A sequence: Gln-Ala-Ile-Glu-Tyr-Asn-Leu-Lys-Arg-Thr-Pro-Arg-Cys. Bold letters indicate peptides that inhibit high-affinity agonist binding. Reprinted with permission from Ref. 7.

should be synthesized as controls. If not done so commercially, peptides should be purified using reversed-phase HPLC and mass spectrometry analysis performed to verify peptide identity. Peptide masses are determined by electrospray mass spectrometry using a Vestec single-quadropole mass spectrometer with electrospray interface by the Protein and Carbohydrate Structure Core Facility at the University of Michigan.

HPLC Purification of Receptor Peptides

Vydac 218TP C_{18} analytical column (0.46 × 25 cm), 1 ml/min flow rate, or Vydac 218TP C_{18} semipreparative column (1.0 × 25 cm), 2.5 ml/min flow rate.

Gradient: 0–15% B over 10 min; 15–30% B over 35 min; 30–90% B over
5 min followed by column reequilibration in Solvent A.
Solvent A: 0.1% (v/v) trifluoroacetic acid (TFA) in HPLC-grade water
(filter and degassed 20 min).
Solvent B: 0.1% TFA in HPLC-grade acetonitrile (filtered and degassed
1 min).
1 μg/μl peptide dissolved in solvent A; monitor 220-nm absorbance

Injections of 10 μg peptide are used to optimize the gradient by making
it more shallow over the region of major peak elution. If the peptide is less
than 95% pure, it is repurified using a semipreparative C$_{18}$ column. Vacuum
centrifugation or lyophilization is used to remove solvent. Peptides are stored
desiccated at −20°C. Peptides are dissolved in distilled water and frozen in
small aliquots to avoid repeated freezing and thawing.

Mass spectrometry analysis is instrumental in determining that the trypto-
phan in peptide Q is essential for activity, because when it is not deprotected
following peptide synthesis, the peptide is inactive (28). Thus, it is essential
to confirm peptide purity and identity.

Ligand Binding

Human platelets are reported to possess only α_{2A}AR (38), whereas NG108-
15 cells express α_{2B}AR (22). Plasma membranes are prepared from human
platelets by discontinuous sucrose density gradient centrifugation as described by Neubig and Szamraj (39). Plasma membranes are also prepared
from CHO-K1 cells expressing α_{2A}AR (an alternative to platelet membranes)
and neuroblastoma–glioma hybrid NG108-15 cells as described above for
CHO-K1 cells. Ligand binding studies are performed in TME buffer (50 mM
Tris-Cl, 10 mM MgCl$_2$, 1 mM EGTA, pH 7.6). Membranes (0.1–0.7 mg/ml
for platelets and 0.5–1.5 mg/mg for NG108-15) are preincubated with peptide,
other drugs, and buffer on ice for 45 min (0.5 ml total volume). Radioligand
(1 nM [^{125}I]PIC, 10 nM [^{3}H]yohimbine, or 1 nM [^{3}H]UK 14,304) is then
added and incubated at 25°C for 60 min. Samples are filtered and the bound
radioactivity determined as described above. Nonspecific binding is defined
using 10 μM oxymetazoline (Schering, Bloomfield, NJ) and 10 μM yohimbine
for agonist and antagonist binding, respectively. The G-protein-dependent
agonist binding is determined using 10 μM Gpp(NH)p. The IC$_{50}$ values are
determined by fitting the data to competition curves using the computer
program InPlot (GraphPAD Software, San Diego, CA).

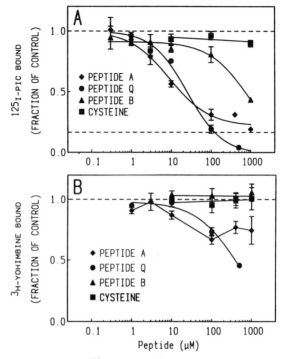

FIG. 4   Effect of peptides on [^{125}I]PIC and [^{3}H]yohimbine binding to platelet membrane α_{2A}AR. Membranes were preincubated with buffer containing cysteine (■) or peptides (A, ◆; Q, ●; B, ▲), and equilibrium binding assays were performed for partial agonist [^{125}I]PIC (A) and antagonist [^{3}H]yohimbine (B). The upper dashed line represents the amount of binding in the absence of peptide, and the lower dashed line represents the fraction of PIC binding remaining when 10 μM Gpp(NH)p was added to the binding assay. Control binding was 138 ± 20 and 315 ± 27 fmol/mg for [^{125}I]PIC and [^{3}H]yohimbine, respectively. Data points represent the mean ± SE of five different experiments. Reprinted with permission from Ref. 7.

Figure 4 shows the effect of peptides on [^{125}I]PIC and [^{3}H]yohimbine binding to platelet membrane α_{2A}AR. Cysteine was used as a control because each peptide was synthesized with a cysteine on either the amino or carboxyl end. Peptides A and Q potently inhibited the binding of partial agonist [^{125}I]PIC with IC$_{50}$ values of 6.7 ± 2.5 ($n = 5$) and 27.2 ± 3.9 ($n = 5$) μM, respectively. Peptide A completely abolished high-affinity agonist binding as defined with 10 μM Gpp(NH)p. In contrast, peptide Q eliminated the specific binding of [^{125}I]PIC. A peptide that disrupts only receptor–G protein interaction should, in theory, only alter high-affinity agonist binding that is

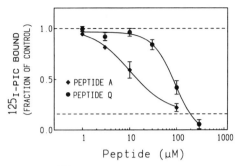

FIG. 5   Effect of peptides on [^{125}I]PIC binding to NG108-15 membrane α_{2B}AR. Membranes were preincubated with either buffer or peptide (A, ♦; Q, ●). The upper dashed line represents the amount of binding in the absence of peptide, and the lower dashed line represents the fraction of PIC binding remaining when 10 μM Gpp(NH)p was added to the binding assay. Control binding was 56 ± 8 fmol/mg for [^{125}I]PIC. Data points represent the mean ± SE of three different experiments. Reprinted with permission from Ref. 7.

dependent on G protein coupling. In contrast to the effects on [^{125}I]PIC binding, peptides A and Q had a diminished effect on antagonist binding (IC$_{50}$ values > 300 μM). The effect of peptide A was not limited to partial agonist [^{125}I]PIC; peptide A had an IC$_{50}$ of 14 μM ($n = 2$, data not shown) for reducing full agonist [^{3}H]UK 14,304 binding. The only other peptide that inhibited [^{125}I]PIC binding was B with an IC$_{50}$ of 900 μM. Peptides C, N, O, and P had no effect on α_{2A}AR ligand binding up to concentrations of 100 μM. Thus, the α_{2A}AR peptides from the second and C-terminal portion of the third cytoplasmic loop are able to block agonist binding selectively.

To evaluate the influence of the active peptides on high-affinity agonist binding to the α_{2B} subtype of the α_2AR, their effects on agonist binding to NG108-15 cell membranes was examined (Fig. 5). Peptides A and Q again inhibited the high-affinity [^{125}I]PIC binding with IC$_{50}$ values of 15 ± 13 μM ($n = 3$) and 71 ± 27 μM ($n = 3$), respectively. In the region of the second cytoplasmic loop (peptide A), there is 67% amino acid identity between α_{2A}AR and α_{2B}AR; in the C-terminal portion of the third cytoplasmic loop (peptide Q) there is 62% amino acid identity. Thus, the primary sequence is probably not the sole determinant of the ability of a peptide to decrease agonist binding. The exact structural determinants of a peptide to modulate G protein function have not been established, although amphiphilicity (40) as well as basic and hydrophobic residues appear to be important (16, 18, 40, 41).

To strengthen the hypothesis that the peptides are acting at the receptor–G protein interface and not the receptor ligand binding site, the effects of

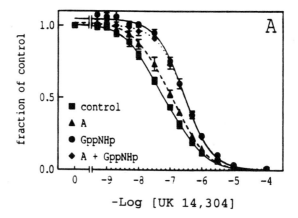

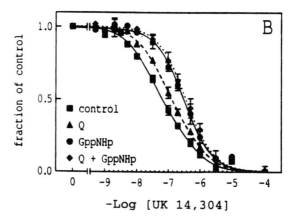

FIG. 6   Effect of peptides A and Q on UK 14,304 competition for [³H]yohimbine binding. Membranes were preincubated with either buffer (■), 100 μM peptide A (A) or 100 μM peptide Q (B) (▲), 100 μM Gpp(NH)p (●), or both (◆). Averaged data points from all experiments are shown as mean ± SD ($n = 2$) or SE ($n = 4$). Reprinted with permission from Ref. 7.

peptides are assessed in α_2AR binding competition assays using radiolabeled antagonist [³H]yohimbine. Membranes are preincubated with 100 μM peptide as described above. Following the preincubation with peptide, increasing concentrations of UK 14,304 (0.1 nM–100 μM), ±100 μM Gpp(NH)p, are added to the tubes. The binding assay is initiated by addition of 10 nM [³H]yohimbine. After a 60-min incubation, the samples are filtered as described above. Figure 6 shows the effect of peptides A and Q on UK 14,304

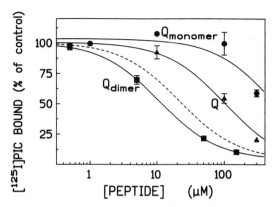

FIG. 7   Effect of monomer and dimer Q peptide on [^{125}I]PIC binding to α_{2A}AR in CHO-K1-transfected cells. Peptides Q (mixture of monomer and dimer (△), Q$_{monomer}$ (●), and Q$_{dimer}$ (■) were preincubated with membranes 15 min on ice and the specific binding of [^{125}I]PIC was determined. The dashed line shows the relative position of the Q$_{dimer}$ data if the Q$_{dimer}$ concentrations were calculated by mass rather than by molar concentrations. Control binding was 720 fmol/mg. Data points represent the mean ± SD of triplicates from a representative experiment that was repeated with similar results. Reprinted with permission from Ref. 28, S. M. Wade, H. M. Dalman, S.-Y. Yang, and R. R. Neubig, *Mol. Pharmacol.* **45,** 1191.

competition with [^{3}H]yohimbine binding. At a concentration of 100 μM peptide, the IC$_{50}$ for agonist is increased from 62 ± 1 to 107 ± 1 nM for peptide A. For peptide Q, the IC$_{50}$ values are increased from 66 ± 1 to 132 ± 1 nM. When Gpp(NH)p is included with the peptide in the binding assay, there is no significant change in the IC$_{50}$ as compared to Gpp(NH)p alone. These results demonstrate that by uncoupling receptor from G protein, the effects of the peptides are eliminated, suggesting that the peptides are not simply blocking ligand binding to receptor but are functioning to decrease receptor–G protein contact.

The Q peptide, when incubated in 50 mM Tris buffer, pH 8, readily oxidized to form dimers connected by the cysteine sulfhydryl. The dimer peptide was purified by HPLC as described above using a gradient of 15–20% Solvent B over 15 min. The Q peptide dimer was more potent than the Q monomer to block α_2AR agonist binding. The ability of the Q peptide to inhibit high-affinity α_2AR agonist binding could be largely accounted for by the activity of the small amount of dimeric peptide present in stock solutions of Q peptide. Figure 7 shows that peptide Q$_{dimer}$ was approximately 10 times more potent in inhibiting [^{125}I]PIC binding that was unpurified Q peptide, whereas the Q$_{monomer}$ was an additional order of magnitude less potent. Because multiple intracellular portions of the G-protein-coupled receptor have been implicated in the functional activation of G proteins, it is possible that the peptide Q_{dimer} may be mimicking the effect of two intracellular regions (28).

To determine if the peptide effects on ligand binding are reversible, a dilution experiment is performed wherein the membranes were preincubated with a concentration of peptide that effectively blocked ligand binding, and then the sample was diluted and a binding assay was performed as usual. The results showed that inhibition of binding was reversible and not due to receptor oxidation or another nonspecific form of receptor inactivation (28).

Peptide Effects on Receptor-Stimulated GTPase

In order to assess the functional significance of the inhibition of high-affinity agonist binding by the $\alpha_2 AR$ peptides, their effects on GTPase activity can be determined. The peptides could simply function as receptor antagonists and not alter the basal GTPase activity, or they could activate (as would activate receptor) or inhibit GTP hydrolysis. The assay for low-K_m GTPase activity in brain membranes has been described (3, 42) and is based on the hydrolysis of [γ-^{32}P]GTP in the presence of magnesium and an ATP-regenerating system. The effects of peptides on GTPase activity were assessed here in platelet membranes.

The GTPase reaction mixture contains [γ-^{32}P]GTP (0.1–0.2 μCi/tube), 1 μM GTP, 180 μM ATP, 100 mM NaCl, 1 mM MgCl$_2$, 0.1 mM EGTA, 1 mM dithiothreitol, 2.2 mM phosphocreatine, 0.2 mg creatine kinase/ml, 25 mM Tris-Cl, pH 7.5, 1 mM cAMP, 0.1 mM isobutylmethylxanthine, and 10 μM propranolol in a total volume of 0.1 ml. The last three components listed are not necessary for the GTPase assay but are included in these experiments so that the conditions are identical to those used for adenylylcyclase inhibition in our laboratory. For GTPase in brain membranes, ATPase inhibitors are included in the assay as well [1 mM App(NH)p and 1 mM ouabain]. In addition, an increase in the GTP concentration to 2 μM will saturate the enzyme and contribute to more consistent results.

Approximately 10 μg of platelet membranes is preincubated with peptide on ice for 15–30 min. The GTPase assay is initiated by addition of membranes to the reaction cocktail, where some tubes contain peptide and/or 10 μM UK 14,304. (This concentration of agonist maximally stimulated GTPase activity. A good prerequisite to these experiments is to determine the concentration of agonist that promotes maximal GTPase stimulation and if the observed increase is blocked by antagonist.) After 15 min at 30°C, the reaction is terminated by addition of 1 ml of ice-cold 25% (w/v) activated charcoal (Sigma), pH 2.3 (using phosphoric acid to adjust the pH). After incubation

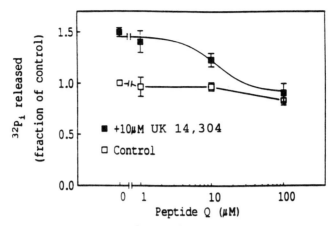

FIG. 8   Effect of peptide Q on GTPase activity in platelet membranes. Membranes were preincubated with either buffer or peptide Q, and GTPase assays were performed in the absence (□) or presence (■) of agonist 10 μM UK 14,304. Reprinted with permission from Ref. 7.

on ice for 30 min to allow the charcoal to bind the nonhydrolyzed nucleotide, the tubes are centrifuged, and a 0.3-ml aliquot of supernatant is added to 10 ml ScintiVerse liquid scintillation cocktail. Low-affinity GTPase activity, which is not affected by agonist or peptides, is subtracted from the total cpm and is determined by the GTPase activity in the presence of 50 μM GTP. Additional controls that are routinely included are tubes containing no membrane protein and tubes that contain all the assay components, but the reaction is stopped at $t = 0$. These additional conditions determine if a great deal of the [γ-^{32}P]GTP has been degraded (to [^{32}P]P$_i$ and GDP) and if the charcoal is able to bind all of the unhydrolyzed guanine nucleotide. Fractional stimulation of GTPase activity is determined from Eq. (1)

$$F = (S - L)/(B - L), \qquad (1)$$

where S equals the stimulated GTPase in the presence of agonist and/or peptide, B is the basal GTPase activity in the absence of agonist or peptide, and L is the low-affinity GTPase activity. This normalization permits averaging of all experiments conducted using different membrane preparations (it is not unusual for maximal receptor-stimulated GTPase to vary between membrane preparations).

Figure 8 shows the effect of peptide Q on α_2AR-stimulated GTPase activity in platelet membranes. Agonist routinely caused a 30–50% increase in GT-

Pase activity and the increase was blocked by 10 μM yohimbine. Peptide Q inhibited agonist-stimulated GTPase activity with an IC_{50} of 15.7 $\pm$ 1.1 μM ($n = 3$) but did not significantly affect basal GTPase level. In contrast, peptide A did not inhibit receptor-stimulated GTPase activity but did slightly increase basal GTPase activity by 10% (data not shown). Peptides B, P, and O did not alter receptor-stimulated GTPase up to peptide concentrations of 30 μM. In platelet membranes, it appears that peptide Q is inhibiting receptor binding to G protein. In addition, Peptide Q has also been shown to stimulate directly G_o/G_i GTPase in lipid vesicles (28).

These methods may provide a basis from which additional receptor–G protein interactions can be characterized as well as the critical G protein interaction regions of the receptor identified.

Acknowledgments

The experiments reviewed in this chapter were performed while I was a member of the laboratory of Dr. Rick Neubig and were supported by Grants HL/GM46417 and GM39561. I thank Susan Wade, Hiroko Dalman, Shang-Zhao Yang, and Rick Neubig for a copy of their manuscript prior to publication (28), as well as Rick Neubig for helpful comments.

References

1. M. Freissmuth, P. J. Casey, and A. G. Gilman, *FASEB J.* **3,** 2125 (1989).
2. M. Rodbell, L. Birnbaumer, S. L. Pohl, and H. M. J. Krans, *J. Biol. Chem.* **246,** 1877 (1994).
3. D. Cassell and Z. Selinger, *Biochim. Biophys. Acta* **452,** 538 (1976).
4. M. I. Simon, M. P. Strathmann, and N. Gautam, *Science* **252,** 802 (1991).
5. L. E. Limbird, *FASEB J.* **2,** 2686 (1988).
6. W. J. Thomsen and R. R. Neubig, *FASEB J.* **2,** A1139 (abstract) (1988).
7. H. M. Dalman and R. R. Neubig, *J. Biol. Chem.* **266,** 11025 (1991).
8. M. A. Gerhardt and R. R. Neubig, *Mol. Pharmacol.* **40,** 707 (1991).
9. R. A. Cerione, J. W. Regan, H. Nakata, *et al., J. Biol. Chem.* **261,** 3901 (1986).
10. R. R. Neubig, R. D. Gantzos, and R. S. Brasier, *Mol. Pharmacol.* **28,** 475 (1985).
11. D. A. Ewald, P. C. Sternweis, and R. J. Miller, *Proc. Natl. Acad. Sci. U.S.A.* **85,** 3633 (1988).
12. R. A. Cerione and E. M. Ross, *in* "Methods in Enzymology" (R. A. Johnson and J. D. Corbin, eds.), Vol. 195, p. 329. Academic Press, San Diego, 1991.
13. A. M. Spiegel, *in* "ADP-Ribosylating Toxins and G Proteins" (J. Moss and M. Vaughan, eds.) pp. 207–224. American Society of Microbiology, Washington, D.C., 1990.

14. F. R. McKenzie, I. Mullaney, C. G. Unson, A. M. Spiegel, and G. Milligan, *Biochem. Soc. Trans.* **16,** 434 (1988).
15. G. Münch, C. Dees, M. Hekman, and D. Palm, *Eur. J. Biochem.* **198,** 357 (1991).
16. T. Okamoto and I. Nishimoto, *J. Biol. Chem.* **267,** 8342 (1992).
17. D. Malek, G. Münch, and D. Palm, *FEBS Lett.* **325,** 215 (1993).
18. T. Ikezu, T. Okamoto, E. Ogata, and I. Nishimoto, *FEBS Lett.* **311,** 29 (1992).
19. B. Konig, A. Arendt, J. H. McDowel, H. Kahlert, P. A. Hargrave, and K. P. Hofmann, *Proc. Natl. Acad. Sci. U.S.A.* **86,** 6878 (1989).
20. C. Kleuss, H. Scherubl, J. Hescheler, G. Schultz, and B. Wittig, *Nature (London)* **358,** 424 (1992).
21. C. Kleuss, J. Hescheler, C. Ewel, W. Rosenthal, G. Schultz, and B. Wittig, *Nature (London)* **353,** 43 (1991).
22. H. Kurose, T. Katada, T. Amano, and M. Ui, *J. Biol. Chem.* **258,** 4870 (1983).
23. R.-R. Huang, C., R. N. DeHaven, A. H. Cheung, R. E. Diehl, R. A. F. Dixon, and C. D. Strader, *Mol. Pharmacol.* **37,** 304 (1990).
24. F. A. P. Ribeiro-Neto and M. Rodbell, *Proc. Natl. Acad. Sci. U.S.A.* **86,** 2577 (1989).
25. B. D. Carter and F. Medzihradsky, *Proc. Natl. Acad. Sci. U.S.A.* **90,** 4062 (1993).
26. D. Graeser and R. R. Neubig, *in* "G Protein-Mediated Signal Transduction" (G. Milligan, ed.), pp. 1–30. IRL Press, Oxford, UK, 1992.
27. O. H. Lowry, N. J. Rosebrough, A. L. Farr, and R. J. Randall, *J. Biol. Chem.* **193,** 265 (1951).
28. S. M. Wade, H. M. Dalman, S.-Y. Yang, and R. R. Neubig, *Mol. Pharmacol.* (1994) **45,** 1191.
29. U. K. Laemmli, *Nature (London)* **227,** 680 (1970).
30. F. R. McKenzie and G. Milligan, *Biochem. J.* **267,** 391 (1990).
31. D. J. Loftus, J. M. Stolk, and D. C. U'Prichard, *Life Sci.* **33,** 61 (1984).
32. J. W. Regan, *in* "The Alpha-2 Adrenergic Receptors" (L. E. Limbird, ed.), pp. 15–74. Humana Press, Clifton, New Jersey, 1988.
33. Y. Salomon, C. Londos, and M. Rodbell, *Anal. Biochem.* **58,** 541 (1974).
34. B. D. Carter and F. Medzihradsky, *Mol. Pharmacol.* **43,** 465 (1993).
35. W. F. Simonds, P. K. Goldsmith, J. Codina, C. G. Unson, and A. M. Spiegel, *Proc. Natl. Acad. Sci. U.S.A.* **86,** 7809 (1989).
36. C. Kleuss, H. Scherubl, J. Hescheler, G. Schultz, and B. Wittig, *Science* **259,** 832 (1993).
37. O. Kisselev and N. Gautam, *J. Biol. Chem.* **268,** 24519 (1993).
38. D. B. Bylund, C. Ray-Prenger, and T. J. Murphy, *J. Pharmacol. Exp. Ther.* **245,** 600 (1988).
39. R. R. Neubig and O. Szamraj, *Biochim. Biophys. Acta* **854,** 67 (1986).
40. T. Higashijima, J. Burnier, and E. M. Ross, *J. Biol. Chem.* **265,** 14176 (1990).
41. H. Mukai, E. Munekata, and T. Higashijima, *J. Biol. Chem.* **267,** 16237 (1992).
42. M. J. Clark and F. Medzihradsky, *Neuropharmacology* **26,** 1763 (1987).
43. C. A. Guyer, D. A. Horstman, A. L. Wilson, J. D. Clark, E. J. Cragoe, Jr., and L. E. Limbird, *J. Biol. Chem.* **265,** 17307 (1990).
44. P. Goldsmith, K. Rossiter, A. Carter, *et al., J. Biol. Chem.* **263,** 6476 (1988).
45. P. Goldsmith, P. Gierschik, G. Milligan, *et al., J. Biol. Chem.* **262,** 14683 (1987).

[10] Somatostatin Receptor Coupling to G Proteins

John R. Hadcock* and Joann Strnad

Introduction

Classification of Somatostatin Receptor Subtypes

Somatostatin-14 (S-14) and the N-terminally extended S-28 are potent inhibitors of many central and peripheral physiological responses. The diverse biological effects of somatostatin are mediated by five distinct somatostatin receptor (SSTR) subtypes (1–9). The five SSTR subtypes are products of different genes, and as members of the G-protein-coupled receptor superfamily, each contains seven putative membrane-spanning domains (1–9). Messenger RNA for all five subtypes has been detected in discrete but overlapping regions of the brain (10). SSTR2 and SSTR5 are the predominant subtypes in peripheral tissues. On the basis of sequence homology among the five subtypes and their diverse pharmacological properties, the SSTR subtypes have been classified as $SRIF_1$ and $SRIF_2$ receptors (11, 12). SSTR2, SSTR3, and SSTR5 belong to the $SRIF_1$ receptor family. SSTR1 and SSTR4 belong to the $SRIF_2$ receptor family.

G Proteins Known to Couple to Somatostatin Receptors

Heterotrimeric guanine nucleotide-binding (G) proteins consisting of α, β, and γ subunits transduce most, if not all, of the signaling mediated by somatostatin. In the inactive state GDP is bound to the α subunit of G proteins. On activation of a G-protein-coupled receptor by agonist, GTP displaces GDP and the heterotrimer dissociates into a GTP-bound α subunit and a $\beta\gamma$ complex (13). When the stimulus is removed, GTP is hydrolyzed to GDP and P_i and the heterotrimer reassociates into an $\alpha\beta\gamma$ complex. It is generally accepted that both α and β subunits can transduce the signals of G proteins. At least 20 α subunits, 5 β subunits, and 7 γ subunits have been identified (13). Several lines of evidence implicate the pertussis toxin-sensitive G pro-

* To whom correspondence should be addressed.

Methods in Neurosciences, Volume 29

tein family [$G_{i\alpha1,2,3}G_{o\alpha a,b}$] in signaling by the somatostatin receptors. Several methods have been utilized to determine somatostatin receptor–G protein–effector coupling. This includes copurification of somatostatin receptor–G protein complexes (14–16), antisense RNA or oligonucleotides to knock out specific G protein subunits (17–19), and immunoprecipitation and blockade of SSTR function using anti-G protein antibodies (20–25). As with many other G-protein-coupled receptors, copurification and coimmunoprecipitation studies of somatostatin receptors indicate that a single somatostatin receptor subtype can couple to more than one G protein (14–16, 20–24). All three $G_{i\alpha}$ subunits and $G_{o\alpha2}$ have been shown to functionally associate with somatostatin receptors.

Effectors Coupled to Somatostatin Receptors

Somatostatin receptors have been shown to modulate the activity of several different classes of effector molecules primarily through pertussis toxin-sensitive G proteins. Pertussis toxin inactivates $G_{i\alpha}$ and $G_{o\alpha}$ by catalyzing the ADP-ribosylation of the C-terminal cysteine of the G_i and G_o α subunits (25). These receptors have been shown to activate potassium channels (26, 27), stimulate tyrosine phosphatases (28), and inhibit calcium channels (29, 30) and adenylylcyclases (31–34). A notable exception is the modulation of the ubiquitous Na^+/H^+-antiporter NHE1(35). The modulating activity of NHE1 by SSTRs is insensitive to pertussis toxin treatment (35). With the exception of adenylylcyclase activity, it is not known whether each effector activity is altered by one or more SSTR subtype. In line with somatostatin receptors coupling to more than one G protein, specific subtypes can functionally couple to more than one G protein. All five subtypes cloned to date can mediate the inhibition of adenylylcyclase activity by somatostatin and its analogs. SSTR1 mediates the activity of NHE1 whereas SSTR2 has no observable effect (35).

This chapter details several pharmacological, biochemical, and immunological methods for studying the coupling of somatostatin receptor subtypes to their cognate G proteins (Fig. 1). We have chosen the rat somatostatin receptor SSTR1 subtype as a model for this chapter.

Pharmacological and Biochemical Methods

Cloning and Expression of Full-Length SSTR1 Subtype

The details of cloning and expression of the full-length rat SSTR1 subtype cDNA are described in the paper by Hadcock et al. (24). Briefly, the full-length SSTR1 cDNA was inserted into the HindIII/XbaI sites of the expres-

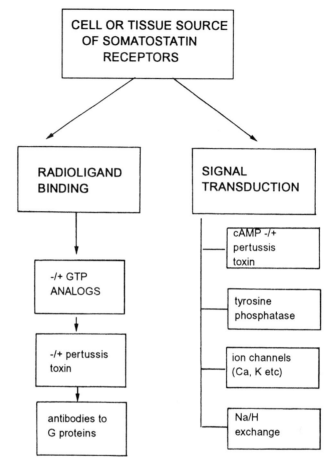

FIG. 1   Flow chart for the analysis of somatostatin receptor–G protein coupling to determine the biochemical and pharmacological characteristics.

sion vector pRC/CMV (Invitrogen). Transfections of Chinese hamster ovary (CHO-K1) cells were performed using a Stratagene (La Jolla, CA) kit for calcium phosphate transfections. We use stable transfections and select six individual G-418-resistant (Gibco/BRL, 300 μg/ml) clones from transfections of receptor cDNAs. Individual clones should be selected for functional assays (i.e., cyclic AMP accumulation assays) to ensure receptor expression in all cells.

Preparation of Membranes from Cells in Culture and Tissues

Crude plasma membranes are prepared according to Eppler *et al.* (14). Cells are detached from plates with phosphate-buffered saline (PBS), pH 7.4, 2 mM EDTA, and centrifuged at 1500 g for 10 min at 4°C. Cells are then resuspended in 10 ml homogenization buffer (1 mM sodium bicarbonate, pH 7.4, 1 mM EDTA, 1 mM EGTA, 5 μg/ml leupeptin, 5 μg/ml aprotinin, 100 μg/ml bacitracin, 100 μg/ml benzamidine) for 10 min on ice. The cells are lysed in a glass/glass Dounce homogenizer (20 strokes). The lysate is centrifuged at 1500 g for 10 min at 4°C. The supernatant is transferred to a new tube and centrifuged at 40,000 g for 20 min at 4°C. This step is repeated twice. The pellet is then resuspended at 1–5 mg/ml in 25 mM Tris–HCl, pH 7.4, containing the same protease inhibitors as above and stored at −80°C for at least 1 year.

To prepare membranes from tissues such as brain we use the method described by Eppler *et al.* (36).

Radioligand Binding Assays

Most G-protein-coupled receptors exhibit two agonist affinity states, high and low. High-affinity agonist binding is dependent on the functional association of receptor with a heterotrimeric G protein. If the receptor does not associate with or is uncoupled from the G protein, agonist binding will be of low affinity and undetectable in radiolabeled agonist saturation binding assays or shifted to the right in displacement curves. For many G-protein-coupled receptors, high-affinity agonist binding can be examined by the sensitivity of binding to nonhydrolyzable GTP analogs (GTPγS or GppNHp) and bacterial toxin-catalyzed ADP-ribosylation. The high-affinity binding of somatostatin agonists such as [^{125}I]Tyr^{11} S-14 to all somatostatin receptor subtypes is decreased on treatment of cells with pertussis toxin or nonhydrolyzable analogs of GTP. Pertussis toxin catalyzes the ADP-ribosylation of a cysteine at the C terminus of the G_i/G_o family of G protein α subunits (25). Nonhydrolyzable analogs of GTP decrease the percentage of receptors found in the high-affinity state and gives and provides an indication of the population of receptors in a high-affinity receptor/G protein complex (37).

All radioligand binding assays are performed in 96-well microtiter plates using binding buffer [50 mM HEPES, pH 7.4, 5 mM $MgCl_2$, 0.25% (w/v) bovine serum albumin (BSA)] containing protease inhibitors (5 μg/ml leupeptin, 5 μg/ml aprotinin, 100 μg/ml bacitracin, and 100 μg/ml benzamidine). We use [^{125}I]Tyr^{11} S-14 (Amersham) because [^{125}I]S-28 displays very high

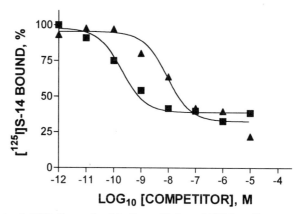

FIG. 2   Effect of GTPγS on the binding affinity of [^{125}I]Tyr^{11} S-14 to the SSTR1 expressed in CHO cells. The ability of increasing concentrations of S-14 (10^{-11} to 10^{-6} M) to displace [^{125}I]Tyr^{11} S-14 (50 fmol, 250 pM) was examined in membranes prepared from CHO cells expressing the SSTR1 subtype that were treated with vehicle (■) and 100 μM GTPγS (▲). The specific binding of [^{125}I]Tyr^{11} S-14 was 15,000 cpm in control membranes and 3600 cpm in the membranes treated with GTPγS.

nonspecific binding in filtration assays (>75%). All components are diluted in binding buffer containing protease inhibitors and added to the microtiter plate wells in the following order: 20 μl binding buffer (for no competitor well) or 20 μl nonradiolabeled S-14 (1×10^{-12} to 1×10^{-5} M final concentration), then 10 μl [^{125}I]Tyr^{11} S-14 (150,000 cpm at 2000 Ci/mmol, 250 pM final concentration). The binding reaction is initiated by adding 2–60 μg of membrane protein in a 170-μl volume. Final reaction volume is 200 μl/well. All incubations are carried out at room temperature for 2 hr. Free radioligand is separated from bound ligand by rapid filtration through a glass fiber filter (IH-201-HA) using an Inotech cell harvester. The filter is then washed several times with cold (4°C) binding buffer lacking BSA and protease inhibitors prior to counting in an LKB gamma master counter (78% efficiency). Also, we have found that Whatman (Clifton, NJ) GFC filters presoaked in 0.3% polyethyleneimine (to decrease background) work well.

[^{125}I]Tyr^{11} S-14 binding to the rat SSTR1 subtype is reduced by 75% when membranes are coincubated with 100 μM GTPγS. In the absence of GTP, greater than 90% of the [^{125}I]Tyr^{11} S-14 binding sites to SSTR1 are of high affinity (K_i = 150 pM). The remaining sites display a K_i = 10 nM (Fig. 2). GTPγS decreases the percentage of high-affinity sites from 90 to 20%. The K_i values for S-14 displacement of [^{125}I]Tyr^{11} S-14 in the presence of GTPγS

(100 μM) are 100 pM (20%) and 11 nM (80%) for the higher and lower affinity sites, respectively. Interestingly, a small percentage of binding sites are high affinity in the presence of 100 μM GTPγS. These data indicate that GTPγS promotes uncoupling of most but not all of the SSTR1–G protein complex. Pertussis toxin treatment (50 ng/ml, 24 hr) of CHO cells transfected with the SSTR1 subtype abolishes 80% of the specific binding of [^{125}I]Tyr^{11} S-14 to membranes prepared from these cells (24). These observations are consistent with a shifting of the receptor to a low-affinity state by uncoupling G_i and/ or G_o from the receptor. Interestingly, SSTR1 also couples to pertussis toxin-insensitive G proteins (35). SSTR1 mediates the activity of the ubiquitous Na^+/H^+-exchanger NHE1 via these pertussis toxin-insensitive G proteins.

Cyclic AMP Accumulation Assay

We use cyclic AMP accumulation assays to determine functional coupling of somatostatin receptors to effectors. All five cloned somatostatin receptors inhibit cyclic AMP accumulation and adenylylcyclase activity (5–7, 24, 38). Cyclic AMP accumulations offer a reliable assay for the transmembrane signaling mediated by receptors coupled to G_s and G_i. There are several excellent methods for measuring cyclic AMP and cyclic AMP accumulation in intact cells and adenylylcyclase activity in membranes (39). These include RIAs, competition assays using adrenal binding protein, and separation of cyclic AMP using Dowex resin and alumina (39). All methods are reliable and vary in cost and difficulty. We will describe the measurement of cyclic AMP accumulation from intact cells using the adrenal binding protein method. This method is inexpensive and reliable and can also be used for the measurement of adenylylcyclase activity in membranes prepared from cells or tissues (39, 40).

Preparation of Adrenal Binding Protein

Bovine adrenal gland extracts are prepared according to the method of Brown *et al.* (40) with minor modifications. Bovine adrenal glands are purchased from Pel Freez and shipped on dry ice. Thaw two to four adrenals and separate the cortices on ice. Chop into small pieces. Resuspend in 1.5 volumes of buffer containing 0.25 M sucrose, 50 mM Tris-HCl, pH 7.4, 25 mM KCl, 5 mM MgCl$_2$, and protease inhibitors (5 μg/ml leupeptin, 5 μg/ml aprotinin, 100 μg/ml bacitracin, 100 μg/ml benzamidine). Homogenize with a Polytron and centrifuge at 2000 g for 5 min at 4°C. Save the supernatant and centrifuge at 5000 g for 15 min at 4°C. Save the supernatant and store as 1-ml aliquots at −80°C. Four adrenal glands are usually sufficient to prepare 30–50 1-ml aliquots. Each preparation should be tested with standards at different dilutions (1:10, 1:15, 1:20) before use in experiments.

Preparation of Cells for Cyclic AMP Accumulations

Somatostatin-mediated inhibition of cyclic AMP accumulation is measured in intact cells coincubated with the diterpene forskolin (10 μM), which activates adenylylcyclase directly and stimulates cyclic AMP accumulation in cells. Aspirate off medium from two 100-mm plates of cells. Add 5 ml phosphate-buffered saline containing 2 mM EDTA to each plate. Wash the cells off the plate. Do not scrape! Pellet the cells and resuspend in Krebs–Ringer phosphate (KRP) containing 2 mM CaCl$_2$ (added fresh). Wash once and resuspend in 2 ml of KRP with CaCl$_2$ containing 100 μM IBMX. Count cells and adjust the concentration of cells to 50,000/70 μl. Incubate 15 min at room temperature before starting the assay.

Generation of Cyclic AMP

Set up and label 12 × 75 polypropylene tubes (samples should be done in triplicate including controls). Add 10 μl activators (in triplicate). For example, we set up a series of samples containing the following components: no added activators (basal), forskolin alone (10 μM, final), and forskolin plus various amounts of S-14 (1 × 10^{-10} to 1 × 10^{-6} M, final). The first two steps can be done on the bench top. Adjust to equivalent volumes with KRP. Add 10 μl 1 mM IBMX to each tube. Place tubes in a shaking 37°C water bath (concentration of cells is 25–100,000 cells per tube) after adding 70 μl cells (to make a final volume of 100 μl per tube) at 15-sec intervals. Incubate for 15 min at 37°C. Add 10 μl of 1.0 N HCl at 10-sec intervals and transfer tubes to ice. Boil for 3 min (tubes may be stored overnight). Neutralize with 10 μl 1.0 N NaOH containing 250 mM Tris, pH 7.4.

Cyclic AMP Assay

Set up a large chest of ice and bury the test tube racks half-way down. In order to set up the tubes for the standard curve, aliquot 50 μl of cyclic AMP standards (Sigma) to each tube in triplicate (e.g., 0, 0.5, 1.0, 2.0, 4.0, 8.0, 16.0 pmol tube), then add 70 μl KRP to each tube of standard. Set up one set of tubes containing 220 μl KRP for the background. Add 50 μl of 10 nM

[^{3}H]cyclic AMP (Amersham) to all tubes (standards and samples). Add 100 μl of adrenal binding protein (diluted 1:15 with ice-cold 20 mM KH$_2$PO$_4$, pH 6.0) to all tubes except the background tubes. The binding protein suspension should be stirring. Incubate for 120 min on ice. Add 100 μl charcoal/BSA solution to each tube (90 mg BSA, 180 mg Norit A, 30 ml 20 mM KH$_2$PO$_4$, pH 6.0). The charcoal/BSA suspension should be stirring in an ice bath. Vortex each tube for 10 sec. Spin 3 min at 2000 g in a refrigerated centrifuge at 4°C. Decant the supernatant into scintillation vials containing 5 ml of Ecoscint (shake each vial well so that it forms a uniform mixture). Quantify by liquid scintillation spectrometry. To generate a standard curve, plot the concentration of the cAMP standards along the x axis and the log of the counts per minute (cpm) that were generated by each concentration along the y axis. When generating the standard curve the "0" competitor sample should account for greater than 40% [^{3}H]cAMP bound. Calculate individual sample results from the standard curve, which must be generated during each experiment. To normalize the data, we express values as picomoles/ 10^6 cells.

Effects of Pertussis Toxin on Somatostatin-Mediated Inhibition of Cyclic AMP Accumulation

Pertussis toxin inactivates the $G_{i\alpha}$ and $G_{o\alpha}$ family of G proteins by catalyzing the ADP-ribosylation of a C-terminal cysteine. To test the effects of inactivation of these G proteins on somatostatin-mediated inhibition of cyclic AMP accumulation, cells are treated overnight with 50 ng/ml pertussis toxin or vehicle. Cyclic AMP accumulations are performed as described above. CHO cells transfected with the somatostatin receptor cDNA are treated for 24 hr with pertussis toxin (holotoxin, 50 ng/ml) or vehicle. This concentration of pertussis toxin has been shown to catalyze completely the ADP-ribosylation of most pertussis toxin-sensitive G proteins (41). As shown in Table I and Fig. 3, both the potency and the efficacy of the somatostatin-mediated inhibition are decreased by incubation of cells with pertussis toxin. This indicates that the functional coupling of somatostatin receptors to adenylylcyclase is transduced predominately by pertussis toxin-sensitive G proteins.

Immunological Approach to Determine SSTR1–G Protein Coupling

The methods described above are used to determine the particular subclass of G proteins (e.g., G_i/G_o) that couple to somatostatin receptor subtypes. Immunoprecipitation of receptor–G protein complexes with antisera selec-

TABLE I   Somatostatin-Medicated Inhibition of Cyclic AMP
Accumulations in CHO Cells Expressing SSTR1
Subtype: Effect of Pertussis Toxin Treatment[a]

Treatment	Maximal inhibition	ED_{50} (nM)	n
None	50%	1.1 ± 0.35	4
PTX	20%	41 ± 5	4

[a] CHO cells expressing the SSTR1 subtype were treated with vehicle (control) or 50 ng/ml pertussis toxin (PTX) for 24 hr. The ability of S-14 to inhibit forskolin-stimulated (10 μM) cyclic AMP accumulation in intact CHO cells expressing the SSTR1 subtype was examined. Cyclic AMP accumulations were then performed. Displayed are the means $\pm$ SEM of four separate experiments. Forskolin-stimulated cyclic AMP accumulation was 59 ± 10 pmol/10^6 cells in control cells and 53 ± 17 pmol/10^6 cells in pertussis toxin-treated cells.

tive for individual G proteins can determine with reasonable certainty the actual G protein α subunits that are associated with a particular somatostatin receptor subtype. This methodology has been used to analyze somatostatin receptor–G protein complexes in brain, in cultured cells (20–22), and in transfected cells expressing the SSTR1 subtype (24). Many different G protein antisera have been used to immunoprecipitate receptor–G protein complexes. Many of the anti-G protein antisera (raised against the same epitope) used in the references cited above are commercially available from several sources, including DuPont/New England Nuclear (Boston, MA) and Calbiochem (San Diego, CA).

Preparation of G Protein Antisera for Immunoprecipitations

Antisera corresponding to sequences 346–355 of rat $G_{i\alpha1,2}$ (SB-04) and 384–394 of $G_{s\alpha}$ (SB-07) were used for immunoprecipitation studies (a gift from Dr. Suleiman Bahouth, University of Tennessee, Memphis). The G protein antisera are prebound to the goat antirabbit IgG-Sepharose at a ratio of 1:10 to 1:20 (v/v, antiserum:goat antirabbit IgG-Sepharose) as described below (24, 42).

Take 600 μl of goat antirabbit IgG-Sepharose (BRL, Gaithersburg, MD) slurry and spin down. Add 300 μl 1× immunoprecipitation buffer (IPA buffer, 20 mM Tris-HCl, pH 7.4, 150 mM NaCl, 5 mM EDTA, 0.5% deoxycholate) containing the following protease inhibitors: 5 μg/ml leupeptin, 5 μg/ml aprotinen, 100 μg/ml bacitracin, 100 μg/ml benzamidine. IPA buffer can be prepared as a 5× stock. Repeat the washing step. Split the goat antirabbit IgG-Sepharose into six equal aliquots. To each aliquot add one antiserum or

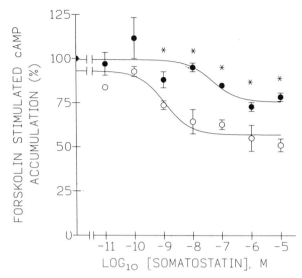

FIG. 3   Somatostatin-mediated inhibition of cyclic AMP accumulations in CHO cells expressing the SSTR1 subtype: effect of pertussis toxin treatment. CHO cells expressing the SSTR1 subtype were treated with vehicle [control ($\bigcirc$)] or 50 ng/ml pertussis toxin ($\bullet$) for 24 hr. The ability of S-14 to inhibit forskolin-stimulated (10 μM) cyclic AMP accumulation in intact CHO cells expressing the SSTR1 subtype was examined. Cyclic AMP accumulations were performed as described in this chapter (see *Experimental Procedures*). Displayed are the means ± SEM of four separate experiments. Forskolin-stimulated cyclic AMP accumulation was 59 ± 10 pmol/10^6 cells in control cells and 53 ± 17 pmol/10^6 cells in pertussis toxin-treated cells. The asterisks denote a significant difference from control cells ($p < 0.05$). Reprinted with permission from Ref. 24, J. R. Hadcock, J. Strnad, and C. M. Eppler, Rat somatostatin subtype 1 couples to G proteins and inhibition of cyclic AMP accumulation. *Mol. Pharmacol.* **45**, 410–416.

preimmune serum (10 μl). Incubate 37°C for 2 hr. Wash three times with IPA buffer. *Note:* the choice of detergent is very important because the [^{125}I]Tyr^{11} S-14, the receptor, and the G protein must remain in a physical association with each other.

Immunoprecipitation of Performed [^{125}I]Tyr^{11} S-14–Receptor–G Protein Complex with Anti-G-Protein Antisera

Immunoprecipitation of the [^{125}I]Tyr^{11} S-14–receptor–G protein (R : L) complex is performed as described by Law *et al.* (20) with several modifications. On the basis of a strategy employed for the purification of the GH$_4$C$_1$ cell

TABLE II   Immunoprecipitation of SSTR1–G Protein Complexes with G Protein α
Subunit Antisera[a]

CHO cell type	Antiserum	Bound [^{125}I]S-14 (cpm)		Specific cpm	n
		$-$S-14	$+$ S-14		
CHO wild type	$G_{i\alpha1,2}$	1060 ± 185	999 ± 213	61 ± 269	4
SSTR1	$G_{i\alpha1,2}$	3530 ± 471	1148 ± 246	2382 ± 401	6
SSTR1–GTPγS[b]	$G_{i\alpha1,2}$	1463 ± 69	998 ± 97	465 ± 69	4
SSTR1	$G_{s\alpha}$	1305 ± 679	1273 ± 751	32 ± 137	6

[a] Membranes were prepared from CHO cells expressing the SSTR1 subtype. Immunoprecipitation of [^{125}I]S-
14–receptor–G protein complex with antisera selective for $G_{i\alpha1,2}$ and $G_{s\alpha}$ was performed as described
above. The "+ S-14" column is [^{125}I]Tyr^{11} S-14 binding in the presence of a saturating concentration of
cold S-14 (100 nM) and accounts for nonspecific binding.

[b] The [^{125}I]S-14/receptor–G protein complex was immunoprecipitated with antisera selective for $G_{i\alpha1,2}$ as
described above. The complex was then incubated with 10 μM GTPγS for 1 hr prior to washing and
quantifying. The data are the means (±SC) of four to six independent experiments. Reprinted with permission from Ref. 24, J. R. Hadcock, J. Strnad, and C. M. Eppler, Rat somatostatin subtype 1 couples to G
proteins and inhibition of cyclic AMP accumulation. *Mol. Pharmacol.* **45**, 410–416.

somatostatin receptor, we choose to prebind the receptor with [^{125}I]Tyr^{11} S-14
before a solubilization step (14). This procedure has been shown to enhance
greatly the stability of receptor–G protein complexes when compared to
solubilization followed by a ligand binding step (14). The half-life of receptor–G protein complex is ~18 hr when ligand is prebound to receptor prior
to solubilization. The half-life of receptor–G protein complex is ~2 hr when
the ligand binding step is performed after solubilization of membranes (14).
Membranes (50 μg) from wild-type CHO cells or cells transfected with SSTR1
are suspended in binding buffer (50 mM HEPES, pH 7.4, 5 mM MgCl$_2$, 0.25%
BSA) containing protease inhibitors plus 500 pM [^{125}I]Tyr^{11} S-14 (~300,000
cpm) in the presence and absence of cold 100 nM S-14 (final volume of 150 μl).
After prebinding of [^{125}I]Tyr^{11} S-14 to membranes for 1 hr, the samples are
pelleted and then solubilized in binding buffer containing 0.15% deoxycholate : lysophosphatidylcholine (D : L). The insoluble fraction is removed by centrifugation and 25 μl of anti-G-protein antiserum prebound to goat antirabbit
IgG-Sepharose is added. The samples are incubated overnight, washed three
times, and quantified by gamma spectroscopy in an LKB gamma counter. The
concentration of D : L used has been shown to solubilize ~50% of the R : L
complex with a half-life for the complex of 18 hr (14). Therefore, approximately
25% of the original specifically bound ligand (R : L complex) remains intact
after the solubilization and incubation step. In a typical assay this represents
a decline from 50,000 cpm specifically bound to 12,500 cpm bound.

The ability of antisera generated against the C-terminal sequence of $G_{i\alpha1,2}$
and $G_{s\alpha}$ to immunoprecipitate [^{125}I]Tyr^{11} S-14–SSTR1–G protein complexes
is displayed in Table II. The $G_{i\alpha1,2}$ antiserum immunoprecipitates ~20% of

the total specifically bound $[^{125}I]Tyr^{11}$ S-14 remaining after solubilization and overnight incubation. In agreement with the GTPγS-dependent loss of binding, a substantial decrease is observed in the amount of $[^{125}I]Tyr^{11}$ S-14 that immunoprecipitates with the anti-$G_{i\alpha1,2}$ antisera. Incubation of the complex with 10 μM GTPγS decreases the specific counts in the immunoprecipitate by 80%. In contrast, no specific binding is observed with either the C-terminal antiserum of $G_{s\alpha}$ or in wild-type CHO cell membranes. Immunoprecipitation of the $[^{125}I]Tyr^{11}$ S-14–SSTR1–G protein complex accounts for 20–30% of the bound $[^{125}I]Tyr^{11}$ S-14 remaining after solubilization and overnight incubation. Initially, 50,000 cpm $[^{125}I]Tyr^{11}$ S-14 specifically bound to membranes.

Summary

This chapter has focused on the examination of the coupling of G proteins to the somatostatin receptor SSTR1. With the molecular cloning of five distinct subtypes, analysis of each individual receptor is possible. It is now apparent that in many tissues and cells more than one subtype is expressed, making it difficult to assign G proteins and effectors to individual receptor subtypes. Expression of individual receptor subtypes in cell lines that have little or no other somatostatin receptor provides a system for the analysis of specific receptor–G protein coupling that is easier to interpret.

References

1. Y. Yamada, S. R. Post, K. Wang, H. S. Tager, G. I. Bell and S. Seino, *Proc. Natl. Acad. Sci. U.S.A.* **89,** 251–255 (1992).
2. W. Meyerhof, H.-J. Paust, C. Schonrock, and D. Richter, *DNA Cell Biol.* **10,** 689–694 (1991).
3. F. W. Kluxen, C. Bruns, and H. Lubbert, *Proc. Natl. Acad. Sci. U.S.A.* **89,** 4618–4622 (1992).
4. X.-J. Li, M. Forte, R. A. North, C. A. Ross, and S. Snyder, *J. Biol. Chem.* **267,** 21307–21312 (1992).
5. K. Yasuda, S. Rens-Damiano, C. D. Breder, S. F. Law, C. B. Saper, T. Reisine, and G. I. Bell, *J. Biol. Chem.* **267,** 20422–20428 (1992).
6. J. F. Bruno, Y. Xu, J. Song, and M. Berelowitz, *Proc. Natl. Acad. Sci. U.S.A.* **89,** 11151–11155 (1992).
7. A.-M. O'Carrol, S. J. Lolait, M. König, and L. Mahan, *Mol. Pharmacol.* **42,** 939–946 (1992).
8. M. Vanetti and V. Höllt, *FEBS Lett.* **331,** 260–266 (1993).
9. J. D. Corness, L. L. Demchyshyn, P. Seeman, H. H. M. Van Tol, C. B. Srikant, G. Kent, Y. C. Patel, and H. B. Niznik, *FEBS Lett.* **321,** 279–284 (1993).

10. J.-F. Bruno, Y. Xu, J. Song, and M. Berelowitz, *Endocrinology* **133,** 2561–2567.
11. K. Raynor, W. A. Murphy, D. H. Coy, J. E. Taylor, J.-P. Moreau, K. Yasuda, G. I. Bell, and T. Reisine, *Mol. Pharmacol.* **43,** 838–844 (1993).
12. K. Raynor, A. M. O'Carroll, H. Kong, K. Yasuda, L. C. Mahan, G. I. Bell, and T. Reisine, *Mol. Pharmacol.* **44,** 385–392 (1993).
13. E. J. Neer, *Protein Sci.* **3,** 3–14 (1994).
14. C. M. Eppler, J. R. Zysk, M. Corbett, and H. M. Shieh, *J. Biol. Chem.* **267,** 15603–15612 (1992).
15. D. R. Luthin, C. M. Eppler, and J. Linden, *J. Biol. Chem.* **268,** 5990–5996 (1993).
16. P. J. Brown and A. Schonbrunn, *J. Biol. Chem.* **268,** 6668–6676 (1993).
17. C. M. Moxham, Y. Hod, and C. C. Malbon, *Science* **260,** 991–995 (1993).
18. Y. F. Liu, K. H. Jacobs, M. M. Rasenick, and P. R. Albert, *J. Biol. Chem.* **269,** 13880–13886 (1994).
19. C. Kleuss, H. Scherubl, J. Hescheler, G. Scultz, and B. Wittig, *Science* **259,** 832–834 (1993).
20. S. F. Law, D. Manning, and T. Reisine, *J. Biol. Chem.* **266,** 17885–17897 (1991).
21. M. Tallent and T. Reisine, *Mol. Pharmacol.* **41,** 452–455 (1992).
22. S. F. Law, K. Yasuda, G. I. Bell, and T. Reisine, *J. Biol. Chem.* **268,** 10721–10727.
23. R. Murray-Whelan and W. Schlegel, *J. Biol. Chem.* **267,** 2960–2965 (1992).
24. J. R. Hadcock, J. Strnad, and C. M. Eppler, *Mol. Pharmacol.* **45,** 410–416 (1994).
25. L. Birnbaumer, J. Abramson, and A. M. Brown, *Biochem. Biophys. Acta* **1031,** 163–224.
26. S. Mihara, R. North, and A. Suprenant, *J. Physiol.* **390,** 335–355 (1987).
27. H. Wang, C. Bogen, T. Reisne, and M. Dichter, *Proc. Natl. Acad. Sci. U.S.A.* **86,** 9616–9620.
28. M. Pan, T. Florio, and P. Stork, *Science* **256,** 1215–1217 (1992).
29. T. Reisine, *J. Pharmacol. Exp. Therap.* **254,** 646–651 (1990).
30. T. Reisine and S. Guild, *J. Pharmacol. Exp. Therap.* **235,** 551–557 (1985).
31. B. D. Koch and A. Schonbrunn, *Endocrinology* **114,** 1784–1790 (1984).
32. K. Jacobs, K. Aktories, and G. Schultz, *Nature (London)* **303,** 177–178 (1983).
33. N. Mahy, M. Wookails, D. Manning, and T. Reisine, *J. Pharmacol. Exp. Therap.* **247,** 390–396.
34. B. Koch, J. Blaylock, and A. Schonbrunn, *J. Biol. Chem.* **263,** 216–225 (1988).
35. C. Hou, R. L. Gilbert, and D. Barber, *J. Biol. Chem.* **269,** 10357–10362 (1994).
36. C. M. Eppler, J. D. Hulmes, J.-B. Wang, B. Johnson, M. Corbett, D. R. Luthin, G. R. Uhl, and J. Linden, *J. Biol. Chem.* **268,** 26447–26451.
37. R. S. Kent, A. DeLean, and R. J. Lefkowitz, *Mol. Pharmacol.* **17,** 14–23 (1980).
38. J. Strnad, C. M. Eppler, M. Corbett, and J. R. Hadcock, *Biochem. Biophys. Res. Commun.* **191,** 968–976 (1993).
39. R. W. Farndale, L. M. Allan, and R. B. Martin, *in* "Signal Transduction, A Practical Approach" (G. Milligan, ed.). IRL Press, Oxford, England, 1992.
40. B. L. Brown, J. D. M. Albano, R. P. Ekins, and A. M. Sgherzi, *Biochem. J.* **121,** 561–562 (1971).
41. J. K. Liao and C. J. Homcy, *J. Biol. Chem.* **268,** 19528–19533 (1993).
42. J. R. Hadcock, M. Ros, D. C. Watkins, and C. C. Malbon, *J. Biol. Chem.* **265,** 14784–14790 (1990).

[11] Galanin Receptor–G Protein Interactions and Stimulation of GTPase Activity

Susan L. Gillison, Gunnar Skoglund, and
Geoffrey W. G. Sharp

Introduction

Galanin, a peptide of 29 or 30 amino acids, depending on the species (1, 2), has a variety of effects in the central and peripheral nervous systems, as well as in nonneuronal tissues (3). These effects are mediated by receptor-activated heterotrimeric G proteins. Accordingly, the first galanin receptor to be cloned and sequenced (4) has a structure similar to those in the super-family of seven transmembrane-spanning receptors. One can anticipate that other galanin receptors will be cloned and sequenced in the future. Already there is evidence in favor of multiple galanin receptors. This evidence comes from structure–activity and antagonist studies, and from the fact that three distinct types of signal transduction mechanisms seem to be operative in the diverse actions of galanin. This implies receptor interactions with different types of heterotrimeric G proteins. Some of the evidence for multiple galanin receptor subtypes is as follows: First, in many tissues studied, the N-terminal portion of the galanin molecule is important for activation of the receptor, and 3–29 galanin is inactive. However, in hypothalamus and anterior pituitary it appears that 3–29 galanin is fully active (5). Second, the receptor antagonists galantide and M-15 act as antagonists of galanin in the CNS and pancreatic β-cells, but are full agonists on gastrointestinal smooth muscle (6). Third, galanin receptors use different signal transduction systems in different cells. Those recognized so far fall into three common groups: (a) Inhibition of adenylylcyclase along with effects on K^+ and Ca^{2+} channels (7) and, in secretory cells, inhibition of exocytosis; these effects are mediated by one or more of the pertussis toxin (PTX)-sensitive G_{i1}, G_{i2}, G_{i3} proteins and the G_o proteins; (b) mobilization of intracellular Ca^{2+} (8) presumably mediated by G_q; and (c) stimulation of adenylylcyclase (6) presumably by G_s. By analogy with receptors for other ligands, e.g., the adrenergic or muscarinic receptors, these data would indicate that a family of galanin receptors is responsible for the diversity.

In the inactive state, heterotrimeric G proteins exist in their $\alpha\beta\gamma$ conformation, with GDP bound to the α subunit. On stimulation by a ligand-bound receptor, the α subunit exchanges GDP for GTP and the α and $\beta\gamma$ subunits dissociate and separate from the receptor. The activated α subunit (GTP bound) and the liberated $\beta\gamma$ then associate with specific target molecules to exert their effects. Because the α subunit exhibits intrinsic GTPase activity, which may be modulated by the target effector, the αGTP subsequently hydrolyzes to αGDP, which allows the α and $\beta\gamma$ subunits to reassociate and complete the cycle. Although this "activation" cycle proceeds slowly in the basal state, hormonal or peptidergic activation of receptors gives a measurable increase in the rate of GTPase activity. Thus, the measurement of receptor-stimulated GTPase activity can give an estimate of the activity of a population of receptor-coupled G proteins.

Studies in our laboratory have characterized the interaction between the galanin receptor and its cognate G proteins in the insulin-secreting RINm5F β-cell line by examining galanin-stimulated high-affinity GTPase activity. The method that we describe in this chapter is one that we have developed for use on these cells. In the β-cell, galanin inhibits insulin secretion, after high-affinity receptor interaction, by acting on at least four distinct mechanisms. It activates K_{ATP} channels and hyperpolarizes the cell, reduces the activity of L-type Ca^{2+} channels, inhibits adenylylcyclase to lower cyclic AMP levels, and inhibits exocytosis by an, as yet, unknown mechanism, at a distal site in stimulus–secretion coupling (7). The latter could possibly be exerted on the mechanism of exocytosis per se. All these effects are abolished by prior treatment with pertussis toxin and, therefore, are mediated by G_i and G_o proteins (9). It has been established that $G_{i1,2,3}$ proteins interact with the activated galanin receptor (10) and that $G_{i2,3}$ proteins are responsible for the inhibition of adenylylcyclase (11). The identities of the G protein mediators responsible for the other effects, on the channels and on exocytosis, are currently under investigation.

Method

General

The method used in our laboratory is based on that of Cassel and Selinger (12), with some modifications (13, 14). In order that conditions can be changed easily from one experiment to the next, the reagents are made up in four parts, a–d (each at $4\times$ final concentration). Equal volumes (25 μl) are added to each assay tube, on ice, in the following order: (a) nucleotide regenerating system, (b) test agents, dithiothreitol (DTT), ouabain, (c) [^{32}P]GTP, and (d) membrane protein sample.

The reactions are carried out in a final volume of 100 μl in 12 × 75-mm borosilicate glass tubes. Each reaction is started with the addition of the membrane protein, vortexed, and immediately put into a 30°C water bath. Each reaction is stopped with the addition of 0.9 ml of stop solution and the tube is put on ice until centrifuged. The tubes are centrifuged for 10 min at 1800 g and 200-μl aliquots of the supernatant are removed, mixed with 4 ml scintillation fluid, and the [^{32}P]P$_i$ is quantified by liquid scintillation spectrometry.

Assay Stocks

1. 4× Salt buffer: 400 mM NaCl, 20 mM MgCl, 0.4 mM EDTA, 50 mM Tris, pH 7.4.
2. 4× Regenerating system: 4 mM ATP, 40 mM creatine phosphate, 20 U/ml creatine phosphokinase, 4 mM APP(NH)P. The 4× regenerating system should be made up in 4× salt buffer. Once prepared, small aliquots should be stored at −20°C so they are only thawed once.
3. GTP stocks (make up in distilled H$_2$O). (i) 2×10^5 cpm/ml [γ-^{32}P]GTP: It is recommended to dilute each batch of radiolabeled GTP to approximately this activity so that aliquots can be frozen and stored at −20°. For each individual assay, thaw a fresh aliquot. Before using each batch of radiolabeled nucleotide, it is advisable to test the GTP stock for purity (% hydrolysis), as described later. (ii) Unlabeled GTP stocks of 20 μM and 20 mM: As above, these should be frozen in small aliquots at −20°C. These stocks provide convenient concentrations to dilute into the assay for high- and low-affinity determinations, respectively.
4. 100 mM DTT (make up in distilled H$_2$O): Freeze in small aliquots at −20°C.

200 mM H$_3$PO$_4$ (pH 2.3); store at 4°C protected from light.

Preparation of Assay Components

For each component (a–d) make more than a sufficient volume for 25 μl/ tube in the assay. Each condition should be tested in triplicate. The final concentration of each reagent in the assay will be as follows:

12.5 mM Tris, pH 7.4
100 mM NaCl
5 mM MgCl$_2$
0.1 mM EGTA
1 mM ATP
10 mM creatine phosphate

5 U/reaction creatine phosphokinase
1 mM APP(NH)P
50,000 cpm/reaction [γ-^{32}P]GTP
Unlabeled GTP at appropriate concentrations.
1 mM Ouabain
2 mM DTT
2–10 μg/reaction protein

Regenerating System (Component a)
Thaw stock and keep on ice.

Test Agents, DTT, Ouabain (Component b)
For each experimental condition, make up (in Tris buffer) the test agent, DTT, and ouabain at 4× the final concentration: 4 mM ouabain (which should be made fresh on the day of the assay in distilled H_2O and diluted into the assay stocks), 8 mM DTT, 4× test agent (or vehicle).

[γ-^{32}P]GTP (Component c)
For each [GTP] used in the assay, dilute unlabeled GTP stock to 4× the final concentration with 2×10^3 cpm/ml [γ-^{32}P]GTP to yield 5×10^4 cpm/25 μl. The optimal [GTP] to use to detect high-affinity GTPase activity in various cell membrane preparations may vary, and should be determined for each case (see below). In purified RINm5F plasma membranes, high-affinity GTPase activity was easily detectable at 0.5 μM GTP, after subtraction of low-affinity activity in the presence of 50 μM GTP.

Membrane Protein (Component d)
Each membrane preparation, either from frozen stock or a fresh preparation, should be diluted to 4× the desired concentration in 12.5 mM Tris buffer, pH 7.4. The optimal protein concentration will vary according to the method of membrane purification, and should be determined in each case. The galanin receptor–G protein interactions in this laboratory were studied in purified RINm5F plasma membranes, which were stored at −70°C. Galanin-stimulated GTPase activity was not detectable in less purified fractions.

Stop Solution
The stop solution consists of 20 mM H_3PO_4, prepared from 200 mM stock solution, containing 5% activated charcoal (e.g., Norit). The stop solution should be stirred for at least 30 min before use. Total volume should be more than sufficient for 0.9 ml/tube.

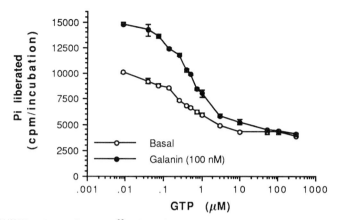

FIG. 1 [^{32}P]P$_i$ release from [γ-^{32}P]GTP incubated with purified RINm5F cell membranes, at various concentrations of GTP, in the absence and presence of 100 nM galanin. The amount of [γ-^{32}P]GTP used in each incubation tube provided 62,000 cpm. GTP concentrations ranged from 0.01 to 300 μM. The results shown are typical for such an experiment performed in triplicate. From Ref. 7, with permission.

Characterization of Basal and Galanin-Stimulated GTPase Activity

High-Affinity GTPase Activity

Isolated membrane preparations contain both low- and high-affinity specific GTPases, as well as nonspecific nucleoside triphosphatases (NTPases). Therefore, conditions must be optimized to isolate the high-affinity GTPase activity characteristic of the heterotrimeric guanine nucleotide-binding proteins. As shown by Cassel and Selinger (12), the addition of APP(NH)P and ATP, in the presence of an ATP-regenerating system, will decrease GTP hydrolysis by nonspecific NTPases. Low-affinity GTPase activity is determined by the addition of unlabeled GTP. As shown in Fig. 1, increasing the concentration of GTP decreased the liberation of [^{32}P]P$_i$ from [γ-^{32}P]GTP up to 50 μM. Thus, this concentration of GTP (50 μM) was used to determine the low-affinity GTPase activity present in the isolated plasma membranes. Below this concentration, galanin-stimulated [^{32}P]P$_i$ liberation was easily detectable. For subsequent experiments, each condition assayed was tested in the presence of both a low (0.5 μM) and high (50 μM) concentrations of GTP, and the low-affinity component was subtracted from the total GTPase activity measured at the lower concentration.

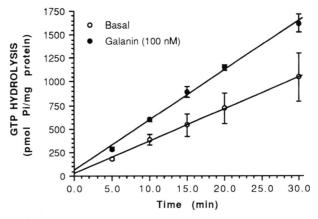

FIG. 2 High-affinity GTPase activity in the absence and presence of 100 n*M* galanin as a function of incubation time. The rates of GTP hydrolysis shown are due to 5 µg of RINm5F cell membrane protein per reaction at 5, 10, 15, 20, and 30 min. From Ref. 7, with permission.

Optimization of Time and Protein Concentration

As shown in Fig. 2, high-affinity GTPase activity in RINm5F membranes (pmol P_i/mg protein) was linear when the reactions were carried out between 5 and 30 min. As calculated by the slope of the lines under basal and stimulated conditions, 100 n*M* galanin stimulated basal GTPase activity approximately 72%. In other studies (data not shown), similar experiments were performed at shorter time points. When 30 µg/ml membrane protein was used, the high-affinity activity could be determined at as little as 1 min. At higher protein concentrations, activity at time points less than 1 min should be detectable. GTPase activity was proportional to the concentration of membrane protein in the assay under both basal and galanin-stimulated conditions at protein concentrations of 10–160 µg/ml. Similar characterization studies should be carried out on each type of membrane preparation studied, because G protein expression levels may vary markedly in different cell types, and somewhat modified conditions may be required.

Pertussis Toxin Inhibition of Galanin-Stimulated GTPase Activity

The galanin receptors expressed in the RINm5F cells are coupled to multiple isoforms of the PTX-sensitive G proteins G_i and G_o. The ability to block agonist-stimulated-GTPase activity by prior treatment with PTX is indicative

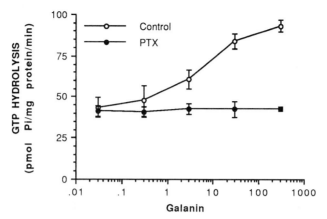

FIG 3   Effects of 0.3 to 300 nM galanin on high-affinity GTPase activity of RINm5F cell membranes with or without PTX treatment. The results are taken from a typical experiment performed in triplicate. From Ref. 7, with permission.

of this coupling. High-affinity GTPase activity in control membrane preparations is dependent on the concentration of galanin in the assay. In this and other assays, half-maximal responses were seen at approximately 5 nM, which is similar to the inhibitory effect of the peptide on insulin secretion. In membranes prepared from cells pretreated with PTX, galanin was without effect as a result of the ADP-ribosylation and functional inactivation of the G proteins, which couple to the galanin receptor. The concentration response characteristics of galanin on GTPase activity and the lack of response in membranes from PTX-treated cells are both shown in Fig. 3. To pretreat the cells, 30 ng/ml PTX is added to the culture media for 18–20 hr prior to membrane preparation. Cells plated at the same time and density are grown in parallel with the appropriate control cells. We have found treatment with the toxin at this concentration and duration to completely ADP-ribosylate the PTX-sensitive G proteins present in the membranes, as assessed by blockade of subsequent *in vitro* [^{32}P]ADP-ribosylation. Shorter treatments may be sufficient if higher concentrations of toxin are used.

Troubleshooting

Plasma Membrane Purification

Two methods of membrane preparations were tested for use in this assay: a crude particulate fraction and a purified plasma membrane fraction. Crude membrane preparations were prepared as follows. Confluent cell cultures were washed twice in ice-cold buffered saline and harvested. The cells were lysed by 30 min exposure to hypotonic Tris buffer, pH 7.4, containing phenyl-methylsulfonyl fluoride (PMSF), followed by Dounce homogenization. They were centrifuged at 1000 g for 10 min at 4°C and the pellet discarded. The supernatant was centrifuged for 30 min at 96,000 g at 4°C and the final pellet suspended at 5–10 mg/ml protein in 20 mM Tris, pH 7.4, containing PMSF. This preparation was homogenized by several strokes in a Teflon–glass Dounce homogenizer, aliquoted, and stored at −70°C until use.

Purified plasma membranes were prepared according to the method of Ullrich and Wollheim (15). The monolayers were washed in homogenization buffer (250 mM sucrose, 5 mM HEPES, pH 7.4, 0.5 mM EDTA, and 0.5 mM PMSF), and harvested in the same buffer. The cell suspension was pressurized in a Parr tissue homogenizer to 130 psi at 4°C for 30 min. The homogenate was centrifuged at 700 g for 15 min and the supernatant mixed with 90% isosmotic Percoll to yield a 15% solution. A self-generating Percoll gradient was formed by centrifugation at 48,000 g for 25 min in a Ti 70 fixed-angle rotor. The membranes were collected at a density of approximately 1.04 g/ml, as determined by density marker beads in paired gradients. The membranes were then diluted 1 : 1 with homogenization buffet and centrifuged at 70,000 g for 90 min in a swinging bucket rotor (SW50.1). The final pellets were collected from the Percoll cushion and diluted to approximately 2 mg/ml with 20 mM Tris, pH 7.4, containing PMSF. The preparation was homogenized by 10 strokes in a Teflon–glass Dounce homogenizer and stored in small aliquots at −70°C until use. Protein concentrations were determined using the Bradford method and bovine serum albumin standards. This method gives an eightfold purification of the plasma membranes as assessed by 5′-nucleotidase activity. Although the first (simpler) method of membrane preparation was suitable for galanin receptor binding studies, and showed the effects of G protein interactions on binding affinity (9, 16), high-affinity GTPase activity was not detectable due to the large low-affinity component.

$[\gamma\text{-}^{32}P]GTP$ Stock Hydrolysis

We have found up to a 10-fold variation in the purity of $[\gamma\text{-}^{32}P]GTP$ stocks supplied by various sources. This background $[^{32}P]P_i$ (due to low-level hydrolysis) needs to be minimized so that it does not mask the $[^{32}P]P_i$ product of the GTPase activity present in the membranes. In a typical assay, less than 20% of the labeled $[\gamma\text{-}^{32}P]GTP$ is hydrolyzed by the low-affinity GTPases in the isolated membranes. If background hydrolysis of the stock preparations approaches this level, the sensitivity of the assay is undermined. In our hands, at protein concentrations of 3–5 μg/reaction, the best results were obtained from batches with very low degradation, i.e., 1–2% $^{32}P_i$ in the [γ-^{32}P]GTP source. When degradation exceeds this level, the protein concentration in the assay should be increased to improve the signal-to-noise ratio.

To test the purity and suitability of the stock for the assay, an aliquot of the nucleotide is diluted to 50,000 cpm in 100 μl of distilled H_2O on ice, mixed with 0.9 ml stop solution, and an aliquot of the supernatant is analyzed by liquid scintillation counting after centrifugation. Percent "hydrolysis" can be calculated by cpm supernatant/total cpm.

Reproducibility

When purified plasma membranes were prepared using carefully controlled conditions, we found little difference in the fold stimulation of high affinity activity by maximal concentrations of galanin. Our experiments were all performed on membranes prepared from cells of similar passage number, from approximately 58 to 70. Actual absolute values of high-affinity GTPase activity (expressed as pmol/min/mg) were also similar enough from one preparation to another, as well as from one experiment to another, to allow the raw data to be analyzed by either analysis of variance or Student's t-test, when appropriate. For example, in eight independent experiments, basal GTPase varied by approximately 11% (mean and SEM = 36 ± 4.0 pmol/min/mg). Similarly, GTPase activity in the presence of 100 nM galanin varied by only 6% (105 ± 6.2 pmol/min/mg). Because basal activity is ill-defined and the stimulated activity is due to specific heterotrimeric G protein activity, using the absolute values is likely a more accurate representation of basal and stimulated activity than the terminology of percent (%) or fold stimulation.

References

1. H. F. Evans and J. Shine, *Endocrinology* **129,** 1682 (1991).
2. G. L. McKnight, A. E. Karlsen, S. Kowalyk, S. L. Mathewes, P. O. Sheppard, P. J. O'Hara, and G. J. Taborsky, Jr., *Diabetes* **41,** 82 (1992).
3. T. Bartfai, G. Fisone, and U. Langel, *Trends Pharmacol. Sci.* **13,** 312 (1992).
4. E. Habert-Ortoli, B. Amiranoff, I. Loquet, M. Laburthe, and J.-F. Mayaux, *Proc. Natl. Acad. Sci. U.S.A.* **91,** 9780 (1994).
5. D. Wynick, D. M. Smith, M. Ghatei, K. Akinsanya, R. Bhogal, P. Purkiss, P. Byfield, N. Yanaihara, and S. R. Bloom, *Proc. Natl. Acad. Sci. U.S.A.* **90,** 4231 (1993).
6. Z. F. Gu, W. J. Rossowski, D. H. Coy, T. K. Pradhan, and R. T. Jensen, *J. Pharmacol. Exp. Therap.* **266,** 912 (1993).

7. A. M. McDermott and G. W. G. Sharp, *Cell. Signalling* **5,** 229 (1993).
8. T. Sethi and E. Rozengurt, *Cancer Res.* **51,** 1674 (1991).
9. G. W. G. Sharp, Y. LeMarchand-Brustel, T. Yada, L. L. Russo, C. R. Bliss, M. Cormont, L. Monge, and E. Van Obberghen, *J. Biol. Chem.* **264,** 7302 (1989).
10. S. L. Gillison and G. W. G. Sharp, *Diabetes* **43,** 24 (1994).
11. A. M. McDermott and G. W. G. Sharp, *Diabetes* **44,** 453 (1995).
12. D. Cassel and Z. Selinger, *Biochem. Biophys. Acta* **452,** 538 (1976).
13. G. Kloski and W. A. Klee, *Proc. Natl. Acad. Sci. U.S.A.* **78,** 4185 (1981).
14. G. Skoglund, C. R. Bliss, and G. W. G. Sharp, *Diabetes* **42,** 74 (1993).
15. S. Ullrich and C. B. Wollheim, *FEBS Lett.* **247,** 401 (1989).
16. M. Cormont, Y. LeMarchand-Brustel, E. Van Obberghen, A. M. Spiegel, and G. W. G. Sharp, *Diabetes* **40,** 1170 (1991).

[12] Cloning and Site-Directed Mutagenesis Studies of Gonadotropin-Releasing Hormone Receptor

Stuart C. Sealfon, Wei Zhou, Niva Almaula, and Vladimir Rodic

Introduction

During the past few years the primary sequences of a great number of G-protein-coupled receptors (GPCRs) have been elucidated by molecular cloning (1). The availability of these clones facilitates the study of the structure and function of these receptors using a variety of molecular biological and pharmacological approaches. In this chapter, the techniques utilized to clone the gonadotropin-releasing hormone receptor (GnRHR) and to study its functional domains are presented. The GnRHR is a seven-transmembrane GPCR that is coupled to phospholipase C via a G_q-subtype G protein. The biology of the GnRHR system and its regulation and coupling have been reviewed (2–4). The following sections focus on elucidating the principles we have followed in designing experiments. Specific protocols that are followed for various procedures can be found at the end of the chapter.

Gonadotropin-Releasing Hormone Receptor Cloning Strategy

A variety of approaches have been utilized for molecular cloning of membrane receptors. These include receptor purification and amino acid sequencing followed by library screening with oligonucleotides, expression cloning in *Xenopus laevis* oocytes or mammalian cells, and cloning strategies that rely on sequence homology, such as polymerase chain reaction (PCR) cloning with highly degenerate oligonucleotides. Many of these methods are described in detail in an earlier volume in this series (5–9).

In cloning the GnRHR, a sequence homology-based approach was pursued (10). The strategy, which is summarized in Fig. 1, relied on isolation of potential GnRHR clones utilizing PCR with highly degenerate oligomers and efficient screening of the partial length clones identified using a *Xenopus* oocyte-based hybrid arrest assay (11).

In any receptor cloning project, it is usually advantageous to use a cell

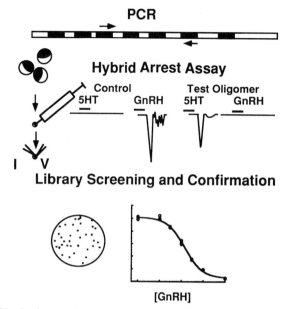

FIG. 1 GnRHR cloning strategy. The mouse GnRHR was cloned by RT-PCR using degenerate oligomers designed against the conserved transmembrane domains of GPCRs. After subcloning the PCR fragments and sequencing, the insert containing a partial-length GnRHR clone was identified using a *Xenopus* oocyte-based hybrid-arrest assay. Clones containing the entire coding region were isolated by cDNA library screening and their identity confirmed by functional expression.

line that expresses high levels of the target receptor, because this both increases the concentration of the target mRNA and decreases the complexity of the mRNA population to be screened. The GnRHR was cloned from the αT3-1 gonadotroph cell line. αT3-1 cells, which express high levels of the GnRHR, were developed by Mellon and co-workers (12). Heterologous expression of the mammalian GnRHR in oocytes using αT3-1 RNA suggested that this cell line would be a suitable source for cloning the receptor (13).

There are several caveats to bear in mind when pursuing a PCR-based homology cloning strategy. First, the GPCR to be cloned may not prove to be a member of the family of homologous proteins used to design the cloning oligonucleotides. Three distinct families of G-protein-coupled receptors have been identified by molecular cloning. The three classes are the metabotropic glutamate receptors (14), the secretin–calcitonin–parathyroid hormone class (15–18), and the large GPCR superfamily (1), of which the GnRHR proved to be a member. The primary sequences of all the clones, when analyzed for hydrophobicity, contain seven putative transmembrane domains. However, the three classes do not share any discernible sequence homology. If the receptor sought is not a member of the sequence family used for oligonucleotide design, homology-based cloning will fail.

Another potential shortcoming of a cloning strategy that relies on sequence homology and PCR is that it may be difficult to bias this approach toward the receptor of interest. The PCR reaction is likely to generate a large number of novel partial-length sequences that could represent any GPCRs in addition to other unrelated sequences that happen to be amplified in the reaction. In fact, PCR has led to the cloning of a large number of "orphan" receptors for which the ligand identity has not been determined (e.g., Refs. 19 and 20). Because significant effort is entailed in isolating, expressing, and characterizing a full-length clone after cloning the PCR fragment, it may not be feasible to target cloning to a particular desired receptor. PCR-based cloning has often been utilized to isolate any new GPCR from a particular target tissue, rather than a particular receptor type (21–23). In order to adapt this approach to the cloning of a specific receptor such as the GnRHR, an efficient screening test must be developed to determine which cDNA fragment isolated by PCR is likely to represent the desired receptor. In cloning the GnRHR, the rapid testing of partial-length clones was accomplished by an oligomer hybrid-arrest assay in *X. laevis* oocytes. PCR products were subcloned and sequenced, and antisense oligonucleotides were synthesized. The ability of these oligomers to inhibit receptor expression induced by GnRHR mRNA containing RNA in *Xenopus* oocytes was evaluated (Fig. 1; see also Ref. 11). This provided a reliable and rapid testing procedure that allowed definitive assessment of short DNA sequences isolated by PCR. Following identification of a positive partial sequence, the full-length sequence could then be isolated from a library and its identity confirmed by receptor expression. Using a cell line as the RNA source for cloning also helps limit the number of GPCR sequences that are likely to be expressed and identified in the PCR reactions. A different approach to use PCR for targeted cloning has been to design oligonucleotides against a GPCR subfamily, for example, the opioid receptors, and to use these more selective primers to isolate novel subfamily members (see Ref. 8 for a description of this approach).

Oligonucleotide Design and Polymerase Chain Reaction

In order to design cloning primers, it is necessary to generate an alignment of the GPCR family. We originally generated an alignment by hand, based on alignment of invariant GPCR residues (1). A computer algorithm-generated alignment is available at the EMBL server at Heidelberg and instructions on receiving it can be requested by sending the message GET HELP:GEN-ERAL.HELP to tm7@embl-heidelberg.de. The cloning of oligonucleotides can be highly degenerate (including up to several thousand individual species). We have not utilized inosines in the oligonucleotides to decrease the degeneracy, although that has been used successfully in degenerate oligomer design for cloning by many groups.

The design of oligomers is a highly empirical process and any given set can run into unpredictable difficulties due to the oligomer–oligomer interactions to which PCR reactions are prone. Highly degenerate oligomers, which are needed for homology cloning, increase the chances of encountering these difficulties. Furthermore, the sequence assumptions made in designing the oligomer may not be valid for the as-yet unknown sequence one desires to isolate, and therefore any given set of primers may actually fail to recognize the target. Therefore, several regions of sequence homology should be targeted by the degenerate primers, a process limited only by the funds available and the enthusiasm of the investigators.

The PCR oligomers used to clone the first GnRH receptor identified were designed against helix 3 and helix 6 of the GPCR family and were 4096- and 256-fold degenerate, respectively (10). The oligomers with which we were ultimately successful were one pair out of many pairs that failed to identify the receptor sought.

The process of oligomer design will be illustrated with a specific example in which we attempt to isolate a novel GnRHR in a species not yet cloned. We would start with an alignment of the known GnRHR amino acid sequences, presented in part in Fig. 2. Several regions of high homology are noted, particularly in the transmembrane domains. Two approaches would be possible to limit the yield of non-GnRHRs: design the oligomers to the non-transmembrane domain (which, although conserved among GnRHRs, are not at all similar among different GPCRs) or design the oligomers against a unique but conserved TMH feature of the GnRHRs. A sequence motif of this type that would be suitable for oligomer design is the loss of the TMH 2 Asp found in nearly every other GPCR (see Fig. 2 and Ref. 1. Thus oligomers that recognize the TMH 2 Asn-87 and the TMH 7 Asp-318–319 should be highly selective for GnRHRs. Mismatches are tolerated in proportion to their distance from the 3′ end of each oligomer. Therefore placing the 3′ end of each oligomer at Asn-78 and Asp-318–319 should provide maximal specificity.

Thus the 5′ oligomer would be directed against the sequence KHLTLAN (single-letter amino acid coding is used) and, if all degeneracies are included, the oligomer would be 5′-AA(A,G)CA(T,C)CTNACNCTNGCNAA-3′ (N, all four nucleotides; total degeneracy, 1024). The 3′ region of the oligomer is more important in determining specificity and for this reason the oligomer

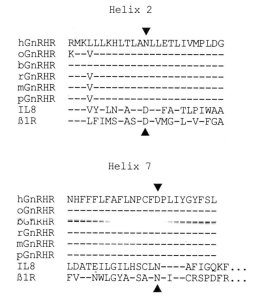

FIG. 2 Amino acid sequence alignment. The sequences of all known GnRHRs (human, ovine, bovine, rat, mouse, and pig) in the helix 2 and helix 7 domain have been aligned. For comparison, the sequences of the interleukin-8 receptor and the β_1-adrenergic receptor are also shown. Dashes indicate residues identical to the human GnRHR at that position. The positions of the Asp in helix 2 and Asn in helix 7, which are conserved in most other GPCRs, are denoted by arrowheads.

ends after the second codon for Asn. The 3' oligomer, directed against the sequence DPLIYGY, would be 5'-TANCC(G,A)TA(T,G,A)AT(C,T)AANG-G(G,A)TC (degeneracy, 384). Desired cloning sites are added to the 5' end of each oligomer. The precaution of having someone else check the oligomer design is worthwhile, because it is easy to make a costly error in recording the oligomer. Degeneracy can be limited by selecting regions for design with less degenerate amino acids; this is particularly desirable at the 3' end of each oligomer. Codon usage frequency tables can also be consulted in designing degenerate oligomers, and rare codon usage can be neglected.

RNA from an appropriate tissue source is isolated and first-strand DNA is synthesized for reverse transcriptase (RT)-PCR. It is preferable to use poly(A) RNA isolated by oligo(dT)-cellulose and to prime the cDNA synthesis with random oligomer mixtures. If there is a long 3' untranslated region the sequences recognized by the degenerate oligomers may be poorly represented. PCR conditions can be found later in this chapter (see *Detailed Methods*).

Evaluation of Partial Sequences

The PCR reaction is run on an agarose gel and bands in the anticipated size range are subcloned and sequenced. In cloning the GnRHR, the sequences isolated were run against the Genbank using the BLAST program (24, 25). Known sequences are eliminated from analysis and novel sequences showing homology to other GPCR members are selected for further analysis.

The partial-length clones isolated were tested by a *Xenopus* oocyte expression-based hybrid-arrest assay. The approach will be briefly outlined here. A full description of the methodology of *Xenopus* oocyte expression of heterologous receptors and of hybrid-arrest assays is beyond the scope of this chapter and can be found in an earlier volume in the series (7, 11). The GnRHR has been expressed in *Xenopus* oocytes using both rat pituitary RNA and αT3-1 cell RNA (13, 26). The expressed receptor can be detected by the calcium-dependent chloride current, which is recorded by two-electrode voltage clamp on exposure of the cells to GnRH (see Fig. 1). *Xenopus laevis* oocytes possess an efficient RNase H-like activity that leads to degradation of complementary RNA–oligomer hybrids (27, 28). This serves as the basis of the hybrid-arrest assay. Because the response in oocytes can be variable, a serotonin 5-HT$_{2C}$ receptor internal control was utilized. Oocytes were injected with a combination of αT3-1 and rat brain RNA, leading to the functional expression of both GnRH and 5-HT$_{2C}$ receptors. The RNA was then mixed with an oligomer that was antisense to either the 5-HT$_{2C}$ receptor or to the test sequence and the response to GnRH and serotonin was evaluated in groups of oocytes injected with the mixtures. The investigator performing the electrophysiological recording did not know beforehand the experimental group to which a given set of cells belonged. This approach provided a reliable and rapid test of the ability for the test sequences to recognize the GnRHR mRNA. Protocols for probe labeling and cDNA library screening can be found at the end of this chapter.

Following isolation and sequencing of a full-length clone, the identity of the receptor must be confirmed by functional expression. The GnRHR was expressed both in *Xenopus* oocytes (10, 13) and in COS-1 cells (29) and was shown to have appropriate pharmacology and coupling to signal transduction. The methods utilized for oocyte expression of the cloned receptor can be found elsewhere (10, 13) and the protocols followed for COS-1 transfection and radioligand receptor binding are discussed later (see also Ref. 30).

Overview of Functional Domain Studies

A driving force of investigations into the structure of the GnRHR is the hope that this insight will provide the basis for the rational design of novel analogs.

Key goals are to define the binding pocket for GnRH and the precise sites of interaction between the receptor and various ligands.

Mutagenesis, deletion, and ligand cross-linking studies suggest variability of the location of the binding pocket and ligand interaction sites among different classes of GPCRs. Three patterns emerge constituting the neurotransmitter, glycoprotein hormone, and peptide receptors. The best studied is the binding pocket of the opsins and the neurotransmitter receptors. The transmembrane domains appear to form a hydrophilic pocket for ligand binding within the transmembrane domains (1, 31, 32). The glycoprotein hormone receptors, including receptors for thyroid-stimulating hormone, follicle-stimulating hormone, and luteinizing hormone, have large amino-terminal domains that encompass the high-affinity ligand-binding sites (33–35). For the peptide GPCRs, the high-affinity binding site appears to require both extracellular and transmembrane residues. High-affinity peptide binding to the substance P receptor, for example, requires residues in the amino terminus (36), extracellular loops (36, 37), and transmembrane domains (38). Given the similar size of GnRH, it is likely that a similar pattern of extracellular and transmembrane domain-binding determinants will be observed, a hypothesis that is supported by mutagenesis studies completed to date.

In mutagenesis experiments, a single amino acid substitution is introduced into the receptor and the resultant effects on receptor function are determined in radioligand binding and/or coupling assays. Although deductions should be and are made on the basis of these experiments, it is essential to recognize that it is impossible to determine unambiguously the structural basis for the alteration in receptor function caused by a particular mutation in a GPCR. A mutation may alter function by alternative mechanisms. For example, binding or coupling may be altered because the locus mutated represents a direct site of ligand (or G protein) interaction. However, the same disruption of function could result indirectly from the mutation, inducing a structural perturbation of the binding pocket or G protein coupling interface. Because the effect of a mutation derives from a complex combination of these potential direct and indirect effects, conclusions must be cautiously developed.

Considerable insight into the difficulties inherent in deducing how particular mutations have altered receptor function can be gleaned from the mutagenesis studies of prokaryotic proteins of known structure. The functional alterations of site-directed mutants have been correlated with the precise structural alterations observed in solved crystal structures of the mutant proteins (39–41). These studies reveal the daunting complexity of the mechanisms by which a mutation alters function. For example, Matthews and co-workers have solved the structures of two second-site revertants of lysozyme that restore function that had been disrupted by a previous mutation. From the mutant and wild-type structures, they infer that the original disruptive mutation of Thr-26 to Gln interferes with enzyme activity, because the mutation perturbs the backbone structure, causing the side chain of Gln-26 to protrude into the active site cleft (41). Thus the mutation, although not at a site of catalytic activity, alters enzyme catalysis indirectly by disturbing the surface of the cleft. In studies performed on GPCRs, there are no precise structural data for the wild-type receptor, much less the mutants. One can predict, however, that many of the reports of loss of binding affinity following a receptor mutation reflect not the loss of a site of ligand interaction (as it is often interpreted) but a perturbation of the surface of the binding cleft. In mutagenesis studies of GPCRs, a large number of offsetting energy alterations underlie the final energetic effect of a particular mutation in enzymes (40).

In designing and interpreting the results of mutational experiments it is crucial to bear these caveats and limitations in mind. However, if the inherent problems in interpreting the results of mutation experiments are kept in the forefront, then one is more likely to design experiments that lead to meaningful conclusions. One set of guidelines has been presented by Ward *et al.* (42). The following guiding principles are followed in our laboratory.

1. Assay mutant constructs using both radioligand binding and functional assays. Any particular mutation is likely to induce a variety of effects potentially involving direct actions (i.e., interaction of the side chain with ligand) and a variety of indirect effects due to interaction with other receptor loci. Evaluations of the mutant receptor using both ligand binding and functional assays of coupling to signal transduction provide more insight into the basis of the various functional effects observed following introduction of a mutation. For the GnRHR we utilize both membrane and whole cell binding assays and agonist-stimulated phosphoinositol hydrolysis assays for mutant receptor characterization.
2. Analyze multiple mutations at a locus designed to test specific hypotheses. The ability to interpret the basis for loss of function is augmented by examining the effects of many different exchanges that test hypotheses about the role of side-chain properties. The principles of selection of the particular amino acid to introduce are discussed below.
3. Evaluate the effects of coordinated and compensating mutations. Although difficult to find, compensating mutations (i.e., mutations that restore a function disrupted by a first mutations) are especially informative in identifying the intramolecular networks of interactions within the receptor that have been disrupted. Potential sites are identified by searching sequence alignments for two loci that show compensatory alteration. For example, the exchange of the Asp and Asn at two positions of the GnRHR led us to explore the effect of mutations at both sites (see Fig. 2, Ref. 45, and below).

4. In studying the binding pocket, evaluate multiple agonists and antagonists selected to evaluate specific hypotheses about the molecular basis of ligand–receptor interaction. The seminal study along these lines was the identification of the putative sites of interaction between the β-adrenergic receptor and the *para-* and *meta-*hydroxyl groups of catecholamine (43). In that study it was demonstrated that the functional effect of mutating particular serine residues in helix 5 of the receptor was equivalent to using ligands lacking the proposed interacting hydroxyl groups.

5. Utilize evolutionary "experiments." A number of studies have utilized evolutionary "mutations," for example, the identification of the basis for pharmacological differences of the human and rat HT1B receptors (44) and the reciprocal mutant of the GnRHR (45). For example, the differences in affinity of Arg-8–mammalian GnRH and nonmammalian GnRHs, which lack a basic residue in the eighth position, provide a focused problem for investigation by mutagenesis (46).

6. Incorporate other approaches to confirm results. When the results of mutagenesis approaches can be confirmed by other experimental approaches, the likelihood of understanding the mechanisms involved is increased. One approach to mapping the binding site cleft of a GPCR that has been described is the systematic introduction of single cysteine residues, which can be tested by chemical reactivity for their location in the binding site (47, 48). This approach, which we have not yet utilized, holds the promise of independently confirming the likelihood that particular side chains are accessible to the ligand. Three-dimensional computational modeling and molecular dynamic simulations of ligand–receptor complexes can also provide insight into the possible molecular mechanisms that underlie the functional effects of particular mutations (49).

7. Evaluate parallel mutations. The effects of a series of exchanges at homologous positions in two different receptors can be extremely informative in understanding the mechanisms by which mutations perturb function. When different residues are tolerated in the same position in two receptors, considerable insight into the microenvironment of the locus in the two receptors can be attained. Examining the effects of mutations at the conserved TMH 2 and 7 loci (indicated by arrowheads in Fig. 2) in both the GnRH and 5-HT receptors has revealed differences in the side chains tolerated at both positions between the two receptor. For example, the residues (in helix 2/7) Asn/Asn are tolerated in the GnRHR (45) but lead to an uncoupled 5-HT receptor (47). In contrast, Asp/Asp eliminates GnRHR function but the 5-HT receptor remains functional with introduction of these residues at the same loci. These results reveal a clear difference in the microenvironment of these loci in the two receptors. Sequence comparison can suggest which residues underlie these differences between

the receptors. Further mutations at other loci may thus identify other side chains that are adjacent in space and contribute to the unique microenvironment at these loci in each receptor.

In addition, we have found that collaboration with laboratories in other disciplines, such as theoretical development of computational models and experimental pharmacology, has been invaluable in improving the design, execution, and interpretation of experimental results. In one study, for example, Harel Weinstein's laboratory evaluated the same mutations in molecular dynamic simulations of a three-dimensional computational model that we were studying experimentally (49).

Selection of Amino Acid Substitutions

The selection of the amino acid to introduce by mutagenesis must be based on the hypothesis to be tested. A simple rule is that any site that appears interesting should be probed with multiple mutations. In general, mutations should be selected to evaluate the specific physiochemical properties of the side chain, such as charge, hydrogen bonding potential, hydrophobicity, volume, and shape. A chart summarizing the relative position of hydrogenbond donors and acceptors is found in Figs. 3 and 4, and the rank order of the volume of residues buried in proteins is found in Table I. A common approach followed in the literature is that of Ala scanning,

TABLE I Rank Order of Volume of Residues Buried in Proteins[a]

Residue	Å	Residue	Å
Gly	66	Gln	161
Ala	92	His	167
Ser	99	Leu	168
Cys-S–S	106	Ile	169
Cys–SH	118	Lys	171
Thr	122	Met	171
Asp	125	Arg	202
Pro	129	Phe	203
Asn	135	Tyr	204
Val	142	Trp	238
Glu	155		

[a] Adapted from Ref. 76.

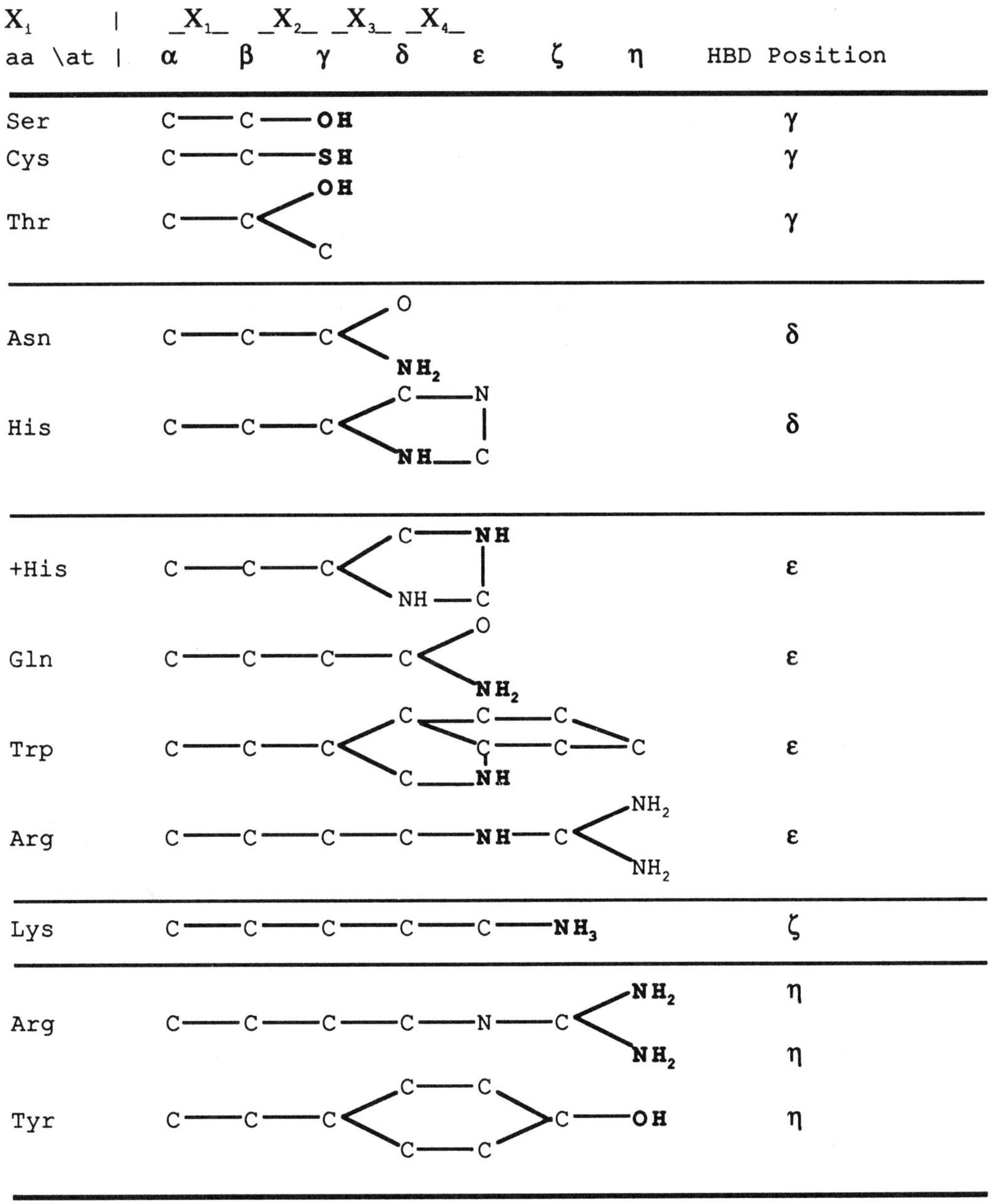

X_i		X_1	X_2	X_3	X_4				
aa \at		α	β	γ	δ	ε	ζ	η	HBD Position

Ser C——C——OH   γ

Cys C——C——SH   γ

Thr C——C〈 OH / C   γ

Asn C——C——C〈 O / NH$_2$ δ

His C——C——C〈 C——N | NH——C   δ

+His C——C——C〈 C——NH | NH——C   ε

Gln C——C——C——C〈 O / NH$_2$ ε

Trp C——C——C〈 C—C—C / C—C=C | C——NH   ε

Arg C——C——C——C——NH——C〈 NH$_2$ / NH$_2$ ε

Lys C——C——C——C——C——NH$_3$ ζ

Arg C——C——C——C——N——C〈 NH$_2$ / NH$_2$ η

Tyr C——C——C〈 C—C / C—C 〉C——OH   η

FIG. 3 Identification of hydrogen-bond donor (HBD) positions along amino acid side chains. The relative positions of the HBDs of all side chains that can donate protons are indicated. Courtesy of Juan A. Ballesteros in the laboratory of Harel Weinstein.

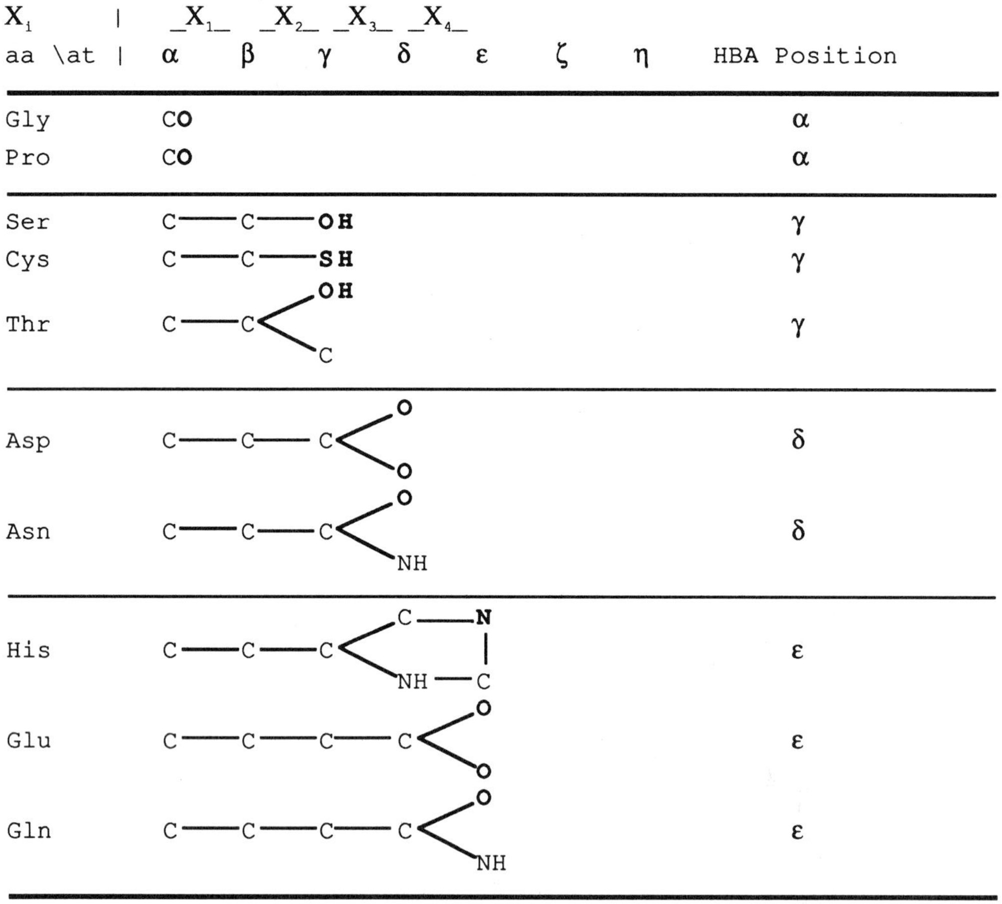

X_i	|	$\underline{X_1}$	$\underline{X_2}$	$\underline{X_3}$	$\underline{X_4}$				
aa \at	|	α	β	γ	δ	ε	ζ	η	HBA Position
Gly		CO							α
Pro		CO							α
Ser		C——C——O H							γ
Cys		C——C——S H							γ
Thr		C——C(—O H)(—C)							γ
Asp		C——C——C(=O)(—O)							δ
Asn		C——C——C(=O)(—NH)							δ
His		C——C——C(—C—N)(—NH—C)							ε
Glu		C——C——C——C(=O)(—O)							ε
Gln		C——C——C——C(=O)(—NH)							ε

FIG. 4   Identification of hydrogen-bond acceptor (HBA) positions along amino acid side chains. The relative positions of the HBAs of all side chains that can accept protons are indicated. Note that the α-carbon HBA listed for Pro and Gly refers to the backbone C=O at the $i - 4$ position when the residue is present in an α helix. The presence of these residues in the helix renders the $i - 4$ carbonyl accessible for hydrogen bonding. Courtesy of Juan A. Ballesteros in the laboratory of Harel Weinstein.

whereby each amino acid is substituted by Ala (42, 50). Ala substitutions introduce changes in both charge or H-bonding capabilities and in size (Table I), potentially leaving a cavity in comparison to other side chains. Although this may be a reasonable initial strategy in many instances, it clearly must be followed up by other substitutions at sites of interest to clarify the basis of the effect found. For example, if Ala at a site of His in the wild-type receptor alters function, additional mutations may be designed to evaluate the likelihood of a hydrogen-bond donor or acceptor role. Uncharged His is usually a hydrogen-bond donor at the δ position and a hydrogen-bond acceptor at the ε carbon. Thus, if a His is involved in hydrogen bonding, Asn would best mirror its hydrogen-bonding capacity, and Gln its hydrogen-bond accepting capacity (Figs. 3 and 4). Therefore, mutations of a His to a Gln and Asn would be likely to provide further insight into the function of that side chain.

Revertant Mutations to Probe Helix Proximities

A particularly important type of mutagenesis experiment has been the identification of second mutations that restore the function disrupted by a first mutation. If a second mutation has a restorative effect, it indicates that two mutation sites involved share a microenvironment and interact. We have been able to identify compensating mutations involving the same two loci of both the GnRHR (45) and the serotonin 5-HT$_{2A}$ receptor (49). In both receptors a mutation in helix 2 leads to a loss of receptor coupling, and a second mutation in helix 7 restores function to the receptor. A related approach is to demonstrate that the mutations at two loci have a comparable and nonadditive effect on receptor function. An example of this approach is mutations of two extracellular Cys residues in the β-adrenergic receptor that were postulated to be involved in forming a disulfide bridge (51) (see also Ref. 42).

Methods for Introducing Site-Directed Mutations

In this section, a variety of methods for performing site-directed mutagenesis are reviewed (see also Ref. 52). Single or multiple base changes introduced at the cDNA level are translated into single or multiple amino acid substitutions at the protein level (53). The first step is the design of the mutation oligonucleotide that has the base change(s) required for the desired amino acid change(s).

The size of oligomer synthesized depends on the number of base substitutions required; typically for a single base change an 18- to 21-mer is sufficient. The oligomer is usually designed so that both ends terminate in a C or G. If mutagenesis is to be carried out using single-stranded DNA the orientation (sense, antisense) of the oligonucleotide must be opposite to that of the single-stranded DNA to allow hybridization.

Production of Single-Stranded DNA (ssDNA) Using Phagemid or Vectors Derived from Filamentous Phage

The gene to be mutated is first subcloned into the phagemid or vector derived from a filamentous phage such as M13 (54, 55). Most commercially available phagemids have an M13 origin of replication for the production of ssDNA via a helper phage. The yield of ssDNA from a phagemid is sensitive to the size of the insert and the site of insertion. ssDNA yields with phagemids are not as reproducible as with vectors derived from filamentous phage (e.g., M13), although the limit for the insert size is larger. Another advantage to the use of phagemid versus M13 is the elimination of an additional subcloning step after mutagenesis, because phagemids can be used directly for expression in eukaryotic cells. M13 is a male-specific bacteriophage. Before infecting its host cell the phage absorbs to the F pilus, which is encoded by the F plasmid of the host cell. Therefore, for M13 replication and ssDNA production to take place, the host cell must carry the F plasmid. Suitable strains include JM109 (56) and TG1 (57).

Introduction of Mutation

The phosphorylated mutation oligomer and ssDNA are annealed. Annealing is accomplished by heating the reaction to 70°C followed by gradual cooling to room temperature. Following the annealing of the mutant oligonucleotide to the ssDNA, the double-stranded plasmid DNA (mutant and wild type) is recovered by extension and ligation reactions. In most protocols the yield of mutant plasmid DNA recovered is increased by a selection procedure, such as digestion of the wild-type DNA by NciI restriction endonuclease (Amersham mutagenesis kit, Arlington Heights, IL). This kit includes thionucleotide (dCTPαS) in the extension reaction. Unlike wild-type DNA strands, all newly synthesized DNA that is primed by the mutation oligomer incorporates the phosphorothionucleotide (58). Thus DNA containing the thionucleotide is resistant to digestion with NciI (59) and digestion selects for plasmid with the desired mutation. The Promega (Madison, WI) altered-sites mutagenesis kit selects for the mutant plasmid DNA by ampicillin selection. The extension reaction is primed by two oligonucleotides simultaneously, the mutation oligomer and a second oligonucleotide that confers ampicillin resistance. The mutant plasmids, therefore, confer ampicillin resistance. The reaction mixture containing the DNA plasmids are then transformed into competent BHS 71-18mutS *Escherichia coli* cells (60), which are deficient in mismatch repair (61). The Kunkel method (62) for increasing the yield of mutant plasmid utilizes wild-type ssDNA containing uracil instead of thymidine synthesized in a *dut⁻*, *ung⁻ E. coli* strain, whereas the *in vitro*-synthesized mutant DNA contains thymidine. The uracil-containing strand is degraded when introduced into an *E. coli* host strain not deficient in these enzymes. The efficiency of mutagenesis using the above methods is fairly high (70–90% in our experience with the Promega system). Accordingly, sequencing a minimum of five of these transformants should result in the isolation of several mutant constructs.

PCR-Based Mutagenesis

PCR mutagenesis requires the use of four oligomers instead of one oligomer to introduce a mutation (63). This technique is especially useful for GC-rich DNA because of the resultant high degree of secondary structure of the template, which requires a high temperature for the DNA to denature in order for the mutation oligomer to anneal. A major drawback of this technique, in addition to the expense of the oligomers, is that the entire protein-coding sequence of the insert must be sequenced after mutagenesis due to the high error rate of *Taq* DNA polymerase.

Avoiding Inadvertent Mutations

A potential concern of any approach to mutagenesis is the introduction of random mutations due to the high tolerance for mismatches in the strains used. This could in theory lead to false conclusions because the effect observed could be due to the presence of an inadvertent mutation in addition to the targeted mutation. Not infrequently, we have sequenced the entire insert or repeated the mutation a second time because of this concern. Another satisfactory approach is to back-mutate the receptor to the wild-type sequence and demonstrate restoration of wild-type properties.

Expression of Mutant Receptors

Following generation of the receptor mutants, their properties are characterized by heterologous expression. The choice of expression system and binding and functional assays is adjusted to the goals of the experiment. As noted above, we routinely assay both ligand binding and signal transduction of the mutant receptors, typically in transiently expressing COS-1 cells.

Expression in Mammalian Cells

The advantages of expression of GPCRs in bacterial and yeast cells have been reviewed elsewhere (64). The ease of manipulation of microbial cells and their lower maintenance costs are significant. However, mammalian cells appear to be the most appropriate model system for structure–function analysis of GPCRs because the biological response can be assayed in a mammalian milieu, and expression in mammalian cells ensures that the appropriate eukaryotic posttranslational modifications take place.

Numerous cell lines are available from American Type Culture Collection (Rockville, MD). The choice of the particular cell line depends on the receptor to be studied. It is preferable to avoid cell lines containing endogenous receptors of similar pharmacological profile and to select a cell line with the appropriate machinery for signal transduction of the receptor system under study.

Transient and Stable Cell Lines

Mammalian cells may be transiently or stably transfected with an appropriate eukaryotic expression vector. For example, COS cells have an integrated copy of the early region of simian virus 40 (SV40) and express the SV40 T antigen. Therefore, vectors that have the SV40 origin of replication, such as pCDNAI/Amp (Invitrogen), are appropriate for transient expression in these cells. Stable transformants incorporate the introduced DNA into their chromosomal DNA and rely on a selectable marker.

Methods of Transfection

Methods of transfection include the use of $CaPO_4$ precipitation, DEAE-dextran mediated transfection, lipopolyamine-mediated gene transfer, protoplast fusion, microinjection, electroporation, and retroviral infection. Currently we are using lipofectamine reagent (GIBCO-BRL), a lipopolyamine reagent to transfect cells with plasmid DNA. Although the reagent is expensive, we find that it gives more reproducible levels of expression than other techniques. The protocol we follow for lipofectamine transfection is given in a later section.

Characterization of Mutant Receptors

The mutant GnRHR constructs are assayed by radioligand binding and by evaluating concentration–response curves for phosphoinositol hydrolysis. In order to correlate the results of functional studies directly with the binding studies, we assay binding both in membrane preparations and in intact cells. Although whole cell binding can be difficult to perform in primary cultures, our results in transfected COS-1 cells have been quite satisfactory (Fig. 5). The protocols we follow for phosphoinositol assays and radioligand binding are given below.

Detailed Methods

Low-Stringency Polymerase Chain Reaction

The PCR reaction conditions are as follows:
    5 μl of dNTP mix containing 2 mM of each nucleotide (final concentration 200 μM).
    2 μg each degenerate oligomer.
    ~1 μg DNA template from reverse transcriptase reaction.
    5 μl 10× PCR buffer as recommended by manufacturer.
    5 units Taq polymerase.
    Water to 50 μl.

In a low-stringency reaction, 40 cycles are performed at 94°C (1 min), 45°C (1 min), and 72°C (1 min) followed by a 5-min extension at 72°C.

Probe Labeling by Random Hexamers

The partial-length cDNA probe identified in the GnRHR cloning by hybrid-arrest assay was excised with $EcoRI/SalI$ and recovered from an agarose gel. The probe was labeled by random hexamer priming. The DNA insert (2 ng + total water to 34 μl) was boiled for 2 min, then chilled on ice for 3 min and briefly spun at 4°C. Add 10 μl hexamer labeling mix (New England Biolabs), 1 μl Klenow (5 U), and 5 μl [α-^{32}P]dCTP. Incubate 30 min at 37°C then boil 2 min. The probe can be used directly for library screening.

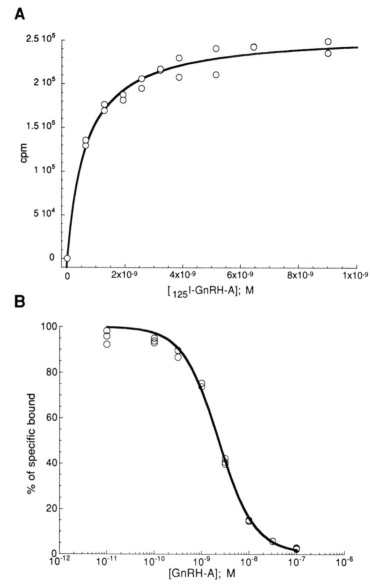

FIG. 5 GnRHR saturation and competition binding performed on intact cells. Intact cell binding assays were done on COS-1 cells transfected with the wild-type mouse GnRHR cDNA as described in the text. (A) Saturation binding. (B) Competition binding. The affinities determined by saturation binding and the K_i values obtained with GnRH-A competition were essentially indistinguishable.

cDNA Library Screening

A lambda-Zap cDNA library (Stratagene, La Jolla, CA) was screened under standard conditions. The primary plates were spread at a density of 40,000 plaques/150-mm plate. After 7 hr of growth at 37°C, duplicate filters of nitrocellulose were applied for 2 and 4 min, respectively and denatured on Whatman (Clifton, NJ) paper previously saturated with 0.2 N NaOH and 1.5 M NaCl. Following denaturation, the filters were neutralized on Whatman paper containing 0.4 M Tris (pH 7.6) and 2× SSC or containing 2× SSC sequentially for 3 min each. Prehybridization buffer consisted of formamide (50 ml), 20× SSC (25 ml), 1 M Tris, pH 7.6 (2 ml), 100× Denhardt's solution (1 ml), ssDNA (5 mg/ml) (1 ml), 20% SDS (0.5 ml), and water to 100 ml.

The filters were prehybridized in 60 ml buffer at 42°C for 3 hr and hybridized with probe at 42°C overnight. Washing was performed sequentially (with change of buffer) in 2× SSC/0.1% SDS at room temperature for 15 min, 42°C for 30 min, and 55°C for 30 min. A final wash was performed in 0.2× SSC/0.1% SDS at 63°C for 30 min. At this point the primary filters should have no counts above background detectable with a hand-held β-counter. After drying, filters were exposed to X-ray film with an intensifying screen at −70°C for 2–3 days. Plaques positive on duplicate filters were excised and purified through ~3 subsequent rounds of screening. Alternate labeling protocols and library screening approaches can be found in any basic molecular biology methodology book (65).

Transient Transfection of COS-1 Cells

We presently use lipofectamine (GIBCO-BRL) to transfect cells with plasmid DNA. The method involves mixing the plasmid and reagent in serum-free medium, allowing DNA–liposome complexes to form, and overlaying cells with this solution.

The method of transfection has a number of variables that need to be optimized for each cell type and plasmid, the most important being (1) ratio of plasmid DNA to lipofectamine reagent, (2) ratio of lipofectamine (which is toxic) to cells, (3) time of exposure of cells to DNA–liposome complexes, and (4) number/density of cells at which to carry out the transfection protocol.

In our hands using COS-1 cells and pCDNAI/Amp encoding GnRHR or 5HT receptors, high levels of expression were achieved using the following protocol:

1. Cells (3 × 10^6) were plated (seeded) on a 10-cm tissue culture dish the day before transfection.

2. For each plate, mix 8 μg DNA plus 0.8 ml serum-free medium and 48 μl lipofectamine plus 0.8 ml serum-free medium. Allow complexes to form for 30 min at room temperature.
3. Add serum-free medium to a final volume of 6 ml/plate and overlay cells that have been rinsed twice with serum-free medium.
4. At 5 hr after transfection add an equal volume (6 ml) of medium containing 20% FBS.
5. Feed the cells with fresh complete medium the following day (18–24 hr after transfection).
6. Harvest cells for radioligand binding studies 48–72 hr after transfection.

Stable Transfection

The method of transfection is the same as for transient transfection except that cotransfection of two plasmids (or gene and selective marker on same plasmid) is carried out. The cells are split 72 hr following transfection, at a densities of 1:10 and 1:20. Stable transformants are selected for by growth of the cells in antibiotic-containing medium, depending on the selective marker present on the plasmid, e.g., pRSVneo (Invitrogen) confers resistance to G418. The concentration of G418 required needs to be determined for each cell line.

Radioligand Binding Assays

A radioligand–receptor assay for gonadotropin-releasing hormone (GnRH) receptors, employing a radioiodinated superagonist analog as the labeled ligand, is used to analyze the receptor-binding properties of wild-type and mutated receptors. This receptor assay involves incubation of ^{125}I-labeled GnRH-A, which is the tracer in the assay, with GnRH receptors present either on intact cells or crude cell membrane preparations. After incubation, separation of bound and free tracer is achieved by washing intact cells or membranes two to three times at 4°C with washing buffer, followed by harvest or filtration of intact cells or membranes, respectively.

Radioiodination of GnRH-A

In our radioligand–receptor assays for pituitary gonadotropin-releasing hormone receptors we use a radioiodinated superagonist analog of GnRH (des-Gly^{10}[D-Ala^6],LH-RHethylamide; GnRH-A) as the labeled ligand. Such analogs bind to a single class of high-affinity sites present in rat anterior pituitary.

Because only a small fraction of [^{125}I]iodo-GnRH binds to high-affinity sites (<25%) and because [^{125}I]iodo-GnRH is inactivated during the incubation with plasma membrane preparations, it is reasonable to use superagonist GnRH-A as the labeled ligand to overcome these problems (66, 67). Radioiodination of GnRH-A as the preferred method of radiolabeling provides a ligand with high specific activity and high affinity. As a relatively inexpensive, easy to perform, and reliable method of iodination we use Iodogen reagent (Pierce Chemical Co., Rockford, IL) to catalyze iodination. Iodogen can be used for a wide range of proteins and peptides, permitting theoretical iodine incorporation with minimal oxidation damage and producing tracer stable for up to 3 months (68).

Plasma Membrane Proteins

Plasma membranes are prepared from COS-1 cells according to our modification of the method graciously provided by Susan Laws (69). Briefly, COS-1 cells (American Type Culture Collection, Rockville, MD) are used as an expression system to obtain GnRH receptor proteins. Cells are grown in Dulbecco's modified Eagle's medium (DMEM) supplemented with fetal bovine serum (10% v/v final) and seeded into 10-cm tissue culture dishes at 3 million cells per dish the day before transfection. Transfection is carried out as described above. Three days after transfection cells are harvested and centrifuged. Membrane pellets are stored at $-80°C$ for up to 2 months. Crude plasma membranes are prepared by homogenizing cell pellets in a glass homogenizer with assay buffer, followed by centrifugation at 15,000 g for 30 min at 4°C. Pellets are resuspended in assay buffer and protein concentration is determined using the Bio-Rad (Richmond, CA) protein assay reagent (70).

Plasma Membrane Receptor Binding Assay

Materials are as follows:

> Assay buffer (10 mM Tris, 1 mM EDTA, 0.1% (w/v) BSA, pH 7.4).
> Washing buffer (10 mM Tris, 1 mM EDTA, 0.1% (w/v) BSA, pH 7.4).
> 12 × 75-mm borosilicate glass tubes.
> [^{125}I]GnRH-A.
> Unlabeled peptide (GnRH-A, GnRH, etc.).

The following solutions are used in all assay tubes:

> 100 μl of plasma membrane preparation (~60–100 μg of proteins per assay tube).

100 μl of [^{125}I]GnRH-A (60,000 cpm/100 μl; 44 pM final concentration in assay tube).
100 μl of assay buffer.
100 μl of unlabeled test peptide.

Each assay tube containing 60,000 cpm [^{125}I]GnRH-A is incubated for 90 min on ice with 60–100 μg membrane proteins in the presence of increasing concentrations of the test peptides. Nonspecific binding is assessed in the presence of 0.1 μM unlabeled GnRH-A or 1 μM GnRH, depending on which peptide is used as a displacing ligand, and total binding is determined in the absence of unlabeled ligand. Following incubation, which is performed in 12 × 75-mm borosilicate glass tubes in a total volume of 0.4 ml assay buffer and terminated by dilution with 3 ml ice cold washing buffer (pH 7.4), the reaction is immediately filtered in a cell harvester under vacuum through glass fiber filters (GF/C Whatman) presoaked in washing buffer. Filters are then rinsed three times with washing buffer, placed in glass tubes, and the retained radioactivity is counted in a gamma counter.

Whole Cell Receptor Binding Assay
Materials are as follows:

Cultured cells in monolayer in a 24-well multiwell dish.
[^{125}I]GnRH-A (60,000 cpm/well).
Unlabeled peptide (serial dilutions starting from 1 μM final concentration in well).
Assay buffer: Krebs–Ringer solution with HEPES (25 mM HEPES, 1% BSA, pH 7.4 at 4°C).
Washing buffer: Krebs–Ringer solution with HEPES (25 mM HEPES, 0.2% BSA, pH 7.4 at 4°C).

Cells from 10-cm dishes are harvested with trypsin 24 hr following transfection and plated into 24-well multiwell tissue culture dishes. The binding assay is performed 72 hr after transfection. Cell monolayers are incubated in a total volume of 0.4 ml, which consists of 0.2 ml assay buffer, 0.1 ml of [^{125}I]GnRH-A at a final concentration of 60,000 cpm/well and either 0.1 ml buffer (for determination of total binding) or 0.1 ml unlabeled peptide at a series of increasing concentrations. The incubation lasts 2 hr at 4°C. The incubation medium is then removed and cells are washed three times with washing buffer at 4°C. Cells are harvested 1 M NaOH and bound radioactivity is counted in a gamma counter. The Lowry method is used for determining the amount of proteins present in each well (71).

Measurement of Phosphoinositol Hydrolysis

Activation of the receptor in the presence of lithium results in the accumulation of inositol monophosphates, which is the basis of the assay (72). The measurement of total inositol phosphates (IP) in the presence of LiCl provides an estimate of the extent of receptor activation.

Various methods for the measurement of inositol lipids and phosphates have been described (see Ref. 73 and references therein). We measure accumulation of total inositol phosphates, by formic acid extraction, from transfected COS-1 cells followed by separation on Dowex columns.

Cells are transfected as described above. After 24 hr, cells in each 10-cm plate are split by trypsin into two 12-well plates; 48 h after transfection cells are prelabeled with myo-[^{3}H]inositol (0.5–2 μCi/ml) for 16–18 hr in inositol-free medium. The medium is then aspirated and cells rinsed with either 1× Hanks' balanced salt solution (HBSS), 20 mM HEPES, pH 7.5, or serum-free DMEM. Add 1× HBSS, 20 mM HEPES, pH 7.5, containing agonist and 20 mM LiCl and incubate cells for the desired time. The medium is aspirated and the reaction is stopped by addition of 10 mM formic acid, 4°C, which also serves to extract cellular lipids. Alternatively, perchloric acid may be used (74).

Inositol phosphates are separated on Dowex columns by ion-exchange chromatography [a variation of the method developed by Berridge (75)].

1. Recharge Dowex columns (column volume, approximately 0.5 ml) using 2.5 ml of 3 M ammonium formate, 0.1 M formic acid.
2. Wash with 10 ml of 10 mM myo-inositol, 10 mM formic acid.
3. Apply samples.
4. Wash columns with 5 ml of 10 mM myo-inositol, 10 mM formic acid.
5. Wash columns with 10 ml of 60 mM sodium formate, 5 mM borax.
6. Elute total inositol phosphates in 4.5 ml of 1 M ammonium formate, 0.1 M formic acid.

Acknowledgments

The authors thank Harel Weinstein, Robert Millar, and their laboratories for collaboration in many studies and for important contributions to the development of the ideas presented. We are grateful to Saul Maayani and Barbara Ebersole for advice and assistance in the development of various pharmacological assays and to Susan Laws for providing protocols for GnRH-A labeling and radioligand binding assays. Juan A. Ballesteros kindly provided the figures summarizing the location of hydrogen bonding. The work on the GnRH receptor is supported by NIH Grant RO1 DK46943.

References

1. W. C. Probst, L. A. Snyder, D. I. Schuster, J. Brosius, and S. C. Sealfon, *DNA Cell Biol.* **11,** 1 (1992).
2. S. C. Sealfon and R. P. Millar, *in* "Oxford Reviews of Reproductive Biology" (H. M. Charlton, ed.), p. 255. Oxford University Press, Oxford, England, 1994.
3. S. C. Sealfon and R. P. Millar, *Cell. Mol. Neurobiol.* **15,** 25–42 (1995).
4. S. S. Stojilkovic, J. Reinhart, and K. J. Catt, *Endocr. Rev.* **15,** 462 (1994).
5. D. I. Schuster and R. B. Murphy, *in* "Methods in Neurosciences: Receptor Molecular Biology" (S. C. Sealfon, ed.), Vol. 25, p. 9. Academic Press, San Diego, 1995.
6. M. J. Walsh, *in* "Methods in Neurosciences: Receptor Molecular Biology" (S. C. Sealfon, ed.), Vol. 25, p. 44. Academic Press, San Diego, 1995.
7. R. D. Zuhlke, H.-J. Zhang, and R. H. Joho, *in* "Methods in Neurosciences: Receptor Molecular Biology" (S. C. Sealfon, ed.), Vol. 25, p. 67. Academic Press, San Diego, 1995.
8. D. K. Grandy, Q.-Y. Zhou, C. Bouvier, C. Saez, and J. R. Bunzow, *in* "Methods in Neurosciences: Receptor Molecular Biology" (S. C. Sealfon, ed.), Vol. 25, p. 90. Academic Press, San Diego, 1995.
9. S. C. Sealfon, *in* "Methods in Neurosciences: Receptor Molecular Biology" (S. C. Sealfon, ed.), Vol. 25, p. 3. Academic Press, San Diego, 1995.
10. M. Tsutsumi, W. Zhou, R. P. Millar, P. L. Mellon, J. L. Roberts, C. A. Flanagan, K. Dong, B. Gillo, and S. C. Sealfon, *Mol. Endocrinol.* **6,** 1163 (1992).
11. M. Tsutsumi and B. Gillo, *in* "Methods in Neurosciences: Receptor Molecular Biology" (S. C. Sealfon, ed.), Vol. 25, p. 105. Academic Press, San Diego, 1995.
12. J. J. Windle, R. I. Weiner, and P. L. Mellon, *Mol. Endocrinol.* **4,** 597 (1990).
13. S. C. Sealfon, B. Gillo, S. Mundamattom, P. L. Mellon, J. J. Windle, E. Landau, and J. L. Roberts, *Mol. Endocrinol.* **4,** 119 (1990).
14. Y. Tanabe, M. Masu, T. Ishii, R. Shigemoto, and S. Nakanishi, *Neuron* **8,** 169 (1992).
15. H. Y. Lin, T. L. Harris, M. S. Flannery, A. Aruffo, E. H. Kaji, A. Gorn, L. J. Kolakowski, H. F. Lodish, and S. R. Goldring, *Science* **254,** 1022 (1991).
16. T. Ishihara, S. Nakamura, Y. Kaziro, T. Takahashi, K. Takahashi, and S. Nagata, *EMBO J.* **10,** 1635 (1991).
17. A. B. Abou-Samra, H. Juppner, T. Force, M. W. Freeman, X. F. Kong, E. Schipani, P. Urena, J. Richards, J. V. Bonventre, J. Potts, Jr., *et al., Proc. Natl. Acad. Sci. U.S.A.* **89,** 2732 (1992).
18. R. Chen, K. A. Lewis, M. H. Perrin, and W. W. Vale, *Proc. Natl. Acad. Sci. U.S.A.* **90,** 8967 (1993).
19. S. N. Schiffmann, F. Libert, G. Vassart, J. E. Dumont, and J. J. Vanderhaeghen, *Brain Res.* **519,** 333 (1990).
20. K. A. Eidne, J. Zabavnik, T. Peters, S. Yoshida, L. Anderson, and P. L. Taylor, *FEBS Lett.* **292,** 243 (1991).
21. M. Parmentier, F. Libert, J. Perret, D. Eggerickx, C. Ledent, S. Schurmans, E. Raspe, J. E. Dumont, and G. Vassart, *Adv. Second Messenger Phosphoprotein-Res.* **28,** 11 (1993).

22. F. Libert, M. Parmentier, A. Lefort, C. Dinsart, J. VanSande, C. Maenhaut, M.-J. Simons, J. E. Dumont, and G. Vassart, *Science* **244,** 569 (1989).
23. L. A. Selbie, N. A. Townsend, T. P. Iismaa, and J. Shine, *Mol. Brain Res.* **13,** 159 (1992).
24. S. F. Altschul, W. Gish, W. Miller, E. W. Myers, and D. J. Lipman, *J. Mol. Biol.* **215,** 403 (1990).
25. W. Gish and D. J. States, *Nat. Genet.* **3,** 266 (1993).
26. K. A. Eidne, A. I. McNiven, P. L. Taylor, S. Plant, C. R. House, D. W. Lincoln, and S. Yoshida, *J. Mol. Endocrinol.* **1,** R9 (1988).
27. P. Dash, I. Lotan, M. Knapp, E. R. Kandel, and P. Goeley, *Proc. Natl. Acad. Sci. U.S.A.* **84,** 7896 (1987).
28. W. Meyerhof and D. Richter, *FEBS Lett.* **266,** 192 (1990).
29. S. C. Sealfon, M. Tsutsumi, W. Zhou, L. Chi, C. Flanagan, J. S. Davidson, M. Golembo, I. Wakefield, P. L. Mellon, J. L. Roberts, and R. P. Millar, *in* "GnRH, GnRH Analogs, Gonadotropins and Gonadal Peptides" (P. Bouchard, A. Caraty, J. J. T. Coelingh Bennink, and S. Pavlou, eds.). Proc., Third Organon Round Table Conf., Paris, 1992.
30. R. P. Millar, J. Davidson, C. Flanagan, and I. Wakefield, *in* "Methods in Neurosciences: Receptor Molecular Biology" (S. C. Sealfon, ed.), Vol. 25, p. 145. Academic Press, San Diego, 1995.
31. C. D. Strader, I. S. Sigal, and R. A. F. Dixon, *FASEB J.* **3,** 1825 (1989).
32. E. C. Hulme, N. J. Birdsall, and N. J. Buckley, *Annu. Rev. Pharmacol. Toxicol.* **30,** 633 (1990).
33. Y. Nagayama, D. Russo, G. D. Chazenbalk, and B. Rapoport, *J. Biol. Chem.* **266,** 14926 (1991).
34. H. Loosfelt, M. Misrahi, M. Atger, R. Salesse, M. Thi, A. Jolivet, A. Guiochon-Mantel, S. Sar, B. Jallal, J. Garnier, and E. Milgrom, *Science* **245,** 525 (1989).
35. C. H. Tsai-Morris, E. Buczko, W. Wand, and M. L. Dufau, *J. Biol. Chem.* **265,** 19385 (1990).
36. T. M. Fong, H. Yu, R. C. Huang, and C. D. Strader, *Soc. Neurosci. (Abstr.)* **18,** 454 (1992).
37. T. M. Fong, R. C. Huang, R. H. Yu, and C. D. Strader, *Regul. Peptides* **46,** 43 (1993).
38. R.-R. C. Huang, H. Yu, C. D. Strader, and T. M. Fong, *Biochemistry* (in press) (1994).
39. J. A. Bell, W. J. Becktel, U. Sauer, W. A. Baase, and B. W. Matthews, *Biochemistry* **31,** 3590 (1992).
40. K. P. Wilson, B. A. Malcolm, and B. W. Matthews, *J. Biol. Chem.* **267,** 10842 (1992).
41. A. R. Poteete, D. P. Sun, H. Nicholson, and B. W. Matthews, *Biochemistry* **30,** 1425 (1991).
42. W. H. Ward, D. Timms, and A. R. Fersht, *Trends Pharmacol. Sci.* **11,** 280 (1990).
43. C. D. Strader, M. R. Candelore, W. S. Hill, I. S. Sigal, and R. A. Dixon, *J. Biol. Chem.* **264,** 13572 (1989).
44. D. Oksenberg, S. A. Marsters, B. F. O'Dowd, H. Jin, S. Havlik, S. J. Peroutka, and A. Ashkenazi, *Nature (London)* **360,** 161 (1992).

45. W. Zhou, C. Flanagan, J. A. Ballesteros, K. Konvicka, J. S. Davidson, H. Weinstein, R. P. Millar, and S. C. Sealfon, *Mol. Pharmacol.* **45**, 165 (1994).
46. C. A. Flanagan, I. I. Becker, J. S. Davidson, I. K. Wakefield, W. Zhou, S. C. Sealfon, and R. P. Millar, *J. Biol. Chem.* **269**, 22636 (1994).
47. J. A. Javitch, X. Li, J. Kaback, and A. Karlin, *Proc. Natl. Acad. Sci. U.S.A.* **91**, 10355 (1994).
48. J. A. Javitch, D. Fu, J. Chen, and A. Karlin, *Neuron* **14**, 825 (1995).
49. S. C. Sealfon, L. Chi, B. Ebersole, V. Rodic, D. Zhang, J. A. Ballesteros, and H. Weinstein, *J. Biol. Chem.* **270**, 16683 (1995).
50. O. Moro, M. S. Shockley, J. Lameh, and W. Sadee, *J. Biol. Chem.* **269**, 6651 (1994).
51. R. A. Dixon, L. S. Sigal, M. R. Candelore, R. B. Register, E. Rands, and C. D. Strader, *EMBO J.* **6**, 3269 (1987).
52. T. M. Fong, M. R. Candelore, and C. D. Strader, *in* "Methods in Neurosciences: Receptor Molecular Biology" (S. C. Sealfon, ed.), Vol. 25, p. 263. Academic Press, San Diego, 1995.
53. M. Smith, *Annu. Rev. Genet.* **19**, 423 (1985).
54. J. Messing, *in* "Methods in Enzymology," Vol. 101, p. 20. Academic Press, New York, 1983.
55. J. Viera and J. Messing, *in* "Methods in Enzymology," Vol. 153, p. 3. Academic Press, New York, 1987.
56. C. Yanisch-Perron and J. Messing, *Gene* **33**, 103 (1985).
57. T. J. Gibson, PhD thesis, Cambridge University, Cambridge, 1984.
58. J. W. Taylor, J. Ott, and F. Eckstein, *Nucl. Acids Res.* **13**, 8764 (1985).
59. K. L. Nakamaye and F. Eckstein, *Nucl. Acids Res.* **14**, 9679 (1986).
60. B. Kramer, W. Kramer, and H.-J. Fritz, *Cell* **38**, 879 (1984).
61. R. Zell and H.-J. Fritz, *EMBO J.* **6**, 1809 (1987).
62. T. A. Kunkel, *PNAS* **82**, 488 (1985).
63. D. H. Jones, R. L. Sakamoto, and B. H. Howard, *Nature (London)* **344**, 793 (1990).
64. D.. Strosberg and S. Marullo, *Trends Pharmacol. Sci.* **13**, 95 (1992).
65. J. Sambrook, E. F. Fritsch, and T. Maniatis, "Molecular Cloning: A Laboratory Manual," 2nd Ed. Cold Spring Harbor Laboratory Press, Cold Spring Harbor, New York, 1989.
66. R. N. Clayton and K. J. Catt, *Endocrinology* **106**, 1154 (1980).
67. R. N. Clayton, R. A. Shakespear, J. A. Duncan, and J. C. Marshall, *Endocrinology* **105**, 1369 (1979).
68. P. R. P. Salacinski, C. McLean, J. E. C. Sykes, V. V. Clement-Jones, and P. J. Lowry, *Anal. Biochem.* **117**, 136 (1981).
69. S. C. Laws, M. J. Beggs, J. C. Webster, and W. L. Miller, *Endocrinology* **127**, 373 (1990).
70. M. Bradford, *Anal. Biochem.* **72**, 248 (1976).
71. O. H. Lowry, N. J. Rosebrough, A. L. Farr, and R. J. Randall, *J. Biol. Chem.* **193**, 265 (1951).
72. M. J. Berridge, P. C. Downes, and R. M. Hanley, *Biochem. J.* **206**, 587 (1982).

73. P. P. Godfrey, *in* "Signal Transduction" (G. Milligan, ed.), p. 105. Oxford University Press, Oxford, England, 1992.
74. E. A. Bone, P. Fretten, S. Palmer, C. J. Kirk, and R. H. Michell, *Biochem. J.* **221,** 803 (1984).
75. M. J. Berridge, M. C. Dawson, C. P. Downes, J. P. Heslop, and R. F. Irvine, *Biochem. J.* **212,** 473 (1983).
76. C. Chothia, *Annu. Rev. Biochem.* **53,** 537 (1984).

[13] Use of *Xenopus laevis* Expression System for Structure–Function Analysis of G-Protein-Coupled Receptor for Parathyroid Hormone and Parathyroid Hormone-Related Protein

Paul R. Turner, Zhengmin Huang, Robert A. Nissenson, and Dolores M. Shoback

Introduction

The receptor for parathyroid hormone (PTH) and parathyroid hormone-related protein (PTHrP) is a member of a small but growing subfamily of G-protein-coupled receptors, which includes receptors for calcitonin, secretin, glucagon, and several other peptides (27). PTH is a major regulator of calcium homeostasis *in vivo*, and receptors for PTH/PTHrP (13) are localized to osteoblasts, kidney cells, and many other cell types. PTH is thought to act in a classical endocrine manner, reaching its target organs through the circulation. In contrast, PTHrP interacts with the same receptors with an identical affinity, but probably via an autocrine or paracrine mechanism (10). Stimulation of PTH/PTHrP receptors by either agonist results in adenylate cyclase activation, through the stimulatory G protein G_s. At higher concentrations of PTH or PTHrP, the same receptors also increase the intracellular free Ca^{2+} concentration ($[Ca^{2+}]_i$) (2), an effect probably mediated via a G protein of the G_q/G_{11} family. G protein subunits in this family can couple to phospholipase C (PLC), which catalyzes the breakdown of phosphatidylinositol 4,5-bisphosphate. This leads to increases in inositol 1,4,5-trisphosphate (1,4,5-InsP$_3$), stimulating mobilization of Ca^{2+} from intracellular storage sites (3). Increases in inositol phosphates induced by PTH or PTHrP are typically smaller and more difficult to detect than are changes in cyclic AMP. This has been demonstrated in studies using COS-1 cells transiently transfected with the PTH/PTHrP receptor cDNA (26). Similar observations have also been made for a variety of other receptors. Such receptors might, therefore, be best studied in a system in which the signals generated by both pathways are more readily measured. The expression of receptor constructs in *Xenopus laevis* oocytes may provide just such a system (8, 9). In contrast to mamma-

Methods in Neurosciences, Volume 29

lian expression systems, oocytes can generate substantial intracellular Ca^{2+} responses that can be detected through electrophysiologic, radioisotopic, and microfluorimetric techniques. It is likely that this quantitative difference in receptor-induced Ca^{2+} mobilization is due to the increased ability of the expressed receptor to couple to the different G protein subunits present in oocytes, compared to the typical mammalian cells used for transient transfections (1).

It has become increasingly clear that receptors interact via different regions and sequences with their cognate G proteins (1). Intracellular domains of the PTH/PTHrP receptor, including the third intracellular loop and the C-terminal tail, are thought to be the ones that most likely interact with G proteins based on evidence from other receptor systems. We have used strategies such as scanning-alanine mutagenesis of specific regions, and tail truncation to begin to examine the roles of these regions on PTH/PTHrP receptor-mediated signal transduction (11). Because of the relative ease of measuring both adenylate cyclase and PLC activation in *Xenopus* oocytes, we have used this system to examine the effects of specific mutations on the functional properties of the PTH/PTHrP receptor. The following text summarizes the experimental techniques we have used to evaluate coupling of wild-type and mutant PTH/PTHrP receptors to stimulation of adenylate cyclase and PLC activity in *Xenopus* oocytes.

Use of *Xenopus laevis* Oocytes

Oocytes present in the ovary of mature *Xenopus* females are arrested at the first meiotic prophase and contain a large nucleus or germinal vesicle (19). The largest of these oocytes (classified as Stage V or VI) are used for expression studies. These cells are typically ≥ 1 mm in diameter, allowing easy isolation and manipulation. The large size of oocytes plus the ease of handling them enables the measurement of changes in $[Ca^{2+}]_i$ and cyclic AMP using single cells or relatively small numbers of cells. The injection of exogenous mRNAs encoding receptors or ion channels into oocytes will often lead to the synthesis of full-length proteins, which may then undergo posttranslational processing. Such proteins are then inserted into the plasma membrane in a functional form and can be screened for using an assay of receptor or ion channel function (9). For all these reasons, expression cloning with *Xenopus* oocytes has proved very successful (29). These properties of oocytes, together with the tools of receptor mutagenesis and chimeric protein construction, have enabled the dissection of the functional roles of specific regions of receptors, ion channels, and other proteins using this system.

Oocyte Removal

Mature gravid females obtained from *Xenopus* One (Ann Arbor, MI) are maintained in the laboratory as described (7). To remove the oocytes, the animal is immersed in a solution of tricaine in distilled water (0.3%) or covered with crushed ice to produce anesthesia. The animal is placed ventral side up and the abdominal skin is pinched. A small incision is made with sterile fine scissors and enlarged to make an incision about 1 cm long. Ovarian sections are removed with blunt forceps, and those containing the largest oocytes (Stage V or VI) are selected. The ovarian segments are immediately placed in modified Barth's saline with HEPES (MBSH) (see *Appendix* at the end of this chapter for solution compositions). After adequate ovarian tissue is removed, the incision is sutured with two to four stitches (cat gut or nylon) and the animal is allowed to recover in tap water (2–4 hr) before returning it to its tank. Oocytes can be stored at 19° or 4°C and used for up to 1 week following isolation. With careful technique, animals can be routinely used at least four times, and in some cases, for many months.

Oocyte Microdissection

The ovarian sections are poured into sterile petri dishes and cut into smaller pieces (~1–2 cm in size). A 2-mm-diameter capillary tube (bent at a 90° angle and fire polished) to used to pull the largest oocytes (≥1 mm diameter) away from the rest of the tissue. Contact must be maintained between the capillary tube and the bottom of the petri dish, so that the oocyte is not crushed during this manual dissection. Healthy oocytes are spherical, with a distinct black animal pole and yellowish vegetal pole. The equator between these poles should be sharp. The oocytes are transferred with a fire-polished Pasteur pipette to a dish and rinsed three or four times with MBSH. To remove follicular cell layers surrounding the oocytes, they are treated with collagenase (Sigma, St. Louis, MO, type 2a) in MBSH (2 mg/ml) with 100 U/ml penicillin G, 100 μg/ml streptomycin, 10 U/ml gentamicin, and 10% fetal calf serum for 2 hr at room temperature with gentle shaking. After this treatment, the oocytes are washed a few times with fresh MBSH and stored overnight. The next day the follicle cells/vitelline layers may be peeled off with fine forceps, and the oocytes are ready for microinjection.

Preparation of cRNA

The general procedure involves the linearization of the cDNA template at its 3' end, preferably using a blunt-cutting restriction enzyme; *in vitro* transcription of the cDNA using the appropriate RNA polymerase to yield cRNA;

and eventually digestion of the cDNA template. Effort should be made to ensure that all equipment and supplies are RNase free. Gloves should be worn at all times. Pipette tips should be autoclaved. Gel apparatus should be washed with soap and water, rinsed with ethanol, soaked in 3% hydrogen peroxide for at least 10 min, rinsed with diethyl pyrocarbonate (DEPC)-treated RNase-free water, and autoclaved before use. All glassware used for the preparation of RNA must be baked.

Template cDNA Production

For these experiments, mutant PTH/PTHrP receptor cDNA constructs were generated using the polymerase chain reaction (PCR). The opossum kidney (OK) PTH/PTHrP receptor (13) in the vector pCDNA1/Amp (Invitrogen Corp.) was used as a template (a gift from H. Jüppner). Tandem-alanine constructs of the OK PTH/PTHrP receptor cDNA were generated using overlap extension PCR (11). All receptor constructs were maintained, transformed, and amplified in the *Escherichia coli* strain Top 10F'. All mutant receptor cDNA constructs were subcloned into the vector pGEM-HE [a gift of E. Liman (17)], which contains 3' and 5' untranslated regions of the *Xenopus* β-globin gene, specifically engineered to increase mRNA expression in oocytes.

The protocol used in our laboratory to prepare the DNA for transcription is as follows:

1. Purified plasmid DNA (OK/pGEM-HE, 10 μg) is digested with the appropriate restriction enzyme (e.g., *Nhe*I for this vector).
2. After digestion at 37°C for at least 2 hr, 1 μl of proteinase K (1 mg/ml) is added for 30 min at 37°C to degrade RNases and restriction enzyme.
3. Linearized template cDNA is extracted twice with phenol–chloroform (1:1 v/v) and once with chloroform, precipitated with 100% ethanol, and resuspended in RNase-free TE buffer.

In Vitro Transcription Reaction

The presence of a 5' cap structure increases the stability of synthesized RNA, thereby improving the translation efficiency of cRNA following microinjection into oocytes (28). We routinely use the mCAP Kit (Stratagene, La Jolla, CA), which initiates with the 7-MeG(5')ppp(5')G cap analog, in about 90–95% of the resulting RNA transcripts. Typical yields range from 1 to 10 μg cRNA μg cDNA template.

1. In a sterile RNase-free microfuge tube, mix 50 μl of 5× transcription buffer, 10 μg linearized cDNA (see above), 10 μl rNTP mix, 10 μl 1 mM rGTP (to improve the yield of full-length transcripts), 25 μl 3 mM mCAP analog, 10 μl 0.75 M DTT, 50 units of appropriate RNA polymerase (e.g., T7), 100 U RNasin (Promega, Madison, WI) or RNase Block (Stratagene), and DEPC-treated water to a final volume of 250 μl. The reaction is initiated by placing the tube at 37°C for 30 min.
2. The cDNA template is then digested by adding 100 U RNase-free DNase I for 5 min at 37°C. The reaction is stopped by adding 740 μl of DEPC-treated water.
3. The resulting cRNA in aqueous solution is extracted once with 1 volume phenol : chloroform (1 : 1) and then precipitated by adding 0.1 vol of 3 M sodium acetate (pH 5.0) and 2 volumes of absolute ethanol at -20°C for at least 30 min or preferably overnight. The precipitate is pelleted in a microfuge for 20 min at 4°C. The pellet is rinsed with 80% ethanol/20% DEPC-treated water (v/v) and dried.
4. The cRNA is resuspended in 25 μl of DEPC-treated water. *Note:* TE buffer should *not* be used because it is toxic to ooctyes.
5. The cRNA concentration is determined by measuring the absorbance at 260 nm of a diluted sample. The integrity of the cRNA is then assessed by denaturing agarose gel electrophoresis.

Problems

The product of the transcription reaction should consist of a single species of cRNA. Minor contaminants such as incomplete transcription products are sometimes also produced. The amount of these may be decreased by either increasing the nucleotide concentrations (rGTP in the mCAP kit) or by lowering the reaction temperature to 30°C.

Microinjection of cRNA

Pulling Pipettes

A two-stage pull using a Narishige micropipette puller (model PB-7) is used. Pipettes are made from 1 × 90-mm borosilicate glass capillary tubes (GD-1, Narishige, Greenvale, NY) with inside diameters of 0.75 mm and outer diameters of 1 mm. The tubes are baked prior to use, and gloves are worn at all stages.

Beveling of Pipette Tips

Each pipette is lowered onto a microgrinder (Narishige model EG-4, motor set to 60 units). The grinder is prewetted with a sterile gauze dipped in DEPC-treated water. The pipette holder is set at an angle of 20°. The moment of contact between pipette tip and grinder results in water entering the pipette tip by capillary action. Pipettes are then immediately lifted off the grinder and sized using a micrometer. Ideally, tips should be about 25 μm in diameter when beveled.

Microinjection Procedure

Stage V or VI oocytes can be injected with up to 50 nl DEPC-treated water (controls) or cRNA solution. The amount of cRNA injected can range from 0.1 to 100 ng per oocyte. A Hamilton syringe mounted on a Nikon picoinjector (Model PLI-188) is used for this purpose. The maximally effective dose of cRNA should be optimized for each receptor construct and perhaps for each cRNA preparation. Excessive quantities of synthetic RNA may be toxic to oocytes, depending on the sequence and nucleotide balance (29). In our studies with the PTH/PTHrP receptor, we found that maximal responses were routinely observed after injecting ≥25 ng cRNA per oocyte (Fig. 1), and routinely chose to microinject ~50 ng per oocyte. All injected oocytes are checked for viability using the dissecting microscope at 24-hr intervals after injection. Any visibly damaged oocytes are discarded, and the remaining oocytes are washed with fresh media. Oocytes are maintained at 19°C after injection. Assays to screen for expression of functional receptor protein may be done as early as 12 to 24 hr after injection. In most cases, however, highest levels of receptor protein expression are observed ≥48 hr after injection.

Problems in Achieving Adequate Expression of cRNA

If the expression of a functional protein does not seem to be occurring using the screening assays chosen, then one can translate the cRNA in an *in vitro* translation system to confirm its translatability. More specifically, one can also document the synthesis of receptor protein in oocytes by Western blotting, if an antibody to the protein is available. Alternatively, binding of a radioactive ligand to oocytes or oocyte membranes can be used to confirm the expression and membrane insertion of the protein of interest. One can also compare the electrophoretic patterns of biosynthetically labeled proteins

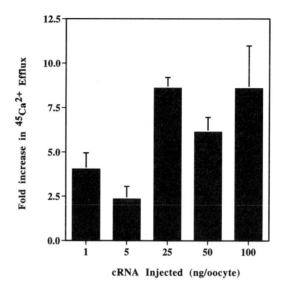

FIG. 1   RNA titration curve. The amount of mRNA microinjected determines the
size of the response to added hormone (PTH). Groups of oocytes were injected with
various amounts of mRNA encoding the opossum kidney PTH/PTHrP receptor (1
to 100 ng). The response measured here was $^{45}Ca^{2+}$ efflux. Data are shown as fold
increases ($\pm$SEM) over the basal prehormone levels of efflux. Each point represents
triplicate assays of five oocytes each.

from injected and control oocytes to determine if a protein of the appropriate
size is being made in cRNA-injected versus control oocytes.

If the protein of interest is not being synthesized, then the following possibilities should be considered:

1. The 5′ cap, which is used to increase mRNA stability, may be modified
   in the oocyte posttranslationally such that it is not functioning.
2. The 5′ and 3′ untranslated regions of the cRNA, if present to any extent,
   may contain sequences that affect mRNA stability. In such a case, the
   removal of such untranslated regions may increase expression levels.
3. The 5′ untranslated region may also have in-frame and out-of-frame ATG
   codons, resulting in incorrect translation initiation sites.
4. A 3′-polyadenylation sequence may need to be inserted to increase the
   stability of a given mRNA. In some cases, the addition of 30 or more
   nucleotides has been demonstrated to increase mRNA half-life in oocytes
   and result in 10- to 20-fold increases in protein expression (29).

If the cRNA is translatable and the cDNA construct does not contain sequences of the types noted above, then it may be important to consider subcloning the cDNA into a vector engineered for high expression in oocytes. These vectors typically include 5' and 3' untranslated sequences from a gene known to be well expressed in oocytes. Two such vectors containing sequences from the *Xenopus* β-globin gene include those developed by Krieg and Melton (15) and the pGEM-HE vector (17), which we have used.

Signal Transduction of Expressed Receptors in Oocytes

Cyclic AMP Responses

Receptors that stimulate adenylate cyclase activity in target cells, such as the β-adrenergic and luteinizing hormone receptors, have been shown to increase adenylate cyclase activity in membranes prepared from receptor cRNA-injected oocytes (8, 13). There has, however, been very little published data on the measurement of basal and hormone-stimulated cyclic AMP levels in intact oocytes. The reliable determination of cyclic AMP in oocytes has been hampered due to assay insensitivity and inherently low basal levels of this second messenger (4). We have successfully detected substantial increases in cyclic AMP content in PTH/PTHrP cRNA-injected oocytes after incubation with PTH. What follows is a description of our experimental and radioimmunoassay protocol (20).

Cyclic AMP Production and Radioimmunoassay
Experiment and Cyclic AMP Extraction
Oocytes are aliquoted into individual wells of a culture plate. The number of eggs per well should occupy no more than the bottom of the plate—for example, 5–10 oocytes per well of a 24-well plate, 10–20 oocytes per well of a 12-well plate, and 20–30 oocytes per well of a 6-well plate. All oocytes are washed and allowed to equilibrate in fresh MBSH at room temperature. The incubation media for a typical experiment would contain various concentrations of agonist such as bPTH(1-34), plus isobutylmethylxanthine (0.4 mM). Incubation media (1 ml per well) are added for 30 min. After the incubation is complete, the solution is carefully and completely removed from each well. Cyclic AMP is extracted by adding 1 ml of ice-cold ethanol to the oocytes. The eggs are gently swirled by tipping and rotating the culture plate. The ethanol is transferred to 12 × 75-mm borosilicate glass tubes, and the extraction is repeated twice more so that there are 3 ml of ethanol in each tube. The tubes are air-dried and covered. The residues can be stored at −20°C until the next part of the assay.

Radioimmunoassay: Day 1

The assay is performed using 12×75-mm glass tubes. Cyclic AMP standards for the assay (typically 0.05–50 pmol per tube) are tested in triplicate. Our laboratory uses a goat anticyclic AMP antiserum (28). The samples are kept on ice and reconstituted with from 300 to 500 μl of ice-cold cyclic AMP assay buffer (0.05 M sodium acetate, pH 6.2). These tubes are vortexed, and 100-μl aliquots are assayed in duplicate; 200 μl of a solution (assay buffer plus 0.2% γ-globulin) containing cyclic AMP antibody and ^{125}I-labeled cyclic AMP (10,000 cpm per 200 μl of solution) is added to each tube. The tubes are vortexed and incubated at 4°C overnight (or longer: we have tried up to ~48 hr).

Bound and free ^{125}I-labeled cyclic AMP are separated after this incubation period by precipitating antibody-bound ^{125}I-labeled cyclic AMP by adding 1.5 ml of a suspension of polyethylene glycol and bovine γ-globulins (see *Appendix*) to each tube. The tubes are vortexed and spun at 3000 rpm for 20 min at 4°C. The supernatant is aspirated and radioactivity in the precipitate is counted in a gamma counter.

Cyclic AMP Generation in Oocytes Injected with cRNA Encoding PTH/PTHrP Receptor

In oocytes injected with OK-PTH/PTHrP receptor cRNA, there is a several-fold increase in cyclic AMP production after exposure to PTH. Mean results from five experiments with PTH/PTHrP receptor cRNA-injected oocytes showed that basal cyclic AMP levels were 0.172 ± 0.057 pmol cAMP per oocyte, compared to cyclic AMP levels of 1.336 ± 0.430 pmol per oocyte after 30 min exposure to bovine PTH (1–34) ($10^{-6} M$). This represents an almost eightfold rise in cyclic AMP content due to bPTH (1–34).

Potential Problems

The size of the sample measured in the assay and the number of oocytes used to generate the sample may need to be varied depending on the sensitivity of the cyclic AMP radioimmunoassay as performed in different laboratories. The radioimmunoassay we use typically allows measurement of samples containing between 0.25 and 25 pmol cyclic AMP. Reliable measurements can be made on samples derived from 15 to 20 oocytes. We have noted substantial variability between batches of oocytes in terms of basal cyclic AMP levels, which could be the result of different stages of maturation of the oocytes (4, 32).

Oocyte Membrane Adenylate Cyclase Assay

This procedure is more laborious than measuring cyclic AMP production in that more oocytes are needed to obtain adequate quantities of membrane protein for the assay.

Membrane Preparation

The method we use is based on previously published techniques (14). All materials and glassware should be kept at 4°C. Typically, a large number of oocytes (200–500) are used to generate membrane fractions. Injected or control oocytes are transferred to an ice-cold Dounce homogenizer. We use a small Dounce homogenizer for <400 oocytes and a medium homogenizer for >400 oocytes. Ice-cold sucrose buffer (see *Appendix*) is added: 500 μl for <400 oocytes, 800 μl for >400 oocytes. The oocytes are homogenized using 8–10 strokes. The homogenate is transferred to ice-cold 15-ml thick-walled plastic centrifuge tubes, and 200 μl sucrose buffer is used to rinse the pestle and Dounce. Tubes are spun at 3000 g for 10 min at 4°C. The supernatant is transferred to clean tubes, taking care to avoid disturbing the large yellowish/greenish pellet. The supernatant is spun at 10,000 g for 10 min at 4°C. The resulting supernatant is transferred to an ice-cold ultracentrifuge tube. The pellet is resuspended in 500–1000 μl of ice-cold sucrose buffer, and recentrifuged at 10,000 g for 10 min. This second supernatant is added to the first, which is spun together in a Ti 50 rotor at 400,000 g for 1 hr at 4°C. The resultant pellet may contain an inner transparent and/or a dark layer, which is typically surrounded by an outer yellowish layer. This outside layer is gently washed and resuspended with 250–500 μl sucrose buffer. Care should be taken not to disturb the inner circles of this pellet. The resulting membrane preparation is stored at −80°C until use.

Adenylate Cyclase Assay

All supplies and solutions are described in published sources (14), and the method we use was originally developed by Salomon *et al.* (24). A cocktail containing assay buffer, ATP-regenerating system, GTP, and [α-^{32}P]ATP is added to 12 × 75-mm glass tubes. The appropriate concentrations of agonists are then added. In the last step, the membranes are thawed, and diluted to ~1 mg/ml in SET buffer (see *Appendix*). At 20-sec intervals, 25 μl of membranes is added to each tube, which is vortexed and incubated at 30°C for 30 min. After incubation, 100 μl of STOP solution (see *Appendix*) is added to each tube, and the tube is placed in a boiling water bath for 3 min. Distilled water is added to adjust the volume of each tube to 1 ml. Cyclic AMP is eluted from each sample using Dowex column chromatography as described (14) and radioactivity is quantified by scintillation spectrometry.

In typical experiments (Fig. 2), we compare the responsiveness of adenylate cyclase activity in membranes from oocytes expressing OK-PTH/PTHrP receptors to maximal concentrations of bPTH (1–34) and forskolin (10^{-5} M). Noninjected oocytes serve as controls (see Fig. 2). Experiments such as this have shown that bPTH (1–34) substantially increases adenylate cyclase activity in membranes prepared from oocytes expressing PTH/PTHrP receptors.

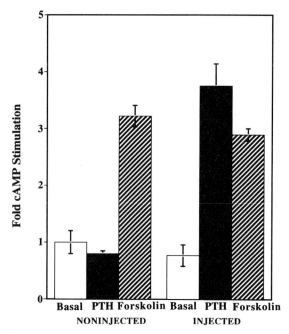

FIG. 2   The change in adenylate cyclase activity in response to added bPTH (1–34) at 10^{-6} M and forskolin at 10^{-5} M are shown for both injected ($n = 200$ oocytes) and noninjected control oocytes ($n = 200$ oocytes). Results are shown relative to the calculated amount of basal cAMP produced per 30-min incubation per tube containing 25 μl of noninjected oocyte membranes (1 mg/ml).

Calcium Mobilization in Xenopus Oocytes

The expression of a variety of receptors in oocytes induces the release of Ca^{2+} from intracellular stores after addition of the appropriate ligands (14). The mechanism responsible for these intracellular Ca^{2+} responses (12, 32) appears to involve PLC activation (3). At present, there are several methods for quantifying intracellular Ca^{2+} release. One of the most popular is the measurement of Ca^{2+}-activated Cl^- currents. Fluorescent Ca^{2+} indicator dyes (31) may also be used as well as ^{45}Ca efflux from prelabeled oocytes (14). All of these methods are outlined briefly below. Our laboratory has had the most extensive experience with the ^{45}Ca efflux assay (5).

Fluorescent Ca^{2+} Indicator Dyes

Fluorescent intracellular Ca^{2+} indicator dyes such as fura-2, indo-1 (31), or fluo-3 (16) have been used to measure $[Ca^{2+}]_i$ in oocytes. This approach may be the method of choice, if the information required is whether expression of a given receptor leads to agonist-dependent increases in $[Ca^{2+}]_i$. If, however, several mutagenized receptors are to be carefully evaluated quantitatively and compared, then a thorough calibration of the dye is required. Because of the variations among oocytes, such a study also necessitates the measurement of $[Ca^{2+}]_i$ in several oocytes to achieve representative results. These requirements can complicate the performance of such studies.

In our experience with microfluorimetric measurements of $[Ca^{2+}]_i$ in oocytes, we typically microinject the free acid of Fura-2 ~30 min before microfluorimetry. This will generally produce a uniform distribution of the dye. Dual excitation of fura-2-injected eggs at 340 and 385 nm is performed using Photon Technology International microfluorimeter (South Brunswick, NJ) and a Nikon fluorescent microscope. In typical experiments, receptor agonists are applied to the bathing chamber after a suitable period of baseline recording of $[Ca^{2+}]_i$. A careful calibration of the fluorescence signals involving the use of ionomycin in the presence of saturating, nominally zero free Ca^{2+}, as well as other free Ca^{2+} concentrations in a physiologic range (1 nM to 1 μM), are often extremely useful.

There are a number of potential pitfalls in this approach (22, 23, 25). First, dyes such as Fura-2 may become compartmentalized, leading to a substantial component of the fluorescence signal due to the presence of indicator, some nonhydrolyzed, in noncytosolic compartments. Second, Fura-2 signals saturate at or near micromolar free Ca^{2+} concentrations (31). Thus, calibration of signals above this level requires the use of dyes with lower affinities for Ca^{2+} such as fluo-3 (16). Third, calibration of fluorescence signals emanating from such a large cell, containing many different intracellular Ca^{2+} stores, can be problematic (23). The pigmentation and density of oocytes make the simple recording of ratio images difficult. This problem can be circumvented by using confocal microscopy and possibly oocytes from albino animals. Those laboratories without access to such expensive microscopes can use other indirect methods of measuring Ca^{2+} release, outlined below.

Ca^{2+}-Activated Outward Cl^- Currents

When Ca^{2+} is released into the cytosol of oocytes, this activates an outward Cl^- conductance. This response is highly sensitive and has been relied on in strategies for expression cloning of several Ca^{2+}-mobilizing receptors (21). Quantification of these currents and comparisons between individual oocytes may, however, be difficult. For example, many groups have assumed that the size of the Cl^- conductance is related to the size of the increase in $[Ca^{2+}]_i$. This may not be true. Rather, the magnitude of the conductance may be related to the rate of increase in $[Ca^{2+}]_i$ (21). An additional problem with this approach is technical. Either incubation with proteases or manual defolliculation of the oocytes is required to prepare them for voltage clamping.

Protease treatment can result in a 30- to 70-fold decrease in Ca^{2+} efflux rates (6), making the size or pattern of the Cl^- current obtained between oocytes more variable. This may make comparisons in estimated absolute Ca^{2+} release difficult. Manual stripping is also quite a tedious procedure and is thought to result in different changes, such as an increase in intracellular free sodium (6).

Generally, Ca^{2+}-activated Cl^- currents are recorded using a two-microelectrode voltage clamp. This is usually straightforward in such a large cell. The procedure requires electrodes that are capable of passing large currents. Electrode tips are usually broken for this purpose and can at best be used for only a few oocytes at a time, due to clogging of the tip with protein, and it therefore will take time to obtain data from large numbers of oocytes. The reader is referred to refs. 21 and 30 for more detailed technical information on this method.

^{45}Ca Efflux

An increase in the rate of ^{45}Ca efflux from prelabeled oocytes can be used as an index of intracellular Ca^{2+} release (14, 18). The advantages of this technique for studying the functional properties of Ca^{2+}-mobilizing receptors include its low cost, the simplicity of assay, and the possibility of quantitative comparisons among groups of oocytes. This latter aspect of the method offers a key advantage over both microfluorimetric and electrophysiologic techniques.

Potential mechanisms underlying the increases in ^{45}Ca efflux observed on activating PLC-coupled receptors include the release of intracellular Ca^{2+} stores and the stimulation of Ca^{2+} extrusion pumps. In other cells, there is also evidence that depletion of intracellular Ca^{2+} stores alone can trigger an increase in plasma membrane Ca^{2+} permeability (3). This would promote increased Ca^{2+} influx and thereby also stimulate Ca^{2+} efflux, if the same mechanism operated in oocytes.

There are different ways to quantify the release of $^{45}Ca^{2+}$ from prelabeled oocytes. For example, one can determine the amount of isotope released as a fraction of the remaining radioactivity in each oocyte. The method we use is adapted from published studies and quantifies the ^{45}Ca effluxed in response to agonist compared to basal (preagonist) levels (14). Typically, five oocytes are studied per experimental point, and all treatments are done in triplicate. Most experiments are performed at 48 or 72 hr after injection.

Oocytes are loaded with ^{45}Ca at 50 $\mu Ci/ml$ in either 0.5 or 1 ml of Ca^{2+}-free MBSH for 2.5 hr at 19°C in a 35-mm tissue culture dish ($\leq$100 oocytes per dish). Loading periods used by different groups vary considerably (from 2 to 20 hr). After the labeling is complete, oocytes are aliquoted into cluster 24-well plates (5–10 eggs per well) and washed at least five times in MBSH containing Ca^{2+} and 0.1% BSA (fatty acid free). This removes any unincorporated radioactivity. After the final wash, fresh medium is added and timing begun. Care must be taken throughout the assay not to damage any of the oocytes when removing or adding fresh media, because this will produce spuriously high $^{45}Ca^{2+}$ efflux.

After loading and the initial washing described above, we use a 70-min washing period, after which basal $^{45}Ca^{2+}$ efflux over either a 5-, 10-, or 20-min interval is determined. Typically, we measure basal efflux in at least three successive timed intervals. During the 70-min washing period, the medium bathing each group of oocytes is removed and replaced every 10 min for a total of seven times. All of these media are discarded. At the 70-min time point, the basal release periods begin. Thereafter, the media samples are collected at the precise intervals chosen and counted. In healthy viable oocytes, basal $^{45}Ca^{2+}$ efflux should be in the range of ~200 cpm over a 5-min period of <500 cpm for a 20-min interval by the end of the basal release period (see Fig. 2). The stimulus (e.g., PTH) is added after all the basal efflux samples are collected. Experimental treatment samples are next taken at the appropriate interval (i.e., 5, 10, or 20 min) for typically three more intervals. Fresh medium containing agonist is added after the removal of each medium sample. The data in Fig. 3 represent pooled efflux over 20-min collection periods. The period of greatest ^{45}Ca efflux for the wild-type OK PTH/PTHrP receptor is within the first 20 min. By the end of 60 min of exposure to bPTH (1–34), ^{45}Ca efflux has returned to basal prestimulus levels (Fig. 3).

Conclusions

Using the methods outlined above, we have started to examine the effects that structural changes in the PTH/PTHrP receptor, such as tail truncation and mutations in intracellular loops, have on signal transduction via both cyclic AMP- and intracellular Ca^{2+}-dependent pathways. The *Xenopus* oocyte expression system may, therefore, provide some evidence as to the location of interaction sites in the receptor for the G proteins that signal via adenylate cyclase and PLC activation.

Appendix: Solutions

1. Modified Barth's saline with HEPES (MBSH): NaCl (88.0 mM), KCl (1.0 mM), $NaHCO_3$ (2.4 mM), $MgSO_4$ (0.82 mM), HEPES (10.0 mM), $Ca(NO_3)_2 \cdot 4H_2O$ (0.33 mM), $CaCl_2 \cdot 2H_2O$ (0.41 mM). Make up a 10× stock, pH 7.6, and sterile filter, store at 4°C. Add antibiotics if necessary.

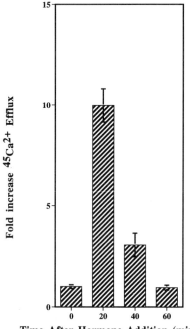

FIG. 3   Time course of the $^{45}Ca^{2+}$ efflux increase following addition of bPTH (1–34) (10^{-6} M) to oocytes expressing the PTH/PTHrP receptor. Oocytes were loaded with $^{45}Ca^{2+}$, washed, and basal efflux determined prior to hormone addition. Basal release was 285 ± 26 cpm per five oocytes over a 20-min period. The $^{45}Ca^{2+}$ efflux in the first, second, and third 20-min periods following hormone addition was then determined [n = 38 experimental points (±SEM) per time point]. No effect of bPTH (1–34) on $^{45}Ca^{2+}$ efflux was seen in noninjected oocytes (not shown).

2. MBSH—Ca free: Same as solution 1, but with no Ca(NO$_3$)$_2$ or CaCl$_2$. Solution is only nominally Ca^{2+} free, because it is not made with EGTA.
3. MBSH/FCS/Pen-Strep-Gent (by volume): 88.9% MBSH, 10% fetal calf serum (FCS), 100 U/ml penicillin, 100 μg/ml streptomycin, and 10 U/ml gentamicin. Adjust to pH 7.6.
4. Sucrose buffer: 75 mM Tris, 12.5 mM MgCl$_2$, 1 mM EDTA, 30% (w/v) sucrose, pH 7.4.
5. PEG milk: Polyethylene glycol 8000 (PEG), 92.5 g; γ-globulin, 0.6 g; assay buffer (0.05 sodium acetate) to 500 ml total volume.
6. SET buffer: 250 mM Sucrose, 1 mM EDTA, 5 mM Tris-HCl.
7. STOP solution: 10 mM ATP, 1 mM [^{3}H]cAMP, 2% sodium dodecyl sulfate (SDS), cAMP (32.9 mg/100 ml), 50 mM Tris-HCl, pH 7.5

Acknowledgments

We gratefully acknowledge the gift of the pGEM-HE vector engineered for high expression in oocytes from Dr. Emily Liman, Department of Neurobiology, Harvard University. We would like to thank Thomas Bambino, Tsui-Hua Chen, Ying Chen, and Stacy Pratt for superb technical assistance. This work was supported by VA Merit Reviews (to RN and DS) and by NIH Grants DK 43400 (to DS) and DK 35323 (to RN), and by a grant from the Academic Senate of the University of California, San Francisco (to DS). Dr. Nissenson is a Research Career Scientist and Dr. Shoback is a Clinical Investigator of the Department of Veterans Affairs. (Address reprint requests to R. A. Nissenson.)

References

1. M. Antonelli, L. Birnbaumer, J. E. Allende, and J. Olate, *FEBS Lett.* **340,** 249–254 (1994).
2. M. Babich, K. L. King, and R. A. Nissenson, *J. Bone Mineral Res.* **4,** 549–556 (1989).
3. M. J. Berridge, *Nature* (*London*) **361,** 315–325 (1993).
4. M. F. Cicirelli and L. D. Smith, *Dev. Biol.* **108,** 254–258 (1985).
5. T-H. Chen, S. A. Pratt, and D. M. Shoback, *J. Bone Mineral Res.* **9**(2), 293–300 (1994).
6. N. Dascal, *CRC Crit. Rev. Biochem.* **22,** 317–387 (1987).
7. A. L. Goldin, *in* "Methods in Enzymology" (B. Rudy and L. E. Iverson, eds.), Vol. 207, pp. 266–279. Academic Press, San Diego, 1992.
8. T. Gudermann, C. Nichols, F. O. Levy, M. Birnbaumer, and L. Birnbaumer, *Mol. Endocrinol.* **6,** 272–278 (1992).
9. J. B. Gurdon, C. D. Lane, H. R. Woodland, and G. Marbaix, *Nature* (*London*) **233,** 177–181 (1971).
10. B. P. Halloran and R. A. Nissenson, "Parathyroid Hormone Related Protein: Normal Physiology and Its Role in Cancer." CRC Press, Boca Raton, Florida, 1992.
11. Z. Huang, Y. Chen, and R. A. Nissenson, *J. Biol. Chem.* **270,** 151–156 (1995).
12. L. F. Jaffe, "Biology of Fertilization," Vol. 3, pp. 127–165. Academic Press, Orlando, 1985.
13. H. Jüppner, A. B. Abou-Samra, M. Freeman, X. F. Kong, E. Schipani, J. Richards, L. F. J. Kolakowski, J. Hock, J. T. Potts, Jr., H. M. Kronenberg, and G. V. Segre, *Science* **254,** 1024–1026 (1991).
14. B. K. Kobilka, C. MacGregor, K. Daniel, T. S. Kobilka, M. G. Caron, and R. J. Lefkowitz, *J. Biol. Chem.* **262,** 15796–15802 (1987).
15. P. A. Krieg and D. A. Melton, *in* "Methods in Enzymology," Vol. 155, pp. 397–415. Academic Press, Orlando, 1987.
16. J. Lechleiter, S. Girard, D. Clapham, and E. Peralta, *Nature* (*London*) **350,** 505–508 (1991).

17. E. R. Liman, J. Tytgat, and P. Hess, *Neuron* **9,** 861–871 (1992).
18. M. Lupu-Meiri, H. Shapira, N. Matus-Leibovitch, and Y. Oron, *Pflugers Archiv.* **417,** 391–397 (1990).
19. Y. Masui and H. J. Clarke, *Int. Rev. Cytol.* **57,** 185–282 (1979).
20. R. A. Nissenson, G. J. Strewler, R. D. Williams, and S. C. Leung, *Cancer Res.* **45,** 5358–5363 (1985).
21. I. Parker and Y. Yao, *Cell Calcium* **15,** 276–288 (1994).
22. M. Poenie, J. Alderton, R. Steinhardt, and T. Tsien, *Science* **233,** 886–889 (1986).
23. M. W. Roe, J. J. Lemasters, and B. Herman, *Cell Calcium* **11,** 63–73 (1990).
24. Y. Salomon, C. Londos, and M. Rodbell, *Analyt. Biochem.* **58,** 541–548 (1974).
25. M. Scanlon, D. A. Williams, and F. S. Fay, *J. Biol. Chem.* **262,** 6308–6312 (1987).
26. H. Schneider, J. H. M. Feyen, and K. Seuwen, *FEBS Lett.* **351,** 281–285 (1994).
27. G. V. Segre and S. R. Goldring, *Trends Endocrinol. Metab.* **4,** 309–314 (1993).
28. D. M. Shoback and J. M. McGhee, *Endocrinology* **122,** 382–389 (1988).
29. H. Soreq and S. Seidman, *in* "Methods in Enzymology," Vol. 207, pp. 225–265. Academic Press, San Diego, 1992.
30. W. Stühmer, *in* "Methods in Enzymology" (B. Rudy and L. E. Iverson, eds.), Vol. 207, pp. 320–339. Academic Press, San Diego, 1992.
31. R. Y. Tsien, T. J. Rink, and M. Poenie, *Cell Calcium* **6,** 145–157 (1985).
32. P. R. Turner and L. A. Jaffe, "The Cell Biology of Fertilization." Academic Press, San Diego, 1989.

[14] Production of G Protein α-Subunit-Specific Monoclonal Antibodies

Xiaohua Li and Richard S. Jope

Introduction

The relatively large number of G protein α subunits and the high homology of their primary sequences make it difficult to distinguish individual subtypes. Although toxin-catalyzed ADP-ribosylation has been used as a method to identify the G_s (cholera toxin-sensitive) and the G_i/G_o (pertussis toxin-sensitive) families of α subunits, neither of these toxins is specific to only one substrate and their actions depend on the association state of the α subunits. The need for assessing individual G protein α subunits in many investigations, including those involving purification, localization, metabolism, modification, and function, has led to the development of antibodies. Polyclonal antisera have been produced against purified G protein α subunits or synthetic peptides corresponding to unique amino acid sequences of α subunits (1–3). These have been of critical importance in many of the studies that have made important advances in the understanding of G proteins. Production of polyclonal antibodies requires a large amount of protein or peptide and the productive lifetime of the animals used is limited. In contrast, monoclonal antibodies have the advantage of offering an unlimited supply as soon as the antibody-producing clones are established. The use of monoclonal antibodies to study G proteins has been much less extensive than that of polyclonal antibodies. Monoclonal antibodies to G protein subunits have been produced against transducin (4–7), $G_{o\alpha}$ (8–10), $G_{i\alpha 1}$ (10), $G_{i\alpha 2}$ (10), and G_{β} (8). Recent work has shown that monoclonal antibodies generated using native proteins as antigens can have high specificity to a G protein subtype, such as $G_{i\alpha 1}$ or $G_{i\alpha 2}$, whereas a similar specificity is difficult to achieve with a polyclonal antibody when native proteins are used as antigens. This chapter describes in detail the methods that have been used for making monoclonal antibodies against G protein α subunits, with emphasis on the methods that have been used successfully in the authors' laboratory. The basic principles of making monoclonal antibodies can be found elsewhere (11) and an outline of the procedures described in this chapter is given in Fig. 1.

Methods in Neurosciences, Volume 29

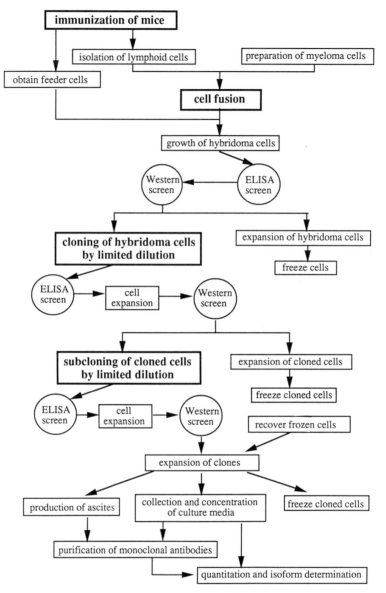

FIG. 1   Flow chart for the preparation of monoclonal antibodies.

Preparation of Antigen

Monoclonal antibodies can be generated successfully using either G proteins purified from mammalian tissues or recombinant G proteins purified from bacterial or eukaryotic expression systems. To make α-subunit-specific antibodies, either the heterotrimer ($\alpha\beta\gamma$) or the α-subunit monomer can be used as the antigen. When using the $\alpha\beta\gamma$ trimer as the antigen, monoclonal antibodies against α subunits rather than $\beta\gamma$ may be more likely to be generated because the $\beta\gamma$ subunits are more hydrophobic and are therefore less immunogenic than α subunits. As with any antibody preparation method, the G protein antigen should be as pure as possible, minimally 90–95% pure. Published purification methods are adequate for this purpose. For example, sequential columns of DEAE-Sephacel, Ultragel AcA34, and heptylamine-Sepharose (12) were found to be adequate to prepare the G_{ou} and G_{iu1} antigens (10), although more highly purified proteins may be optimal for generating fewer nonspecific antibodies. A total of 2–3 mg of purified protein is needed for the entire procedure described below (including immunization and screening).

Immunization of Mice

For each immunization, two female BALB/c mice are injected in order to obtain an adequate amount of tissue for cell culture. The antigen to be injected should be dialyzed into phosphate-buffered saline (PBS) before use to avoid injecting a high concentration of a detergent. For the first injection, antigen (50–100 μg) in PBS (100 μl) is mixed with Freund's complete adjuvant (100 μl) by vortexing until the liquid is viscous. The mixed antigen is injected subcutaneously into two locations on both rear legs (50 μl/site). Every 3 days the mice are injected with 100–200 μg of antigen in 200 μl PBS until five injections have been completed, using the same four injection sites every time. Freund's incomplete adjuvant can be used for the last four injections, but it is not necessary with this procedure. Two days after all five injections have been completed, the mice are ready to be sacrificed for making hybridoma cells.

When compared to systemic injection procedures that are required to obtain a strong immune response in the spleen, the immunization protocol described here takes a shorter period of time and produces high-titer antibodies at lymphoids near the injection sites. Because the time of immunization is relatively short, a high titer of antibodies in the systemic circulation is not expected, so a bleeding titration is not necessary with this procedure. It should be remembered that proteins purified from host animals are not good antigens, so if an antigen purified from mice is to be used, immunization of rats rather than mice may be considered.

Fusion of Mouse Lymphocytes with Myeloma Cells for Hybridoma Production

Myeloma Cells

Many different myeloma cell lines can be used as fusion parents (11). X63Ag8.653 (13) is routinely used in our laboratory to fuse with mouse lymphocytes. Seven days before fusion, myeloma cells are recovered from the freezer and grown in RPMI 1640 medium supplemented with 15% defined fetal bovine serum 100 unit/ml penicillin, 100 μg/ml streptomycin, and 30 μg/ml glutamine. The cells are split 1 day before fusion to ensure optimal viability. Before sacrificing the immunized mice, a test fusion using the myeloma cells and lymphocytes from nonimmunized mice is recommended to check the quality of the myeloma cells, the media, and the serum.

Materials

RPMI 1640 medium
Complete medium (RPMI 1640 medium, 15% defined fetal bovine serum, 100 unit/ml penicillin, 100 μg/ml streptomycin, and 30 μg/ml glutamine)
50× Hypoxanthine, aminopterin, and thymidine (HAT) selective medium
50× Hypoxanthine and thymidine (HT) selective medium
Polyethylene glycol (PEG)
10-ml syringes and 18-gauge needles
50-ml conical tubes
200-ml culture flask
10-cm culture plate
24-well culture plates
1- and 10-ml pipettes
Cell strainer (70-μm nylon)
70% (v/v) Ethanol
Pair of scissors and forceps
Mouse fixing board

Procedures

1. Sterilize all materials before use. Prewarm serum-free RPMI 1640 medium and PEG to 37°C.
2. Sacrifice immunized mice by cervical dislocation.

3. Soak mice in ethanol, then fix on board.

4. Open mouse skin along the midline of the abdomen and gently separate the skin from the peritoneum toward each side of the abdomen. Inject 6–8 ml of complete medium into the abdomen of each mouse (Fig. 2A), then withdraw the medium (containing macrophages to be used as feeder cells) and keep it in the syringe until use (step 13).

5. Add 10 ml RPMI 1640 medium into a 10-cm culture dish. Dissect the lymphoids (Fig. 2B) (lymphoids in immunized mice are bigger than normal lymphoids and are therefore easier to dissect) from the lower body sites, put in the medium, and disrupt the tissue. Pass the torn tissue through a 10-ml pipette several times to make a cell suspension. Transfer with a 10-ml glass pipette all of the cells suspended in the medium through a cell strainer to a 50-ml conical tube. Rinse the dish with 10 ml RPMI 1640 media and filter into the tube.

6. Transfer myeloma cells (the ratio of myeloma cells to lymphocytes is approximately 5:1) to another 50-ml conical tube.

7. Spin both the lymphocytes and the myeloma cells at 1000 g for 5 min at room temperature.

8. Discard the medium. Wash the lymphocytes and the myeloma cells separately with 20 ml RPMI 1640 medium and then spin down the cells and discard the media.

9. Add 20 ml prewarmed RPMI 1640 medium to the myeloma cells, transfer all of the cells with the medium to the pelleted lymphocytes, and mix well. Spin down the cells and discard the medium.

10. Slowly add 1 ml prewarmed PEG to the mixed cells, and pass the cells gently through the pipette several times.

11. Add drop-by-drop prewarmed RPMI 1640 medium up to 5 ml and then bring the volume to 50 ml by slowly increasing the speed of delivery of the medium. Gently shake continuously during addition of the medium.

12. Cap the tube and gently invert it several times. Spin down the fused cells and discard the medium.

13. Add 180 ml complete medium to a 200-ml tissue culture flask. Add 4 ml of 50× HAT selective medium. Mix the feeder cells from step 4 into the flask. Invert to mix well.

14. Add 10 ml of the diluted HAT medium from step 13 to the fused cells, mix well, and transfer back to the flask. Take another 10 ml from the mixture to rinse the tube and transfer back to the mixture. Mix everything well.

15. Add 1 ml of the mixture (containing myeloma cells, lymphocytes, and feeder cells) to each well of a 24-well plate (eight or nine plates are required for 200 ml of mixture).

16. Wrap the plates with plastic wrap to avoid loss of moisture.

17. Grow the mixed cells at 37°C under 95% O_2/5% CO_2 (v/v).

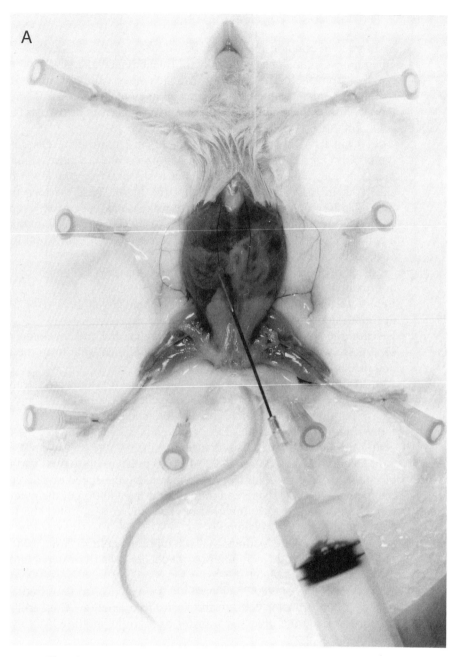

Fɪɢ. 2 Preparation of feeder cells and isolation of lymphoids. (A) Injection and withdrawal of medium from the abdomen of a mouse to obtain feeder cells. (B) Dissection of lymphoids from locations indicated by the arrows.

192

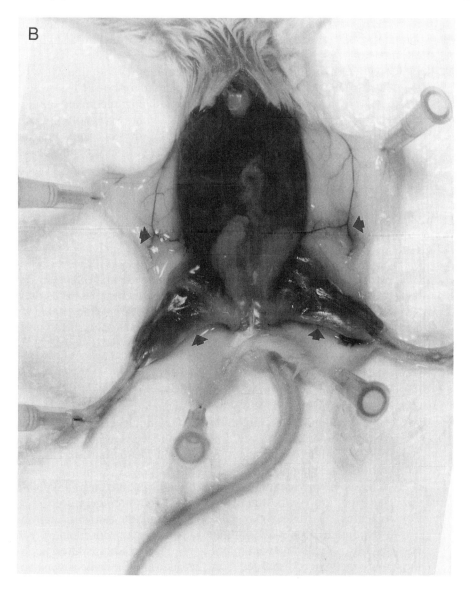

FIG. 2   (continued)

18. After 5 days half of the medium is replaced by complete medium containing HT selective medium (1×) and then 5 days later by complete medium.
19. Growth of hybridoma cells should be monitored continuously. When the cells are 50% confluent (multiple colonies may be seen in each well), they are ready to screen for antibodies with enzyme-linked immunosorbent assay (ELISA) and Western blots. *Note:* Screening the cells too early tends to result in weak immunoreactivity and screening too late may miss the best time for cloning due to cell overgrowth.

Enzyme-Linked Immunosorbent Assay

Screening of Antibody-Producing Hybridomas

ELISA (14) provides a rapid and sensitive assay method to screen for antibody-producing hybridoma cells. The original antigen protein is usually used for screening. The antigen protein is diluted in borate saline (100 mM boric acid, 25 mM sodium borate, 75 mM NaCl) to a final concentration of 1–10 μg/ml, and 50 μl of the antigen solution is added into each well of an ELISA plate using a multichannel pipetter. If the antigen solution contains a high concentration of detergent (as is the case for many G protein preparations), the detergent concentration must be reduced to less than 0.1% to obtain efficient coating of the ELISA plate. After adding the antigen, the plate is covered and incubated overnight at 4°C. One hour before use, the plate is washed with borate saline three times and the wells are blocked by adding 200 μl borate saline containing 1% BSA and incubating for 1 hr at room temperature. The plate is now ready for screening.

Hybridoma culture medium (50 μl) is transferred into an ELISA plate (use sterilized pipette tips) and the plate is left at room temperature for 2–4 hr. To avoid a false positive, a parallel control should be run with each sample. To do this, an ELISA plate without antigen protein but blocked with the BSA solution is incubated with hybridoma culture medium. After incubation, the culture medium is discarded from the plate and the plate is washed three times with borate saline. Goat antimouse immunoglobulin conjugated with alkaline phosphatase is diluted into borate saline containing 1% BSA and 50 μl of this solution is added into each well of the plate. The plate is then incubated with secondary antibody at room temperature for 2 hr. The plate is washed with borate saline three times and 50 μl of alkaline phosphatase substrate solution (1 mM MgCl$_2$, 1 mM diethanolamine, pH 9.8, with 1 mg/ml p-nitrophenylphosphate added immediately before use) is added into each

well and color should develop within 30 min. The absorbance can be determined at 405 nm with an ELISA reader.

Quantitation of Monoclonal Antibody Solution and Immunoglobulin Isoform Determination

In addition to screening the antibody-producing hybridoma cells, ELISA is
also used for immunoglobulin quantitation and isoform determination. These
methods should be applied when a stable subclone has been established and
concentrated or purified antibody has been obtained.

To quantitate the concentration of immunoglobulin in a monoclonal antibody solution, an ELISA plate is coated with goat antimouse immunoglobulin
(4 μg/ml), washed, and blocked with BSA as described above. The coated
plate is incubated with serial dilutions of an immunoglobulin standard (400,
200, 100, 50, 25, 12.5, 6.25, and 0 ng/ml) or monoclonal antibody solution
($1:10^3$ to $1:10^6$ dilutions), washed, and incubated with antimouse immunoglobulin conjugated with alkaline phosphatase as described above. Color is
developed by the method described above and the absorbance is determined
at 405 nm. The concentration of the monoclonal antibody in solution is
calculated according to the standard curve obtained with the immunoglobulin standard.

To determine the immunoglobulin isoform (IgG, IgM, or IgA) of each
monoclonal antibody, an ELISA plate is coated with goat antimouse immuno-
globulin specific for the κ or λ light chain, IgG (γ_1, γ_{2a}, γ_{2b}, and γ_3 heavy
chain), IgM (μ heavy chain), and IgA (α heavy chain), washed, and blocked
with BSA. The monoclonal antibody solution is diluted and incubated with
each antiimmunoglobulin isoform. The plate is further incubated with alkaline
phosphatase-conjugated antimouse immunoglobulin and color is developed.
A monoclonal antibody should contain only one type of light chain and one
type of heavy chain. If more than one type of antimouse immunoglobulin
light chain or heavy chain reacted with the antibody solution being tested,
this antibody is either not completely cloned or is contaminated with another
immunoglobulin.

Western Immunoblotting

Immunoblotting is a sensitive screening method and it provides an initial
estimation of the specificity of an antibody. However, due to the large number
of hybridomas that need to be screened initially, it is usually not practical

for immunoblotting to replace the ELISA for the initial screening. Rather, this method is most useful to confirm antibody-positive hybridoma cells and cloned cells once the ELISA has indicated a positive for the presence of monoclonal antibodies. Purified G protein or membrane proteins are resolved by electrophoresis in 10% (w/v) SDS-polyacrylamide gels. Loading the proteins into individual wells divided by a spacer is not necessary for this purpose. Instead the protein sample can be loaded directly onto the stacking gel. Approximately 1.5–3 μg of purified protein or 1–1.5 mg of membrane protein is needed for one minigel (Bio-Rad). Proteins are transferred to nitrocellulose and the blot is cut into strips (the remaining blots can be stored at −20°C for later screens). Strips are soaked in borate saline containing 5% (w/v) nonfat dry milk and then incubated individually with the hybridoma culture medium (0.5–1 ml) to be tested for 6 hr to overnight. After rinsing with borate saline, rabbit antimouse immunoglobulin (IgG or IgM or both) conjugated with horseradish peroxidase (1 ml, 1 : 3000) is added and the strips are incubated at room temperature for 6 hr. Strips are then washed with borate saline and color is developed within 2 min after immersing in the substrate solution (50 mM Tris-Cl, pH 7.6, 10 mM imidazole, 1 mM 3,3′-diaminobenzidine, activated with 2% (v/v) H_2O_2 before use).

Cloning and Subcloning of Antibody-Containing Hybridoma Cells

ELISA-positive hybridoma cells are likely to contain more than one type of antibody. These antibodies may be specific to the antigen or may be generated by a nonspecific immune response. In addition, more than one isoform of a specific monoclonal antibody may be produced in hybridoma cells. Cloning and subcloning are critical procedures to develop and maintain a single clone producing a single specific monoclonal antibody.

Feeder cells are required for the successful growth of cloned cells. Feeder cells are collected from the peritoneum of a BALB/c mouse by injecting and withdrawing 6–8 ml of complete medium as described in step 4 in the fusion protocol given above. The medium containing feeder cells is diluted in the same medium to 20–50 ml and 100 μl of the diluted feeder cell suspension is added to each well of a 96-well flat-bottom tissue culture plate (9.6 ml/plate). The plate should be incubated for 1 day and examined to make sure there is no contamination before cloning.

Hybridoma cells in an ELISA-positive well of a 24-well tissue culture plate are suspended by passing the cells gently through a 1 ml glass pipette. An aliquot (10 μl) of the hybridoma cell suspension is mixed with 5 ml of complete medium in a sterilized solution basin. Using a multichannel pipetter, 100 μl of the cell suspension in media is added into each well of the first row of a

96-well plate containing feeder cells. Serial dilution is then performed from the first row to the last row of the plate by transferring 100 μl of mixed cell suspension from one row to the next. An additional 100 μl of complete medium is added into each well (except those in the last row) after serial dilution to bring the final volume up to 200 μl in each well. In order to obtain more clones, more than one cloning plate can be made from each ELISA-positive well that contains antibody-positive hybridoma cells. The remaining hybridoma cells from the same well are expanded by transferring into a 25-ml tissue culture flask containing 5 ml complete medium. These expanded cells can be frozen or used for further cloning. The cloning plates are wrapped with plastic wrap and incubated for 1–2 weeks. Cell growth should be monitored and wells containing a single clone identified. These single clones are then screened for antibodies by ELISA. If an inadequate number of wells with a single clone are produced, wells with double clones also can be selected for screening, followed by further cloning later. ELISA-positive clones are expanded into 24-well tissue culture plates, grown in 1 ml complete medium per well to 50% confluency, and screened by Western blotting.

Single clones showing immunoreactivity with the antigen in an ELISA plate or on a Western blot can be identified as monoclonal antibody-producing clones. To confirm the identity of all cells from a "single clone," and to ensure that 100% of the cells from a single clone secrete the monoclonal antibody, repeated subcloning is required. The procedure for subcloning is the same as that for cloning in which cloned antibody-producing cells from 24-well tissue culture plates are diluted sequentially and cultured in 96-well tissue culture plates. ELISA and Western blotting are performed for screening each time. An antibody-positive clone is established when 100% of its subclones produce antibodies.

Subclones producing monoclonal antibodies can be gradually expanded to collect the antibody or frozen for future use. To freeze hybridoma cells or clones, about 10^7 cells are transferred to conical tubes and spun at 1000 g for 5 min at room temperature. The cell pellet is mixed with 92% (v/v) defined fetal bovine serum and 8% (v/v) dimethyl sulfoxide (DMSO) and rapidly frozen and stored at $-80°C$ for use within 6 months to 1 year. For longer storage times the cells should be maintained in liquid nitrogen.

Collection and Purification of Antibodies

Large quantities of monoclonal antibodies can be obtained either by expanding cloned hybridoma cells to collect the culture medium or by injecting cloned hybridoma cells into mice to make ascites. Concentrated culture media containing a monoclonal antibody can be used directly for many purposes, such as immunoblotting, or can be purified further. Making ascites is a less expensive way to obtain a solution containing concentrated monoclonal antibody solution, and purification of ascites is strongly recommended.

To collect culture medium containing a monoclonal antibody, one frozen subclone is thawed at 37°C and mixed with prewarmed complete medium containing a small amount of feeder cells (feeder cells help the recovery and growth of cloned cells). After the cells are growing steadily, the presence of monoclonal antibodies in the cell culture medium should be tested by Western blotting as described above. Hybridoma cells usually secrete less antibody at early times after recovery from the freezer, therefore the culture medium to be tested should not be diluted more than twofold for Western blot analysis. After the production of the monoclonal antibody is confirmed, the cells are grown in RPMI 1640 medium containing 5–10% defined fetal bovine serum, 100 unit/ml penicillin, 100 μg/ml streptomycin, and 30 μg/ml glutamine, and the cells are expanded gradually from a small flask to a large flask. The cell culture medium from each pass is pooled until at least one liter of medium is collected. The pooled medium is then concentrated to about 5% of the original volume with a Millipore (Bedford, MA) protein concentrator. The concentrate is dialyzed against borate saline containing 50% glycerol for 4–6 hr (the solution may be concentrated during dialysis due to the presence of glycerol). Aliquots of the concentrated antibody solution can be stored at −80°C.

To make ascites, an adult female BALB/c mouse is pretreated by injecting 1 ml pristane into the peritoneum. Five days later, the growing hybridoma cells from an antibody-producing clone are collected, spun at 1000 g for 5 min at room temperature, and washed with PBS three times. About 5×10^6 cells are suspended in 1 ml PBS and injected (intraperitoneally) into the pristane-pretreated mouse. After 1–2 weeks, or when the abdomen of the mouse is noticeably enlarged, an 18-gauge needle is inserted into the peritoneum and ascitic fluid is collected into a 10-ml conical tube by gently pressing the abdomen. The collected ascitic fluid is incubated at 37°C for 1 hr, then incubated at 4°C for 6 hr, and then spun at 3000 g for 10 min. The supernatant (containing the monoclonal antibody) is used for further purification. Several batches of ascites can be collected from a single mouse.

Monoclonal antibodies are purified using antimouse immunoglobulin (Ig)–agarose (Sigma). A column (1 cm diameter) is packed with 2–5 ml of antimouse Ig–agarose slurry (1 : 1) and is washed thoroughly with borate saline. Concentrated monoclonal antibody (about 2 mg of total Ig) is diluted in borate saline to 0.1 mg/ml and loaded onto the column. The column is then washed with borate saline until the eluate contains no proteins (proteins that do not bind to immunoglobulin) as monitored by measuring the absorption at 280 nm. The eluate from the borate saline wash is saved for the next purification cycle. The monoclonal antibody is eluted with glycine–NaCl (50 mM glycine, 150 mM NaCl, pH 2.4–2.8), and the solution containing the eluted antibody is immediately neutralized with borate saline. To reuse the column, it is washed with borate saline until the pH of the outflow reaches the pH of borate saline. The column is then ready for the next purification cycle. The eluate containing the purified monoclonal antibody is collected and pooled from each purification cycle until no further protein is eluted by the glycine–NaCl solution. The purified monoclonal antibody is concentrated with an Amicon concentrator through a YM10 Diaflo ultrafilter (Amicon, Danvers, MA). Purified concentrated monoclonal antibody is mixed with an equal volume of glycerol and aliquots are stored at −80°C.

Basic Characterization of Monoclonal Antibodies to G Proteins

Due to the fact that G proteins belong to a superfamily of proteins that share high sequence homology and that also vary among species, both the subtype specificity and the species selectivity of monoclonal antibodies to G proteins need to be determined. Because monoclonal antibodies generated using purified G proteins rather than synthetic peptides react with an unknown amino acid sequence or three-dimensional structure, identification of the epitope will be useful in the application of the antibody. In addition, the reactivity of these monoclonal antibodies to native versus denatured proteins, heterotrimers versus monomers, GDP-bound versus GTP-bound, and ADP-ribosylated or lipid-modified G proteins can be identified to facilitate their use.

Titration of Monoclonal Antibody

Before a monoclonal antibody can be used properly, the optimal concentration of the antibody needs to be determined. To titrate a monoclonal antibody on a Western blot, several serial dilutions (1 : 100 to 1 : 10,000) of a monoclonal antibody solution are incubated with nitrocellulose containing a range of amounts of purified or membrane proteins (1 to 50 μg) from tissues or cultured cell lines to be studied. The range of the concentrations of antibody and of protein that produces a linear increase in band density on the titration curves should be used. Additionally, experimental conditions, such as pH, concentration of salt, and blocking protein, can have a major influence on the binding efficiency of a monoclonal antibody. Therefore, each of these conditions should be varied to determine the optimal conditions for each monoclonal antibody.

Antibody Specificity

To determine G protein α-subunit subtype specificity by Western immunoblotting, the use of purified recombinant G proteins is preferable. Equal amounts of different subtypes of G protein α subunits (50 ng) are run on SDS-polyacrylamide gels, transferred, and probed with the monoclonal antibody to test the specificity of the antibody.

For some applications it is necessary to test the specificity of a monoclonal antibody to a native G protein subtype in solution, which may be different from the specificity obtained with an immobilized protein, such as is encountered with ELISA and Western blots. A competitive ELISA can be used for this purpose. The principle of the competitive ELISA consists of preincubating the monoclonal antibody with a purified G protein to remove the antibody from solution. The pretreated antibody solution is added to an ELISA plate precoated with the G protein antigen. The more antibody that was bound to the G protein subtype in solution, the less will be available to bind to the antigen precoated in the ELISA plate. This determines the reactivity of the monoclonal antibody with the G protein subtype in solution.

Procedure for Competitive ELISA

1. An ELISA plate is precoated with the antigen protein as described in Section V and the plate is rinsed immediately before use.
2. In a conical-bottom microtiter 96-well plate, 50 μl of the monoclonal antibody diluted in phosphate-buffered saline (PBS) containing 0.1% (w/v) gelatin and 0.05% (v/v) Tween 20 is mixed with 50 μl of serial dilutions (0.01–10 μg/ml) of a purified G protein in the same buffer. For this purpose, the use of purified recombinant G proteins is preferable and several highly homologous G protein subtypes can be tested to determine the antibody specificity. For example, recombinant $G_{i\alpha1}$, $G_{i\alpha2}$, and $G_{i\alpha3}$ may be incubated separately with a monoclonal antibody that specifically recognizes $G_{i\alpha1}$ on a Western blot.
3. The antibody–protein mixture is incubated for 1 hr at room temperature in a moist chamber, and 50 μl of the mixture is transferred to individual wells of the precoated ELISA plate.
4. The antibody–protein mixture is incubated in the ELISA plate, washed, and then incubated with alkaline phosphatase-conjugated goat antimouse Ig, and color is developed, as described in Section V. The absorbance is read at 405 nm and data are calculated as percent of the control that contained no G protein in the preincubation and should have the highest absorbance measurement.

Species Selectivity

A monoclonal antibody may demonstrate species selectivity when the monoclonal antibody binds to an amino acid sequence that varies among different species. For example, a monoclonal antibody (1C) specific to $G_{o\alpha}$ generated in our laboratory from G_o/G_i purified from bovine brain reacts strongly with bovine and human $G_{o\alpha}$, but reacts weakly with rat $G_{o\alpha}$ and does not react with murine $G_{o\alpha}$ (10). In contrast, another $G_{o\alpha}$-specific monoclonal antibody (2A) generated from the same preparation showed less species selectivity. A comparison of the amino acid sequences of bovine, human, rat, and murine $G_{o\alpha}$ (15) reveals few differences in the sequences. However, at amino acid residues 93–104, the bovine and human sequences are identical, the rat sequence varies in two amino acids, and the murine sequence has three amino acid alterations, as shown below.

> Bovine: 93 IEYGDKERKADA 104
> Human: IEYGDKERKADA
> Rat: VEYGDKERKADS
> Murine: VEYGDKERKTDS

This sequence comparison indicates that the monoclonal antibody 1C epitope may lie near amino acid residues 93–104, which would cause the different reactivities with bovine, human, rat, and murine $G_{o\alpha}$. To test species selectivity, equal amounts of membrane protein or purified proteins prepared from the same tissue of different species are resolved on a single gel, transferred, and the blot is probed with the monoclonal antibody.

Antibody Epitope

A complete epitope identification can be useful to facilitate the use of the antibody. This can be estimated by limited proteolysis or by competitive binding experiments.

Site-specific proteolysis can help to locate the binding site of an antibody. One example is the cleavage of $G_{o\alpha}$ by trypsin. Trypsin cleaves $G_{o\alpha}$ at amino acids 22 and 208, producing peptides migrating at 37, 25, and 17 kDa (16). In the presence of GTP the 37-kDa fragment is predominantly formed and it is stable, whereas with GDP the 37-kDa peptide is cleaved to the 25- and 17-kDa peptides. The reactivity of a monoclonal antibody specific to $G_{o\alpha}$ on an immunoblot with trypsinized $G_{o\alpha}$ fragments may help to estimate the site of the epitope.

Another method to estimate the antibody binding site is the use of synthetic peptides or G protein fragments for competitive antibody binding experiments. Synthetic peptides corresponding to sequences of the G protein subtype from which the antibody is generated can be used. Alternatively, G protein fragments expressed from defined regions of a G protein cDNA can be used for competitive antibody binding (9). Several concentrations of a synthetic peptide or a G protein fragment are preincubated with the monoclonal antibody, and the reaction mixture is then used to label an immunoblot containing the G protein antigen. If the reaction of the monoclonal antibody to the G protein on the Western blot is blocked by the synthetic peptide or the G protein fragment, the corresponding sequence is likely to contain the epitope.

Monoclonal Antibody Binding to Native versus Denatured Proteins

The reactivity of a monoclonal antibody to a native protein may differ substantially from that to a denatured protein. Testing the reactivity of a monoclonal antibody with a native protein is important to assess the feasibility of using the antibody in experiments that involve nondenaturing conditions, such as immunoprecipitation of G proteins. To compare the reactivity of a monoclonal antibody with native and denatured proteins, a purified G protein is (1) used untreated, (2) denatured by mixing in 1% SDS in PBS, and heating to 85°C for 3 min, and (3) boiled for 5 min. The native and denatured G protein solutions are diluted to several concentrations and reacted with the monoclonal antibody in solution followed by transferring to a precoated ELISA plate for competitive ELISA experiments. The greater the binding of the antibody to the protein in solution, the lower the final ELISA reading will be (Fig. 3). However, the presence of detergents may affect the antibody–protein reaction, so heat denaturing in the presence of SDS may not be representative of the maximal capability of the antibody to bind to the denatured antigen. The effect of SDS or other detergents on antibody binding can be determined by ELISA by diluting the monoclonal antibody with PBS containing different concentration of SDS or other detergents (0.001–0.1%) followed by incubation in an ELISA plate precoated with the G protein antigen.

Comparison of Monoclonal Antibody Reactivity with G Protein Heterotrimers and α-Subunit Monomers

It is often important to know if monoclonal antibodies differentially recognize the αβγ trimer or the dissociated α subunit of a G protein, for example, when the antibody will be used for immunoprecipitation of G proteins after

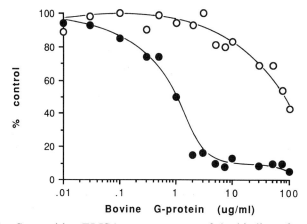

FIG. 3   Competitive ELISA measurements of the binding of native and denatured proteins to a monoclonal antibody. A monoclonal antibody specific for $G_{o\alpha}$ (20 ng/ml) was preincubated with the indicated concentrations of native ($\bigcirc$) or heat-denatured ($\bullet$) purified bovine brain G_o/G_i. The preincubation mixtures were transferred to an ELISA plate for the competitive ELISA assay described in the text. The absorbance was read and the data were calculated as the percent of control (the average absorbance of the controls was 0.480). The data indicate that denatured $G_{o\alpha}$ reacts better than native $G_{o\alpha}$ with the antibody, because less antibody was available for the reaction in the ELISA plate after preincubation with denatured $G_{o\alpha}$ compared with native $G_{o\alpha}$.

receptor activation. To test this, membranes can be preincubated in the absence or presence of various concentrations of GTPγS (to dissociate α from βγ), followed by immunoprecipitation. Immunoprecipitated proteins are identified by Coomassie blue staining or immunoblotting. An example of this experiment is shown in Fig. 4.

Interaction of Monoclonal Antibody with GTP-Binding Activity of G Proteins

Because GTP binding is a common feature of G proteins, the effect of a monoclonal antibody bound to a G protein on its GTP-binding activity may be tested. After preincubation with mouse immunoglobulin (as a control) or a monoclonal antibody, the purified G protein is incubated with 2 μM [^{35}S]GTPγS (0.1 mCi, 1313 Ci/mmol; Dupont, NEN) in binding buffer (50 mM HEPES, pH 8.0, 40 mM MgCl$_2$, 1 mM EDTA, 1 mM DTT, 200 mM NaCl) at 30°C for 1 hr. The binding buffer and incubation conditions can be

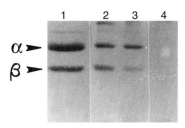

FIG. 4   Reactivity of a monoclonal antibody with G protein heterotrimers and α-subunit monomers. Rat cortical membranes were solubilized with 0.5% CHAPS in 10 mM HEPES, pH 7.4, 1 mM EDTA, 120 mM NaCl, 0.2 mg/ml PMSF, 2 μg/ml aprotinin, and 2 μg/ml leupeptin. $G_{o\alpha}$ was immunoprecipitated using a $G_{o\alpha}$-specific monoclonal antibody (2A) after incubation in the absence or presence of 100 μM GTPγS. Immunoprecipitated proteins were separated in 10% SDS-polyacrylamide gels followed by staining with Coomassie blue. The figure shows purified bovine brain G_o/G_i (1 μg) as a reference (lane 1) and immunoprecipitate using the monoclonal antibody 2A for $G_{o\alpha}$ in the absence (lane 2) or presence (lane 3) of GTPγS, and in the absence of membrane proteins (lane 4). The results indicate that monoclonal antibody 2A immunoprecipitated proteins that comigrated in SDS-polyacrylamide gels with G protein α and β subunits. Preincubation of solubilized rat cortical membranes with GTPγS, which causes dissociation of the α subunit from the $\beta\gamma$ dimer, increased the proportion of the α subunit relative to the β subunit that was immunoprecipitated. Thus, the antibody immunoprecipitated α subunit associated with or dissociated from the $\beta\gamma$ complex.

varied for different G protein subtypes. The reaction mixture is then diluted with ice-cold washing buffer (20 mM Tris-HCl, pH 8.0, 100 mM NaCl, 25 mM MgCl$_2$) and is filtered through 0.45 μm nitrocellulose filters (Whatman). The filters are washed four times, dried, and soaked in liquid scintillation fluid. The radioactivity is counted and GTP binding is calculated as nanomoles/milligram G protein. Reduced GTP binding after monoclonal antibody preincubation indicates blockade of GTP binding by that antibody.

ADP-Ribosylation

Several G proteins are substrates for toxin-catalyzed ADP-ribosylation, such as G_s by cholera toxin and G_i/G_o by pertussis toxin. Determination of the interaction of a monoclonal antibody with ADP-ribosylated G proteins is required prior to using the antibody in experiments that involve the use of these toxins. The ability of an antibody to bind to an ADP-ribosylated G protein can be tested on Western blots. For example, purified G_o or a mem-

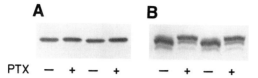

FIG. 5   Reactivity of $G_{o\alpha}$-specific monoclonal antibody to non-ADP-ribosylated and pertussis toxin-catalyzed ADP-ribosylated $G_{o\alpha}$. Rat cortical membranes were incubated for 1 hr at 30°C with NAD in the absence (−) or presence (+) of pertussis toxin (PTX, 10 μg/ml) for ADP-ribosylation. The membrane proteins were resolved on (A) a 10% SDS-polyacrylamide gel or (B) a 9% SDS-polyacrylamide gel with a 4–8 M urea gradient, transferred to nitrocellulose, and the blot was incubated with a $G_{o\alpha}$-specific monoclonal antibody (2A). The result indicates that ADP-ribosylation by pertussis toxin did not alter the reactivity of monoclonal antibody 2A to $G_{o\alpha}$, as shown in (A). The ADP-ribosylation of G_o was confirmed by the slower mobility of $G_{o\alpha}$ on the urea gel (B).

brane preparation containing G_o can be ADP-ribosylated with pertussis toxin with unlabeled NAD as the substrate. Proteins are resolved on a 10% SDS-polyacrylamide gel, transferred to nitrocellulose, and the blot is incubated with the $G_{o\alpha}$-specific monoclonal antibody to be tested. The binding of the monoclonal antibody to the same quantity of $G_{o\alpha}$ before and after ADP-ribosylation is compared. To confirm the ADP-ribosylation of G_o in the presence of unlabeled NAD, G_o can be resolved on a 9% SDS-polyacrylamide gel with a 4–8 M urea gradient, transferred to nitrocellulose, and the blot incubated with the $G_{o\alpha}$-specific monoclonal antibody. Because the mobility of $G_{o\alpha}$ is reduced by ADP-ribosylation (17), ADP-ribosylated $G_{o\alpha}$ can be distinguished from non-ADP-ribosylated $G_{o\alpha}$ (Fig. 5).

In summary, the characterization of several properties of a monoclonal antibody for G protein subunits should contribute to its optimal utilization. Much of the basic knowledge of the properties of the antibody that are required for initial experiments, such as titration of the antibody, specificity, and species selectivity, is described in this chapter. However, novel uses of such antibodies are likely to require further characterization procedures to ensure optimal utilization and interpretation of results. Monoclonal antibodies specific to subtypes of G proteins should prove useful in many studies of the function and expression of these proteins.

Acknowledgments

The authors thank Dr. Susanne Mumby for much useful advice and for collaborating in the characterization of monoclonal antibodies. Supported by NIH Grant MH-38752.

References

1. G. Milligan, *in* "G-Proteins as Mediators of Cellular Signalling Processes" (M. D. Houslay and G. Milligan, eds.), pp. 31–46. Wiley, New York, 1990.

2. A. M. Spiegel, *in* "ADP-Ribosylating Toxins and G-Proteins" (J. Moss and M. Vaughan, eds.), pp. 207–224. 1990.

3. S. M. Mumby and A. G. Gilman, *in* "Methods in Enzymology" (R. A. Johnson and J. D. Corbin, eds.), Vol. 195, p. 215. Academic Press, San Diego, 1991.

4. J. L. Halpern, S.-C. Tsai, R. Adamik, Y. Kanaho, E. Bekesi, H.-F. Kung, J. Moss, and M. Vaughan, *Mol. Pharmacol.* **29,** 515 (1986).

5. H. E. Hamm and D. Bownds, *J. Gen. Physiol.* **84,** 265 (1984).

6. S. E. Navon and B. K.-K. Fung, *J. Biol. Chem.* **262,** 15746 (1987).

7. S.-C. Tsai, R. Adamik, Y. Kanaho, J. L. Halpern, and J. Moss, *Biochem.* **26,** 4728 (1987).

8. S. Peraldi-Roux, P. Mangeat, B. Rouot, J. Oliver, S. Herbute, P. Brabet, V. Homburger, M. Toutant, J. Bockaert, and J. Gabrion, *C.R. Acad. Sci.* (Paris) **312,** 157 (1991).

9. L. van der Voorn, O. Tol., T. M. Hengeveld, and H. L. Ploegh, *J. Biol. Chem.* **268,** 5131 (1993).

10. X. Li, S. M. Mumby, A. Greenwood, and R. S. Jope, *J. Neurochem.* **64,** 1107 (1995).

11. E. Harlow and D. Lane, "Antibodies: A Laboratory Manual." Cold Spring Harbor Laboratory, Cold Spring Harbor, New York, 1988.

12. P. C. Sternweis and J. D. Robishaw, *J. Biol. Chem.* **259,** 13806 (1984).

13. J. F. Kearney, A. Radbruch, B. Liesegang, and K. Rajewsky, *J. Immunol.* **123,** 1548 (1979).

14. E. Engvall and P. Perlmann, *J. Immunol.* **109,** 129 (1972).

15. T. Tsukamoto, R. Toyama, H. Itoh, T. Kozasa, M. Matsuoka, and Y. Kaziro, *Proc. Natl. Acad. Sci. U.S.A.* **88,** 2974 (1991).

16. B. M. Denker, C. J. Schmidt, and E. J. Neer, *J. Biol. Chem.* **267,** 9998 (1992).

17. S. Mumby, I.-H. Pang, A. G. Gilman, and P. C. Sternweis, *J. Biol. Chem.* **263,** 2020 (1988).

[15] Expression and Purification of G Protein $\beta\gamma$ Subunits Using Baculovirus Expression System

Stephen G. Graber, Margaret A. Lindorfer, and
James C. Garrison

Introduction

Guanine nucleotide-binding regulatory proteins (G proteins) play a central role in coupling receptors to effector systems in all eukaryotic organisms examined. Since the appreciation of the roles of G_s in mediating the hormonal stimulation of adenylylcyclase, and of transducin in mediating the electromagnetic stimulation of cyclic GMP phosphodiesterase, models of G protein action have assumed a direct, active role for the α subunit and a passive role for the $\beta\gamma$ subunits as attenuators of α activity (1). In fact, G proteins are classified by the nature of their α subunit, with 21 distinct α subunits grouped into four general classes based on amino acid identities (2). It has become clear from studies with both native and recombinant α subunits that a variety of α subunits are capable of interacting with the same effectors (3); individual receptors are capable of regulating multiple effectors (4); and individual α subunits are capable of activating different effectors (5). Therefore the α subunits by themselves do not determine the specificity of the signal transduction pathways in which they are involved.

Although the existence of β and γ subunits has been known for some time, the traditional view has been that $\beta\gamma$ subunits are functionally interchangeable among α subunits (1). It has become clear that the G protein $\beta\gamma$ subunits play a wider role in signal transduction. Although early reports suggesting a direct modulation of effector function by $\beta\gamma$ subunits (6–8) were largely overlooked, $\beta\gamma$ subunits have now been shown to affect directly certain types of adenylylcyclase (9, 10), K^+ channels (11, 12), and phospholipase C (13–15) and to mediate the response to mating factors in yeast (16). Molecular cloning techniques have revealed the existence in mammals of five distinct β and seven distinct γ subunits, whereas peptide sequences predict the existence of additional γ subunits (2, 17). Examination, by protease mapping and immune recognition, of $\beta\gamma$ subunits, purified from a variety of tissues, revealed complex mixtures of at least two β and six γ subunits (18). A recent attempt to purify $\beta\gamma$ dimers of defined composition from tissue sources met

Methods in Neurosciences, Volume 29

with only partial success, and further documented the complex mixture of βγ dimers found in native sources (17). The interaction between G protein β and γ subunits to form dimers has been shown to be selective for specific combinations (19, 20), further underscoring a likely important role for βγ subunits in determining the selectivity of signaling pathways. Thus the current situation with regard to the βγ subunits is analogous to the earlier situation with the α subunits, in which assignment of function was impeded by the inability to purify subunits of specified subtype. Clearly the ability to prepare functional recombinant βγ subunits of defined subtype will lead to progress in understanding their contribution to selective activation of signal transduction pathways. Several laboratories have used the baculovirus expression system to express recombinant βγ subunits of defined composition, and have shown that the recombinant βγ subunits exhibit the known posttranslational modifications and functional activities of βγ subunits purified from mammalian cells (21–25). Although this approach is more complex than other methods, it is currently the best means of preparing reasonable amounts of βγ subunits of defined composition.

Construction of Baculovirus Transfer Vectors

General Considerations

The high levels of protein expression typically achieved with the baculovirus expression system is due to the use of the powerful promoter of the polyhedrin gene from the *Autographica californica* nuclear polyhedrosis virus (AcMNPV). This promoter regulates the gene encoding the polyhedrin protein that encapsulates the virus (26). If the cDNA for another protein replaces the sequences encoding the polyhedrin protein, a high level of expression of the foreign protein will occur in insect cells infected with the virus. Because the large size of the baculovirus genome (130 kbp) makes direct manipulations of the DNA impractical, recombinant viruses are usually produced by cotransfection of Sf9 cells with wild-type viral DNA and a transfer vector containing the foreign cDNA sequence. The transfer vectors are modified bacterial plasmids that contain the polyhedrin promoter adjacent to a multiple cloning site and flanked by wild-type viral sequences. Homologous recombination *in vivo* substitutes the foreign cDNA sequence (inserted in the multiple cloning site of the transfer vector) for the wild-type polyhedrin sequences, producing a recombinant virus. The recombinant virus can be isolated by visual screening or other methods (see below).

Selection of a particular transfer vector is dependent on individual requirements, such as whether a fusion or nonfusion protein product is desired, and the specific cloning sites available in the cDNA to be inserted into the vector. A variety of vectors are available with and without markers to facilitate selection of the recombinant virus (27–29). At least one vendor (PharMingen) produces vectors that incorporate sequences coding for affinity tags to facilitate purification of the protein product. The simplest subcloning strategies are those that involve directional ligation into existing sites in the polylinker, but constructs may also be produced through the use of oligonucleotide adaptors and linkers, or polymerase chain reaction (PCR)-effected mutagenesis of the specific cDNA clone to be inserted. Subcloning strategies are of necessity specific to each individual cDNA insert and the transfer vector selected, but certain general rules apply. No modifications should be made that alter the sequences that lie downstream from the polyhedrin promoter and upstream from the multiple cloning site, because this region possesses elements that affect transcription (30, 31). Furthermore, the 5' noncoding sequence upstream of the insert start codon should be minimized or eliminated if possible, because the level of protein expression is decreased by increasing distance between the polyhedrin promoter sequences and the start codon of the inserted cDNA (30). Noncoding sequences and modifications (i.e., affinity tags) placed at the 3' end of the construct do not appear to affect gene expression. The identity of the constructs should always be confirmed prior to transfection by either restriction mapping or better, by DNA sequencing using universal primers made for the baculovirus DNA. An excellent description of these techniques can be found in two laboratory manuals (32, 33).

Subcloning Strategy for the G Protein $\beta\gamma$ Subunits

The baculovirus transfer vectors pVL1392 and pVL1393 used in the construction of recombinant baculoviruses were kindly provided by Dr. Max Summers (Texas A & M University), and are now commercially available from several vendors. Obviously, each different β and γ DNA sequence requires different subcloning strategies. The preparation of transfer vectors for different β and γ subunits have been described (21–25). The preparation of the transfer vector for production of the β_1, β_2, and the γ_2 viruses are described here as examples. The coding region of a β_1 cDNA was obtained as a product of a polymerase chain reaction from a human liver cDNA library with an EcoRI cohesive end three nucleotides before the ATG start codon. This sequence was subcloned into the EcoRI/BamHI sites of pVL1392. The 1051-bp ApaI fragment of a β_2 cDNA (34) was subcloned into the BamHI site of pVL1393 using a single-stranded adaptor. The 560-bp BsmI/EcoRI fragment of a γ_2 cDNA (35) was subcloned into the SmaI/EcoRI sites of pVL1393 after filling in the *Bsm*I overhang with T4 DNA polymerase. All constructs have been sequenced to ensure that no changes were made in the amino acid sequences of the proteins.

Production and Purification of Recombinant Baculovirus

Recombinant baculoviruses are obtained by transfecting Sf9 (*Spodoptera frugiperda,* fall armyworm ovary) cells with a 5:1 (mass:mass) mixture of transfer vector and circular wild-type viral DNA using a calcium phosphate precipitation technique modified for insect cells (32). Plasmid and viral DNA are mixed in 950 μl HEBS/CT (137 mM NaCl, 6 mM D(+)-glucose, 5 mM KCl, 0.7 mM Na$_2$HPO$_4$ · 7H$_2$O, 20 mM HEPES with 15 μg/ml calf thymus DNA, pH 7.05) and precipitated with 125 mM CaCl$_2$ for 30 min at room temperature. This mixture is then pipetted onto a preattached Sf9 cell monolayer (2.5 × 10^6 cells in a 25 cm^2 tissue culture flask covered with 2 ml of 1× Grace's medium) and absorbed for 4 hr. The mixture is then removed with a Pasteur pipette; the monolayer is rinsed carefully with TNM-FH medium supplemented with 10% fetal bovine serum, 50 μg/ml gentamicin, and 2.5 μg/ml amphotericin B, and covered with an additional 5 ml of medium. The flasks are then incubated for 4–6 days and checked for signs of infection using an inverted microscope. The anticipated frequency of recombinants utilizing this technique is 0.1–3.0% (32). Recombinants can be selected by visually screening for the *occ*⁻ phenotype. Typically, the *occ*⁻ phenotype is distinguished from the normal phenotype by a gap in the monolayer associated with swollen cells having a grainy cytoplasm, trypan blue-stained cells and cell debris, and the absence of inclusion bodies. Recombinant viruses have been produced in Sf9 cells by transfection using linear viral DNA (31). The use of linear DNA improves the frequency of recombinants to greater than 30%. An additional strategy that can further improve the frequency of recombinants is the use of viral DNA, which requires sequences in the transfer vector to produce a functional virus (36). Some vendors, including InVitrogen, Stratagene, and PharMingen, offer kits for preparing recombinant viruses.

Media from transfections are harvested by centrifugation at 2000 g and titered by plaque assay. Plaque assays may be performed according to the method of Summers and Smith (32). Briefly, medium from a transfection is diluted 10⁴-fold, 10⁵-fold, and 10⁶-fold in TNM-FH medium supplemented with 10% fetal bovine serum, 50 μg/ml gentamicin, and 2.5 μg/ml amphotericin B, and 1 ml is placed on cells plated at 2.0 × 10⁶ cells/plate in 60-mm polystyrene tissue culture plates (Corning or Costar). The plates are incubated for 1 hr at 27°C, after which the medium is aspirated and the plates covered with 4 ml of 50% low-melt agarose (3%, w/v)/50% 2× Grace's medium supplemented with 20% fetal bovine serum and the above antibiotics. After 2–3 days, an additional 2 ml of low-melt agarose overlay containing 0.04% (w/v) trypan blue may be added to assist in the identification of recombinant plaques. Following titer of the transfection medium, 25–50 plates are set up at a density of 125–150 plaques/plate to screen for recombinants. Plaques may be identified based on the observation of the occ^- phenotype. Recombinant viruses identified should be purified through four rounds of plaque purification. Due to the possibility of false recombinant viruses and the fact that different recombinant clones from a single transfection may express protein at different levels, several recombinants should be picked and followed through the initial screening process. Analysis of recombinant proteins can be performed by solubilization of infected Sf9 cells with Laemmli sample buffer (37), resolution of the proteins on an SDS-polyacrylamide gel, transfer of the proteins to nitrocellulose, and immunoblotting with peptide antisera for the G protein subunit. One useful antibody is the NEI-808 antibody (New England Nuclear) that recognizes a common sequence in the β subunits. If an antibody is not available, screening can also be done by dot hybridization using the cDNA clone inserted in the transfer vector as a probe (32). The utility of the polymerase chain reaction for screening recombinant clones has also been demonstrated (38).

Sf9 Cell Culture

Maintenance of Sf9 Cells

Spodoptera frugiperda (Sf9) cells may be obtained from the American Type Culture Collection (Rockville, MD ATCC No., CRL 1711). Cells are maintained in logarithmic growth in TNM-FH medium, which is Grace's medium supplemented with yeastolate (3.33 g/liter) and lactalbumin hydrolyzate (3.33 g/liter), containing 10% fetal bovine serum (32). Antibiotics are not routinely employed in maintenance cultures. Normally, cells are maintained at 27°C as 75-ml suspension cultures in spinner flasks stirred at a rate of 60 rpm in an atmosphere of 50% O_2/50% N_2 (v/v) or 50% O_2/50% air (v/v). For normal passage, spinners are seeded at a density of $(0.8–1.2) \times 10^6$ cells/ml. The doubling time for these cells is typically 18–22 hr. Aliquots from each spinner are diluted 1:1 with 0.4% (w/v) trypan blue in 0.85% (w/v) normal saline and counted using a Neubauer hemocytometer. Routinely, subculturing is done by removing cell suspension and replacing it with fresh medium. Periodically, to prevent the accumulation of toxic by-products, the cells are pelleted by gentle centrifugation (600 g for 6 min) and resuspended in fresh medium.

In our experience subculturing is required every 3 days. Because the health of the cells is essential for high levels of protein expression, care should be taken so that cell viability remains greater than 95% and the cell borders appear cleanly rounded when examined with a microscope.

Scale Up of Sf9 Cell Culture

To obtain enough starting material to prepare a G protein α-subunit affinity matrix of sufficient capacity (see below), suspension cultures have been scaled up to 3-liter spinner flasks. The limiting factor for Sf9 cell growth in suspension culture is dissolved oxygen (39). However, Sf9 cells are fragile and susceptible to rupture on bubble surfaces. Pluronic F68, a block copolymer of polyethylene and polypropylene, protects Sf9 cells from mechanical stress and also controls foaming (40). Thus, in the presence of 0.1% Pluronic F68, the cell suspension may be sparged without excessive cell damage. The antifoam property of Pluronic F68 also allows the use of TNM-FH medium with 10% serum, which supports vigorous cell growth and protein production.

Several large-scale Sf9 cell culture systems have been described, including spinner flasks equipped with oxygen electrodes to measure dissolved oxygen, and airlift fermenters (33, 41). However, the design described below is less costly and is easier to assemble and maintain, and produces cell densities in the range of 3×10^6 cells/ml. Figure 1 illustrates a typical arrangement for large-scale suspension culture of Sf9 cells. A standard 3-liter microcarrier spinner flask (Bellco PN 1965-03000) is configured with an optional, adjustable paddle (Bellco PN 1965-40003) on the central shaft above the magnetic stirring bar. The level of the paddle is adjusted to break the surface of the suspension culture and thereby facilitate oxygen delivery from the head space. The simple configuration of the center port allows for medium introduction by pouring from flamed medium containers. One side arm is fitted with a two-port assembly (Bellco PN 1965-2000). This provides one gas outlet and one medium/cell suspension outlet. The medium port is fitted with a female Luer-lock connector and plugged with a 5 ml Luer-lock syringe when not in use. Cell culture-grade silicone or Pharmed tubing is used for all liquid transfers. The other side arm is fitted with a two-port assembly with a shorter inlet tube (Bellco PN 1965-20001). A pyrex $\frac{1}{8}$-inch inner diameter (i.d.) tube with a 12C frit is connected to the stainless-steel pipe on this port with a short length of silicone tubing. The incoming sparging gas (50% O_2/50% N_2, or simply O_2) is filtered through an Acro 50 NPT filter (Gelman, 0.2 μm, PTFE). Gas delivery through the frit may be controlled either via a floating ball flow meter or a peristaltic pump. The flow should be set to attain a slow but steady stream of gas bubbles, with an effective rate being roughly 10 ml/

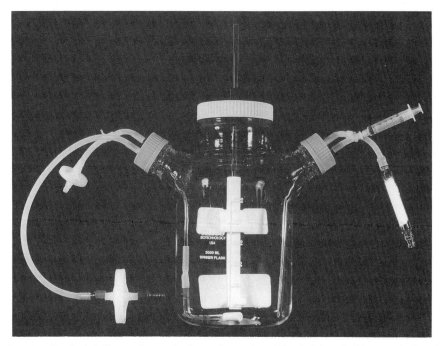

FIG. 1   Typical 3-liter spinner flask configuration. The left-hand side arm is fitted for a gas inlet through a Pyrex frit (on bottom of the vessel) and one for a gas outlet. The right-hand side arm is also fitted for two ports, one for a medium/cell outlet and one for a gas outlet. The suspended stir bar shaft is fitted with an optional, auxiliary, adjustable paddle. All components are commercially available from Bellco Glass, Vineland, NJ.

min for a 3-liter culture. Alternatively, the frit may be removed and gas delivered directly through the $\frac{1}{8}$-inch i.d. pipe. However, the delivery rate must be increased to roughly 225 ml/min for a 3-liter culture, to achieve comparable growth rates and densities. Consequently, the optimal balance among oxygenation, foaming, and cell growth must be sought empirically.

Spinners are seeded at $0.5-0.8 \times 10^6$ cells/ml, and stirred at 50–70 rpm at 27°C. TNM-FH medium supplemented with 10% fetal bovine serum, 10 μg/ml gentamicin sulfate, 2.5 μg/ml fungizone, and 0.1% Pluronic is used for all 1- to 3-liter suspension cultures. After 72 hr, cell density typically reaches $(2.5-3) \times 10^6$. This represents a slightly slower growth rate and lower ultimate cell density than normally attained in 75-ml spinners in 50% O_2/50% N_2 (v/v) atmosphere [doubling time, 22–24 hr; cell density, $(4-6) \times 10^6$ cells/ml]. Cells are transferred to sterile centrifuge bottles by placing the spinners under positive pressure via one of the gas outlet filters, clamping off the gas inlet and auxiliary gas outlet and attaching a sterile transfer line with filling bell to the medium outlet port. This can be conveniently accomplished in a laminar flow hood, using the sparging gas to pressurize the vessel. A pressure of 5 psi is sufficient for a reasonable flow rate and does not cause significant damage to the Sf9 cells.

Infection and Harvest of Sf9 Cells

Sf9 cells may be successfully infected in culture volumes ranging from 75 ml to 3 liters, depending on the amount of final product that is required. Generally, a 75-ml culture will yield 1–1.25 g (wet weight) of infected cells, whereas the 3-liter cultures will yield 40–45 g (wet weight) of cells. For infections, the appropriate number of cells are spun at 600 g at room temperature for 5 min with the centrifuge brake off. Spent medium is removed, and the cell pellet is gently resuspended in the appropriate viral stock using a multiplicity of infection of 1–5 plaque-forming units (pfu) per cell. A small volume of additional medium may be used to aid in cell suspension, but the cell density should be at least 1×10^7 cells/ml during the infection process. After 1 hr at 27°C, infected cells are transferred back to spinner flasks and suspended at $(2.5–3.0) \times 10^6$ cells/ml with fresh medium. Infected cells are harvested between 48 and 72 hr, depending on their appearance and density, usually when trypan blue staining reaches 20–25%. Because infectivity and expression levels vary depending on culture conditions and the particular recombinant virus used, an immunological or functional time course should be done on a case by case basis to determine accurately the infection period yielding the highest level of protein expression. This will also allow determination of the optimum multiplicity of infection and assess the degree of proteolysis that may occur as the infection progresses. Infected cells may be harvested by centrifugation at 800 g for 5 min at 4°C with the centrifuge brake off. The supernatant is aspirated and the cell pellet washed three times in ice-cold phosphate-buffered saline (PBS: 7.3 mM NaH$_2$PO$_4$, 58 mM KCl, 47 mM NaCl, 5 mM CaCl$_2$, pH 6.2). After rinsing, the cells are resuspended in 1 ml homogenization buffer/gram (wet weight) of cell pellet, snap-frozen in liquid nitrogen, and stored at −70°C. The homogenization buffer for α-subunit cell pellets consists of 10 mM Tris-Cl, 25 mM NaCl, 10 mM MgCl$_2$, 1 mM EGTA, 1 mM dithiothreitol (DTT), 0.1 mM phenylmethylsulfonyl fluoride (PMSF), 20 μg/ml benzamidine, and 2 μg/ml each of aprotinin, leupeptin, and pepstatin, 10 mM NaF, 10 AlCl$_3$, pH 8.0, at 4°C. Homogenization buffer for $\beta\gamma$ cell pellets omits

NaF and AlCl$_3$ in an attempt to keep the endogenous Sf9 $\beta\gamma$ subunits complexed with endogenous α subunits.

Preparation of G Protein α-Subunit Affinity Column

Although $\beta\gamma$ subunits can be purified using conventional chromatographic methods (17, 42), preparation of functional protein is potentially enhanced by using a G protein α subunit as an affinity matrix because properly modified α and $\beta\gamma$ subunits interact with each other with higher affinity than their unprocessed counterparts (23, 43). Initial experiments using a small column containing 4–5 mg of α subunit showed the utility of the approach, while revealing that the binding capacity for $\beta\gamma$ subunits of this size column is insufficient for use in routine purifications of multiple combinations of $\beta\gamma$ subunits. It was estimated that an affinity matrix containing 30 mg of an α subunit would allow the purification of 50–150 μg of $\beta\gamma$ dimers from 300 to 500 ml of infected cells. In order to obtain 30 mg of an α subunit, it became necessary to develop the capacity to produce and process amounts of infected Sf9 cells at least an order of magnitude greater than originally reported (44).

Large-Scale Purification of G Protein α Subunits

The scaled-up purification of G protein α subunits follows the previously described protocols (45) with the modifications described below. These modifications have primarily involved the use of larger columns with increased flow rates. All chromatography steps utilize a Waters Model 650 Advanced Protein Purification System (controller and pump), an Isco Model UA5 Detector with 5-mm path length preparative flow cell, and an Isco Foxy fraction collector. The standard large-scale preparation begins with 40–45 g (wet weight) of Sf9 cells infected with a recombinant virus expressing a G$_{i\alpha}$ subunit and yields 10–12 mg of purified α subunit.

The volume of the DEAE column has been increased from 200 to 1000 ml. One liter of Waters 40HR Protein-Pak DEAE (40-μm particle size) is slurry packed into an Amicon (Danvers, MA) Vantage column (9 cm $\times$ 50 cm), resulting in a final bed height of 15 cm. Using 0.062-inch (i.d.) tubing from the pump outlet to the flow cell inlet, and 0.04-inch (i.d.) tubing from the column outlet to the flow cell, results in a total system backpressure of 60 psi at the flow rate of 40 ml/min used to load and elute the column. Although the volumes are increased, the gradient shape and elution profile of the larger column are essentially identical to those reported for the smaller column (45). The GTPγS binding activity is typically contained in 10–12 40-ml fractions. These are pooled, diluted fivefold with 10 mM KPO$_4$, 10 mM Tris-Cl, pH 8.0, at 4°C, and loaded onto a hydroxyapatite column (3 × 15 cm, Mitsui Toatsu Chemical, Tokyo) at a flow rate of 20 ml/min. This column is eluted as previously described (45). The size of the Mono P chromatofocusing (Pharmacia, Piscataway, NJ) column has been increased from an HR 5/20 to an HR 10/30 and the flow rate has been increased to 2 ml/min. The elution conditions remain as previously described (45). Both the hydroxyapatite and chromatofocusing columns are plumbed with 0.04-inch (i.d.) tubing precolumn and 0.02-inch (i.d.) tubing postcolumn. The GTPγS binding fractions from the chromatofocusing column are pooled and dialyzed twice versus 4 liters of HED (50 mM HEPES, 1 mM EDTA, 1 mM DTT, pH 8.0) to remove Polybuffer. Proteins are concentrated to a final concentration of 1–6 mg/ml using an ultrafiltration apparatus (Amicon, PM30 membrane). Concentrated proteins are aliquoted and stored at −70°C.

Preparation of $G_{i\alpha}$ Affinity Matrix

Synthesis of the $G_{i\alpha}$ affinity matrix using the recombinant α subunits is based on the method of Pang and Sternweis (46) as described (15). The following modifications have been made for the preparation of larger $G_{i\alpha}$ affinity columns. Purified recombinant $G_{i\alpha}$ is concentrated to approximately 5 mg/ml, reduced in the presence of 1 mM DTT and 5 μM GDP, and exchanged into thoroughly degassed 200 mM KPO$_4$, 200 mM NaCl, 5 μM GDP, pH 8, using an HR 10/10 Fast Desalting (Pharmacia) column. This protein is immediately reacted with the MBS-activated ω-aminobutylagarose (Sigma, St. Louis, MO), prepared as described (15), at a concentration of approximately 3 mg $G_{i\alpha}$/ml activated resin for 2 hr at 4°C with gentle tumbling. The resulting $G_{i\alpha}$–agarose is first washed with five volumes of 50 mM NaHPO$_4$, pH 7.0, 1 mM EDTA, 300 mM NaCl, 2 mM GDP, 0.1% (w/v) Genapol C-100, and then with five volumes of the same buffer containing 50 mM 2-mercaptoethanol to "cap" any unreacted sulfhydryl groups. Aliquots of the $G_{i\alpha}$ subunits can be removed from the reaction mix before and after the coupling reaction and electrophoresed on a 12% polyacrylamide gel to determine the efficiency of the coupling to the agarose matrix. Figure 2 presents a typical gel showing that approximately 90% of the α subunit has been immobilized on the agarose matrix. The $G_{i\alpha}$–agarose column can be stored at 4°C in 50 mM HEPES, pH 8.0, 100 mM NaCl, 1 mM EDTA, 0.1% (w/v) Genapol C-100, 30% (v/v) glycerol, 1 mM DTT, and 5 μM GDP. For extended storage, 0.02% sodium azide should be added to the storage buffer.

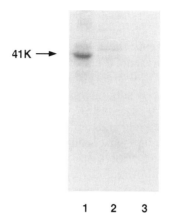

$41K \longrightarrow$

1   2   3

FIG. 2   Coupling of $G_{i\alpha}$ to MBS agarose; 10 mg of $rG_{i\alpha2}$ and 6 mg of $rG_{i\alpha3}$ were reacted with 5 ml of MBS agarose as described in the text. Lanes 1–3 of a 12% acrylamide-SDS gel are used to check the efficiency of coupling of the α subunit to the agarose gel matrix. Lane 1 represents about 5 μg of the $G_{i\alpha2}$ preparation prior to the coupling reaction. Lane 2 represents a 1 : 1 dilution of the supernatant following the reaction and lane 3 represents a 1 : 1 dilution of the first wash of the MBS agarose after the reaction. Note that greater than 90% of the recombinant G protein α subunit present in the starting material has been immobilized on the agarose beads.

Expression and Purification of Recombinant $\beta\gamma$ Subunits

Purification of Recombinant $\beta\gamma$ Subunits

Sf9 cells are infected with the appropriate recombinant baculoviruses, cultured, and harvested as described above. Frozen, harvested cells (typically 2–5 g wet weight) are thawed in 15× their wet weight of ice-cold homogenization buffer and burst by nitrogen cavitation (600 psi, 20 min on ice). Lysed cells are mixed with an equal volume of homogenization buffer containing 0.2% (w/v) Genapol C-100 (Calbiochem) and gently stirred for 1 hr at 4°C. The detergent extract is centrifuged at 100,000 g for 1 hr, and may be either used immediately or stored at -70°C until needed. The Genapol extract (40–60 ml) is passed through a 0.45-μm filter, and applied at a flow rate of 3 ml/min through the pump injection port to a 2 × 10-cm DEAE HR40 (Waters) column equilibrated with TED/CHAPS [50 mM Tris-Cl, 0.02 mM EDTA, 1 mM DTT, 0.6% CHAPS (w/v), pH 8.0, at 4°C]. The column is washed with 36 ml (approximately two column volumes) of TED/CHAPS and eluted with a 96-ml linear gradient from 0 to 300 mM NaCl in TED/CHAPS. The various recombinant $\beta\gamma$ dimers elute at slightly different ionic strengths in the range of 100–200 mM NaCl. Figure 3 depicts the results from the DEAE chromatography step of a $\beta_1\gamma_3$ preparation. In Figure 3, the top panel presents the UV absorbance (280 nm) of the protein eluate; the middle panel presents the proteins in fractions 9–28 resolved on a Coomassie blue-stained SDS polyacrylamide gel, and the bottom panel presents a Western blot of the β subunit made from a duplicate of the gel shown in the middle panel.

To maximize yield and minimize levels of contaminating proteins, a simple dot blot procedure has been developed to monitor the $\beta\gamma$ content of the DEAE fractions prior to pooling them for application to the $G_{i\alpha}$ subunit affinity column. Aliquots (5 μl) of DEAE fractions are denatured by the addition of 1.25 μl 5× SDS sample loading buffer and boiled for 6 min. A 1 × 4-cm strip of nitrocellulose (Schleicher and Schuell, Keene, NH; 0.2 μm pore size) is notched in two places to provide unambiguous orientation. Duplicate 0.3-μl aliquots of each sample are spotted onto the nitrocellulose, allowing the first application to dry completely before reapplication. The "blot" is blocked for 20 min at room temperature in M/TBST (3% powdered milk in 10 mM Tris-Cl, pH 8.0, 150 mM NaCl, 0.05% Tween 20), incubated with 1 : 1000 primary antibody (NEI 808) in M/TBST for 40 min, rinsed three times for 3 mins with M/TBST, incubated with 1 : 750 alkaline phosphatase conjugated secondary antibody (Promega, Madison, WI) in M/TBST for 40 min, rinsed three times for 3 min with M/TBST, rinsed once with deionized water, and developed with the ProtoBlot Alkaline Phosphatase System (Promega). Fractions giving the strongest response in this assay are pooled and applied to the $G_{i\alpha}$–agarose affinity resin.

Prior to use, the $G_{i\alpha}$–agarose affinity resin is washed with 10 volumes of WB1 (20 mM NaHEPES, pH 8.0, 1 mM EDTA, 200 mM NaCl, 0.6% CHAPS, 3 mM DTT, 5 μM GDP). GDP is added to the pooled DEAE fractions to a final concentration of 5 μM, the DEAE pool (normally 15–20 ml) is mixed with the affinity resin, and the mixture is gently tumbled for 1 hr at 4°C. Following a gentle centrifugation (50 g, 5 min), the resin is resuspended in 10 ml of WB1 buffer, and poured into a 1.5 × 12-cm Econo-Pak column (Bio-Rad, Richmond, CA). The bed is washed four times with 2 ml WB1, and twice with 2 ml WB1 with 400 mM NaCl (final concentration). For elution of the recombinant $\beta\gamma$ dimers, 2-ml fractions are collected as follows. Two 2-ml aliquots of elution buffer (20 mM NaHEPES, pH 8.0, 1 mM EDTA, 200 mM NaCl, 50 mM MgCl$_2$, 10 mM NaF, 30 μM AlCl$_3$, 3 mM DTT, 5 μM GDP, 0.6% CHAPS) are applied to the top of the bed, flow is stopped, and an additional 2 ml of elution buffer is placed on the column bed and allowed to stand for 15 min. An additional four 2-ml aliquots of elution buffer are passed over the resin, collected, concentrated to 0.1–0.5 mg/ml with a Centricon 30 concentrator (Amicon), aliquoted, and stored at −70°C until

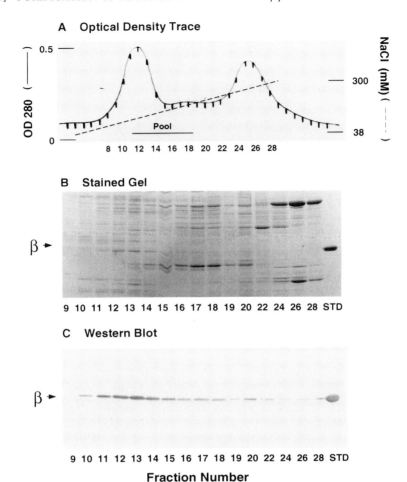

FIG. 3   Representative DEAE chromatography of a typical recombinant $\beta\gamma$ preparation. A 40-ml volume of a 0.1% Genapol C-100 extract of $\beta_1\gamma_3$-infected Sf9 cells was
applied to the DEAE column and eluted with a NaCl gradient; 3-ml (1-min) fractions
were collected. (A) UV absorbance of the protein in the elution fractions. (B) SDS-
gel electrophoresis of the DEAE elution fractions. The proteins were electrophoresed
on a 12% acrylamide-SDS gel and stained with Coomassie Brilliant Blue. A section
of the SDS gel ranges from approximately 50 kDa at the top to 25 kDa at the bottom;
STD, brain $\beta\gamma$. Fractions 11 to 18 contain the $\beta1$ subunit. These fractions were pooled
for application to the G_{ia}–agarose affinity column. (C) Western blot of the DEAE
elution fractions. Proteins from an identical gel to that in β were transferred to
nitrocellulose, blotted with a β common antibody, NEI808 (New England Nuclear),
and developed with alkaline phosphatase (Promega).

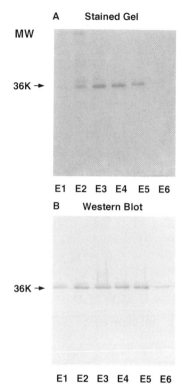

FIG. 4   The $\beta\gamma$ fractions eluted from the $G_{i\alpha}$–agarose affinity column during the purification of $\beta_1\gamma_3$. (A) An SDS-polyacrylamide gel and (B) Western blot of the fractions eluted from the $G_{i\alpha}$–agarose affinity column with $Al^{3+}/Mg^{2+}/F^-$. The proteins in the concentrated, sequential elution fractions, E1–E6, are shown. Each 2-ml fraction was concentrated 5- to 10-fold with a Centricon 30 concentrator (Amicon) prior to electrophoresis; the γ subunits are not detectable with Coomassie blue staining.

use. The resin should then be regenerated by washing with 20 ml of WB1 buffer and placed in storage buffer (see above). Figure 4 depicts the fractions eluted from the affinity resin during the purification of $\beta_1\gamma_3$. The top panel presents a Coomassie blue-stained gel showing that most of the $\beta_1\gamma_3$ elutes in fractions E3 and E4. The Western blot in the lower panel confirms this conclusion.

Estimation of Purity of Recombinant $\beta\gamma$ Subunits

A Pharmacia Phastgel System has been used to assess the purity of the recombinant $\beta\gamma$ subunits and to estimate the concentration of the purified

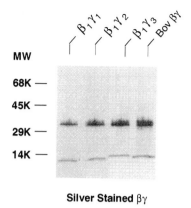

Silver Stained βγ

Fig. 5   Purified recombinant βγ subunits were electrophoresed on 8–25% acrylamide gradient SDS gels and stained with silver. The differences among the mobilities of the three γ subunits are apparent.

protein. A standard curve is generated by loading amounts of purified bovine brain βγ (kindly provided by Dr. Paul Sternweis, Southwestern Medical School) ranging from 5 to 20 ng in individual lanes of an 8–25% polyacrylamide gradient gel. Samples of recombinant βγ preparations are loaded in adjacent lanes. After electrophoresis for 80 V-hr (about 20 min at 250 V), the gels are silver stained according to the manufacturer's procedure for high-sensitivity silver staining, which is based on the method of Heukeshoven and Dernick (47). The amounts of protein in the stained bands are quantitated with the Whole Band software on a Millipore/BioImage Visage 4000 Image Analysis System. Based on this method, the final yield of purified recombinant βγ subunits has been 10–30 μg of protein from 1.5 to 2.0 g (wet weight) of infected Sf9 cell pellet. Figure 5 depicts several preparations of purified recombinant βγ dimers. Note that the preparations are essentially pure as judged by silver staining and that the three different γ subunits have distinct mobilities in this gel system. The γ_1 subunit migrates faster than the γ_2 or the γ_3 subunit.

Assay of βγ Activity

The purification of the βγ dimer can be followed using a number of functional assays, including (a) the ability to support ADP-ribosylation of G protein α subunits by pertussis toxin (48), (b) the ability to activate certain phospholipase C-β isozymes (15), or (c) the ability to activate adenylylcyclase (25).

Additional assays of $\beta\gamma$ activity are described by Higgins and Casey in Chapter 8, this volume. It is suitable to monitor the purification of the recombinant $\beta\gamma$ subunits by Western blotting (using the dot blot assay described above) followed by an activity assay to assess the functionality of the final product. The observation that only properly modified $\beta\gamma$ subunits bind to an α-subunit affinity column (23, 43) is a major advantage of the purification procedure described above. Use of recombinant $G_{i\alpha}$ subunits that are known to be myristoylated (44) for the affinity matrix greatly enhances the probability that the $\beta\gamma$ subunits eluted are active proteins. For routine assay of $\beta\gamma$ activity, ADP-ribosylation of a $G_{i\alpha}$ or $G_{o\alpha}$ subunit is a convenient choice as these α subunits are readily available from preparation of the material used to make the affinity column, and pertussis toxin is commercially available. Another advantage of the ADP-ribosylation assay is that all of the recombinant $\beta\gamma$ combinations tested to date have been active in this assay (21, 23, 25). Thus, the assay should be applicable to new combinations of recombinant $\beta\gamma$ dimers. Accordingly, this assay is described below.

ADP-Ribosylation of Recombinant $G_{i\alpha}$ by Pertussis Toxin

The procedure described by Casey *et al.* (48) has been modified slightly such that the assay is linear with time for the concentrations of Sf9 cell extracts assayed. The recombinant $G_{i\alpha}$ used as substrate may be purified as described above. Each assay tube should contain the following components in a final volume of 40 μl: 75 mM NaHEPES, pH 8.0; 2 mM DTT; 1 mM MgCl$_2$; 1 mM EDTA; 0.5 mM dimyristoylphosphatidylcholine; 10 μM GDP; 5 μM NAD; [^{32}P]NAD ($\sim$5000 cpm/pmol); 400 ng holopertussis toxin (10 μg/ml); 500 ng recombinant $G_{i\alpha}$ ($\sim$ 0.3 μM); and the amount of sample to be assayed. The sample to be assayed is diluted with the above buffer and added in a volume of 5 μl such that the concentration of detergent in the assay does not exceed 0.06%. In practice it is possible to keep the final detergent concentration below 0.01%. The incorporation of ADP-ribose into the $rG_{i\alpha}$ is generally linear for at least 25 min at 30°C for up to 20 ng of purified bovine $\beta\gamma$ under these assay conditions. Therefore, reactions are allowed to proceed for 20 min at 30°C. The holopertussis toxin should be preactivated by incubation at 37°C for 20 min in the presence of 25 mM DTT and 150 μM ATP. The ADP-ribosylation reaction is stopped by the addition of 20 μl 20% SDS (w/v) with 1 mM DTT, and heating for 3 min at 90°C. After alkylation with 20 μl of 40 mM NEM for 15 min at 20°C, the samples are prepared for SDS-PAGE on 12% gels by the addition of 20 μl of 5× Laemmli sample buffer. Following autoradiography with Kodak XK-1 film, the amount of radioactive

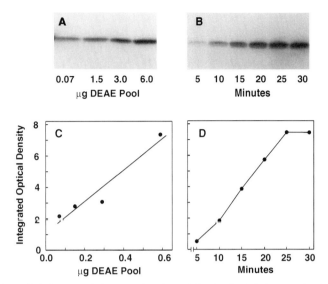

FIG. 6   (A and B) Assay of the $\beta\gamma$ subunit activity in the DEAE pool. (C) $\beta\gamma$ activity detected in varying amounts of the DEAE pool from a $\beta_1\gamma_2$ preparation during a 20-min ADP-ribosylation assay as described in the text. (D) Time course of the ADP-ribosylation assay using the highest dose (0.6 μg) of the DEAE pool depicted in C. The images above the graph (A and B) depict the autoradiogram from which the integrated optical densities were derived.

ADP-ribose incorporated into $rG_{i\alpha}$ is routinely estimated by scanning densitometry of the images using the Whole Band software on a BioImage Visage 4000 Image Analysis system. The standard deviation of triplicate samples is generally 4–5% by this method. Essentially identical results may be obtained by excision of the recombinant $G_{i\alpha}$-containing band from the gel followed by scintillation counting. Figure 6 depicts the activity detected in the DEAE pool from a preparation of recombinant $\beta_1\gamma_2$ subunits. Note from Fig. 6 C and D that the amount of ADP-ribose incorporated into the $G_{i\alpha}$ subunit in this assay is linear with increasing protein and time. Figure 6 A and B show sections of the original autoradiograms used to quantitate the extent of reaction.

Summary

It is possible to obtain reasonable quantities of purified G protein $\beta\gamma$ subunits of defined composition from Sf9 cells infected with recombinant baculoviruses expressing the desired β and γ subunits. The recombinant $\beta\gamma$ dimers contain the known posttranslational modifications of their native counterparts and have been shown to be functional by a variety of criteria. In spite of the complexity of the baculovirus expression system, this is currently the only feasible method of producing reasonable quantities of $\beta\gamma$ dimers of defined subunit composition.

Acknowledgments

The authors thank Dr. Douglas D. Nicoll, formerly of Bellco Glass, Vineland, NJ, for invaluable advice and assistance in developing the large-scale Sf9 cell culture methodologies. We also thank Robert A. Figler for significant contributions to this work and Brian T. Easton for excellent technical assistance.

References

1. A. G. Gilman, *Annu. Rev. Biochem.* **56,** 615–649 (1987).
2. M. I. Simon, M. P. Strathmann, and N. Gautam, *Science* **252,** 804–808 (1991).
3. A. Yatani, R. Mattera, J. Codina, R. Graf, K. Okabe, E. Padrell, R. Iyengar, A. M. Brown, and L. Birnbaumer, *Nature (London)* **336,** 680–682 (1988).
4. A. Ashkenazi, J. W. Winslow, E. G. Peralta, G. L. Peterson, M. I. Schimerlik, D. J. Capon, and J. Ramachandran, *Science* **238,** 672–675 (1987).
5. R. Mattera, M. P. Graziano, A. Yatani, Z. Zhou, R. Graf, J. Codina, L. Birnbaumer, A. G. Gilman, and A. M. Brown, *Science* **243,** 804–807 (1989).
6. D. E. Logothetis, Y. Kurachi, J. Galper, E. J. Neer, and D. E. Clapham, *Nature (London)* **325,** 321–326 (1987).
7. Y. Kurachi, H. Ito, T. Sugimoto, T. Katada, and M. Ui, *Pflugers Archiv.* **413,** 325–327 (1989).
8. T. Katada, K. Kusakabe, M. Oinuma, and M. Ui, *J. Biol. Chem.* **262,** 11897–11900 (1987).
9. W. J. Tang, and A. G. Gilman, *Science* **254,** 1500–1503 (1991).
10. A. D. Federman, B. R. Conklin, K. A. Schrader, R. R. Reed, and H. R. Bourne, *Nature (London)* **356,** 159–161 (1992).
11. J. M. Murphy, S. G. Graber, J. C. Garrison, and G. Szabo, *FASEB J. (Abstr.)* **7,** A1138 (1993).
12. K. D. Wickman, J. A. Iniguez-Lluhl, P. A. Davenport, R. Taussig, G. B. Krapivinsky, M. E. Linder, A. G. Gilman, and D. E. Clapham, *Nature (London)* **368,** 255–257 (1994).
13. M. Camps, C. Hou, D. Sidiropoulos, J. B. Stock, K. H. Jakobs, and P. Gierschik, *Eur. J. Biochem.* **206,** 821–832 (1992).
14. J. L. Boyer, G. L. Waldo, and T. K. Harden, *J. Biol. Chem.* **267,** 25451–25456 (1992).
15. J. L. Boyer, S. G. Graber, G. L. Waldo, T. K. Harden, and J. C. Garrison, *J. Biol. Chem.* **269,** 2814–2819 (1994).

16. K. J. Blumer, and J. Thorner, *Annu. Rev. Phys.* **53**, 37–57 (1991).

17. T. Asano, R. Morishita, T. Matsuda, Y. Fukada, T. Yoshizawa, and K. Kato, *J. Biol. Chem.* **268(27)**, 20512–20519 (1993).

18. H. Tamir, A. B. Fawzi, T. Evans, and J. K. Northup, *Biochemistry* **30**, 3929–3936 (1991).

19. C. J. Schmidt, T. C. Thomas, M. A. Levine, and E. J. Neer, *J. Biol. Chem.* **267**, 13807–13810 (1992).

20. A. N. Pronin, and N. Gautam, *Proc. Natl. Acad. Sci., U.S.A.* **89**, 6220–6224 (1992).

21. S. G. Graber, R. A. Figler, V. K. Kalman-Maltese, J. D. Robishaw, and J. C. Garrison, *J. Biol. Chem.* **267**, 13123–13126 (1992).

22. J. D. Robishaw, V. K. Kalman, and K. L. Proulx, *Biochem. J.* **286**, 677–680 (1992).

23. J. A. Iniguez-Lluhi, M. I. Simon, J. D. Robishaw, and A. G. Gilman, *J. Biol. Chem.* **267**, 23409–23417 (1992).

24. A. Dietrich, M. Meister, K. Spicher, G. Schultz, M. Camps, and P. Gierschik, *FEBS Lett.* **313**, 220–224 (1992).

25. N. Ueda, J. A. Iniguez-Lluhl, E. Lee, A. V. Smrcka, J. D. Robishaw, and A. G. Gilman, *J. Biol. Chem.* **269**, 4388–4395 (1994).

26. L. K. Miller, *Annu. Rev. Microbiol.* **42**, 177–99 (1988).

27. V. A. Luckow, and M. D. Summers, *Bio/Technology* **6**, 47–55 (1988).

28. J. Vialard, M. Lalumiere, T. Vernet, D. Briedis, G. Alkhatib, D. Henning, D. Levin, and C. Richardson, *J. Virol.* **64**, 37–50 (1990).

29. N. R. Webb, and M. D. Summers, *Technique* **2**, 173–188 (1990).

30. V. A. Luckow, and M. D. Summers, *Virology* **167**, 56–71 (1988).

31. B. G. Ooi, C. Rankin, L. K. Miller, P. A. Kitts, M. D. Ayres, and R. D. Possee, *Nucleic Acids Res.* **18**, 5667–5672 (1990).

32. M. D. Summers, and G. E. Smith, *Tex. Agric. Exp. Stn. Bull.* **1555**, 1–56 (1987).

33. D. R. O'Reilly, L. K. Miller, and V. A. Luckow, "Baculovirus Expression Vectors: A Laboratory Manual," pp. 1–347. W. H. Freeman, New York, 1992.

34. B. Gao, A. G. Gilman, and J. D. Robishaw, *Proc. Natl. Acad. Sci. U.S.A.* **84**, 6122–6125 (1987).

35. J. D. Robishaw, V. K. Kalman, C. R. Moomaw, and C. A. Slaughter, *J. Biol. Chem.* **264**, 15758–15761 (1989).

36. P. A. Kitts, and R. D. Possee, *Bio/Technology* **14**, 810–817 (1993).

37. U. K. Laemmli, *Nature (London)* **227**, 680–685 (1970).

38. A. C. Webb, M. K. Bradley, S. A. Phelan, J. Q. Wu, and L. Gehrke, *BioTechniques* **11**, 512–519 (1991).

39. B. Maiorella, D. Inlow, A. Shauger, and D. Harano, *Bio/Technology* **6**, 1406–1410 (1988).

40. D. W. Murhammer, and C. F. Goochee, *BIO/Technology* **6**, 1411–1418 (1988).

41. J. W. Rice, N. B. Rankl, T. M. Gurganus, C. M. Marr, J. B. Barna, M. M. Walters, and D. J. Burns, *BioTechniques* **15**, 1052–1059 (1993).

42. J. K. Northup, P. C. Sternweis, and A. G. Gilman, *J. Biol. Chem.* **258**, 11361–11368 (1983).

43. M. E. Linder, I. Pang, R. J. Duronio, J. I. Gordon, P. Sternweis, and A. G. Gilman, *J. Biol. Chem.* **266,** 4654–4659 (1991).
44. S. G. Graber, R. A. Figler, and J. C. Garrison, *J. Biol. Chem.* **267,** 1271–1278 (1992).
45. S. G. Graber, R. A. Figler, and J. C. Garrison, *in* "Methods in Enzymology" (R. Iyehgar, ed.), Vol. 237, pp. 212–226. Academic Press, San Diego, 1994.
46. I. Pang, and P. C. Sternweis, *Proc. Natl. Acad. Sci. U.S.A.* **86,** 7814–7818 (1989).
47. J. Heukeshoven, and R. Dernick, *Electrophoresis* **9,** 28–33 (1988).
48. P. J. Casey, M. P. Graziano, and A. G. Gilman, *Biochemistry* **28,** 611–616 (1989).

[16] Analysis of G Protein γ Subunits Using Baculovirus Expression System: Requirement for Posttranslational Processing

Janet D. Robishaw

Introduction

The guanine nucleotide-binding (G) proteins play an obligatory role in the signal transduction process by coupling a large number of receptors to a smaller number of effectors. Based on the characterization of G proteins reconstituted with the appropriate receptors and effectors in phospholipid vesicles, the following model has been proposed to describe their role in the signal transduction process (1). Each G protein exists as a complex of $\alpha\beta\gamma$ subunits. In response to a chemical or physical signal acting on the extracellular face of the receptor, activation of the G protein occurs when the α subunit exchanges bound GDP for GTP and dissociates from the tightly associated complex of $\beta\gamma$ subunits. Both the GTP-bound α subunit and the $\beta\gamma$ subunit complex are capable of regulating, either independently or interdependently, a number of downstream effectors, including adenylylcyclases, phospholipase C, phospholipase A_2, various ion channels, and cGMP phosphodiesterase (2–4). Deactivation of the G protein occurs when the α subunit hydrolyzes the bound GTP and reassociates with the $\beta\gamma$ subunit complex.

In mammals, cDNAs have been isolated that encode four distinct β subunits (5) and six distinct γ subunits (6), allowing the formation of several possible $\beta\gamma$ subunit complexes (7, 8). The growing awareness of the multiple roles of the $\beta\gamma$ subunit complexes in the signal transduction process has led to speculation that their molecular heterogeneity would be reflected in their interactions with different α subunits, receptors, and effectors. To address this question, we have expressed and purified 10 different $\beta\gamma$ subunit complexes from the baculovirus system and characterized their interactions with α subunits, adenylylcyclases, and phospholipase C (9, 10). However, because complexes of $\beta\gamma$ subunits can be separated only with strong chaotropic agents, such as urea and sodium dodecyl sulfate (SDS), it has not been possible to determine the individual contributions of the β and γ subunits in terms of their interactions with α subunits, receptors, and effectors. To circumvent this problem, we have recently expressed the β subunits and the γ subunits individually in the baculovirus system (9, 10, 11). We have examined posttranslational processing of the γ subunits in the baculovirus system, revealing both similarities and differences between insect and mammalian cells (12, 13). Finally, we have demonstrated the ability of the appropriately processed γ subunits alone, but not the β subunits alone, to interact with the α subunits (14). Thus, use of the baculovirus system has led to the first demonstration of the individual contribution of the γ subunits in terms of their interactions with the α subunits. In a similar fashion, it is expected that use of the baculovirus system will define the contribution of the γ subunits in terms of their interactions with receptors, receptor kinases, and effectors.

Components of Baculovirus System

The baculovirus system originally developed by Smith and colleagues (15) is useful for the expression of a wide variety of genes, including genes that encode multisubunit proteins. This system is based on the infection of the *Spodoptera frugiperda* (Sf9) insect cells with the *Autographa californica* nuclear polyhedrosis (AcNP) virus. The virus directs the expression of the polyhedrin gene, accounting for 50–75% of the protein expression in infected cells. Because expression of the polyhedrin gene is not essential, a recombinant AcNP virus can be constructed by replacing the polyhedrin gene with the "cDNA of interest," immediately downstream of the polyhedrin promoter. In addition to high-level expression, the baculovirus system has the added advantage of carrying out posttranslational modifications, such as prenylation and carboxymethylation (16, 17), which are required for the biologic activity and membrane localization of the G protein γ subunits (9, 14, 18–20).

Transfer Vector and Viral DNA

The choice of transfer vectors is extensive, although vectors pVL1392, pVL1393, and pVL941 are used most widely (21) (PharMingen, San Diego, CA). All transfer vectors contain pUC8 sequences coding for the origin of replication and ampicillin resistance, and AcNP viral sequences for the polyhedrin promoter, polyadenylation signals, and 5' and 3' untranslated regions necessary for homologous recombination with the wild-type virus. In addition, the pVL1392 and pVL1393 transfer vectors contain a large number of restriction enzyme sites that have proved to be particularly useful for the directed insertion of the cDNA of interest. From earlier results in mammalian and reticulolysate systems (22), it is known that the cDNAs for several γ subunits possess 5' and/or 3' untranslated sequences that significantly inhibit their expression in these systems. Therefore, as a precaution, we routinely remove the 5' and 3' untranslated sequences before inserting the coding regions of these cDNAs into the transfer vector. This can be readily accomplished by polymerase chain reaction (PCR) amplification of the coding regions of these cDNAs, using the appropriate oligonucleotide primers made to the 5' and 3' ends of the coding regions of these γ subunits. If desired, different restriction enzyme sites can be added to ends of the oligonucleotide primers to allow directed insertion of the coding region into the transfer vector.

The production of recombinant virus is accomplished by cotransfection of Sf9 cells with the transfer vector containing the cDNA of interest and the wild-type viral DNA. On homologous recombination, the polyhedrin gene in the wild-type viral DNA is replaced with the cDNA for one of the γ subunits, resulting in the production of the recombinant viral DNA. Until recently, the identification of plaques containing the recombinant viral DNA (<1%), from a high background of plaques containing wild-type viral DNA (>99%), required several rounds of plaque purification, which was a very laborious process. Recently, however, two modifications of the wild-type viral DNA have been shown to greatly speed this process. The first modification involves the use of a linearized form of viral DNA for transfection (23) (Invitrogen, San Diego, CA). Because this linearized form of viral DNA cannot replicate in the absence of recombination, the high background of plaques containing wild-type viral DNA is largely eliminated. The second modification involves the use of a mutationally modified type of viral DNA (PharMingen) that cannot replicate in the absence of recombination. Both the linearized and mutationally modified forms of viral DNA have been successfully used for the production of recombinant viruses encoding the γ subunits (9, 10, 12, 13).

Growth and Transfection of Sf9 Cells

General procedures for growth of Sf9 cells are described elsewhere (24). Briefly, *S. frugiperda* (Sf9) insect cells are grown in monolayer in TNM-FH medium (Grace's medium supplemented with 3.33 g/liter yeastolate and 3.33 g/liter of lactalbumin hydrolyzate) containing 10% fetal bovine serum (Gibco, Gaithersburg, MD). It is important to screen lots of fetal bovine serum for possible toxicity to Sf9 cells prior to use. Likewise, it is important to culture Sf9 cells in an incubator equipped with a cooling coil (Forma, VWR Scientific, Bridgeport, NJ) to maintain a temperature of 27°C. Under optimal growth conditions, Sf9 cells should double every 18 to 20 hr. For transfection, Sf9 cells are seeded at 3×10^6 cells per 60-mm dish. A procedure for calcium phosphate transfection of Sf9 cells is described elsewhere (24). Alternatively, a kit for transfection of Sf9 cells is available commercially (PharMingen). After transfection, Sf9 cells are placed in 3 ml of TNM-FH medium containing 10% (v/v) fetal bovine serum and are cultured at 27°C for 4 days.

Identification and Amplification of Recombinant Virus

After 4 days, the supernatant from transfected cells is cleared of cellular debris by centrifugation and is diluted in Grace's medium. Monolayers of Sf9 cells (2×10^6 cells per 60-mm dish) are infected with diluted virus ($1 : 10^3$ to $1 : 10^6$) for 1 hr, with the aim of producing monolayers with <100 plaques. After aspirating the diluted virus, Sf9 cells are overlayed with 1% (v/v) agarose and incubated at 27°C until plaques become visible (5 to 6 days). The plaques can be used for recombinant virus identification or for determination of the recombinant virus titer. If the mutationally modified form of viral DNA (PharMingen) is used for transfection, >99% of the plaques are expected to contain recombinant virus capable of expressing the cDNA of interest. Nevertheless, it is suggested that 10 well-separated plaques be selected for identification by PCR and immunoblot analysis, because different plaques may vary widely in terms of viral replication and/or protein expression. In order to amplify the recombinant virus, an agarose plug from a single plaque is retrieved with a sterile pipette tip and incubated in 500 μl of TNM-FH medium at 4°C overnight. A 100-μl aliquot of this medium is used to infect Sf9 cells [$(6-8) \times 10^6$ cells per 100-mm dish]. After 3–4 days, the supernatant from infected cells is collected by centrifugation and titered as described above. Two rounds of such amplification routinely yield a high-titer stock of each recombinant virus at 1×10^8 plaque-forming units (pfu)/ml. For short-term use (several months to 1 year), a high-titer stock of each recombinant virus is stored at 4°C. For longer term use, the viral DNA is isolated from each recombinant virus and stored at −80°C. Then, as necessary, the viral DNA is thawed and transfected into Sf9 cells to produce the recombinant virus. Procedures for isolation of viral DNA have been described (24).

Expression of γ Subunits

In contrast to mammalian systems (19), the baculovirus system allows expression of the various γ subunits without coincident expression of β subunits. As shown in Fig. 1, Sf9 cells infected with recombinant viruses for the known γ subunits express the corresponding proteins of the correct size (5- to 7-kDa range), which are readily detectable by the different γ-subunit-specific

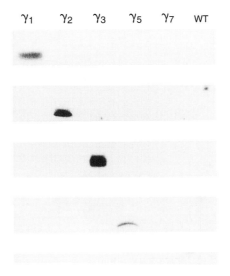

Fig. 1   Baculovirus expression of the various γ subunits. Cholate extracts of particulate fractions from Sf9 cells infected with recombinant viruses encoding the indicated γ subunits or the wild-type virus (100 μg) at a multiplicity of infection of 2 are subjected to SDS-PAGE on a 15% gel followed by immunoblot analysis with the various antibodies. Antibodies used for blotting were the anti-γ₁ antibody, A4; the anti-γ₂ antibody, A75; the anti-γ₃ antibody, B53; the anti-γ₅ antibody, D9; and the anti-γ₇ antibody, A67, as described previously (25); WT, wild type.

antibodies (25). Furthermore, proteins of similar size are not detected by these antibodies in Sf9 cells infected with the wild-type virus. Taken together, these results demonstrate the utility of the baculovirus system for the expression of the γ subunits.

For efficient expression of the γ subunits in the baculovirus system, the Sf9 cells must be appropriately receptive to infection. In this regard, the Sf9 cells must be in log-phase growth and must be >98% viable. For expression on a large scale, the Sf9 cells are cultured in suspension, rather than in monolayer, in spinner flasks. Cells are seeded in spinner flasks at $(0.5–1.0) \times 10^6$ cells/ml and are grown to $(2–2.5) \times 10^6$ cells/ml, with the volumes in spinner flasks determined by the need for aeration, as described by the manufacturer (Bellco Glass Inc., Vineland, NJ). Also, for large-scale expression, the cells are grown in a low-serum medium to contain costs and to facilitate the subsequent purification of the expressed proteins. For this

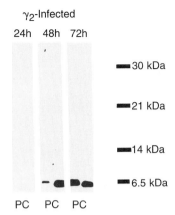

γ_2-Infected

24h   48h   72h

━30 kDa

━21 kDa

━14 kDa

━6.5 kDa

PC   PC   PC

FIG. 2   Time course of expression of the γ subunits. A total of 2.5×10^6 cells were infected with the recombinant virus encoding the γ_2 subunit at a multiplicity of infection of 2. After harvesting at 24, 48, or 72 hr after infection, the cells are fractionated by centrifugation at 100,000 g for 60 min to yield the cytosolic (C) and particulate (P) fractions. To monitor the time course of expression of the γ_2 subunit, equivalent amounts of the C and P fractions are resolved by SDS-PAGE on a 15% gel, transferred to nitrocellulose, and incubated with the anti-γ_2 antibody (A75) at a 1:200 dilution, using a ^{125}I-labeled GAR secondary antibody for detection (25).

purpose, we use IPL41 medium supplemented with 2× lipid concentrate and 1% fetal bovine serum (Gibco, Gaithersburg, MD).

Time Course

The polyhedrin promoter is expressed late in the infection, allowing the protein of interest to be expressed to a high level, even if the protein turns out to be toxic to these cells. Using this promoter, expression levels have ranged from 1 to 10 mg/liter, depending on the characteristics of the γ subunit being expressed. These expression levels are fairly typical of membrane proteins that require extensive posttranslational processing, such as the γ subunits. A representative time course is shown in Fig. 2. Sf9 cells infected with recombinant virus encoding the γ_2 subunit express a protein of the correct size (6-kDa range) for the γ_2 subunit, beginning at 48 hr after infection, similar to the time course reported for polyhedrin expression after infection with wild-type virus. The greatest amount of the γ_2 subunit is found in the cytosolic fraction at 48 hr after infection, whereas roughly equal amounts of this protein are found in the cytosolic and particulate fractions at 72 hr after infection. The marked increase in the amount of the γ_2 subunit in the particulate fraction at 72 hr after infection results from posttranslational processing of this protein (12). Thus, to produce the highest amount of γ subunit that has undergone processing, cells are harvested at 72 hr after infection.

Subcellular Distribution

A common problem encountered when recombinant proteins are expressed in Sf9 insect cells is that the protein of interest does not fold correctly, resulting in the formation of insoluble aggregates of protein that lack biologic activity. As a quick test for this, Sf9 cells infected with recombinant virus are fractionated into cytosolic and particulate fractions by centrifugation at 100,000 g for 60 min. Subsequently, the particulate fraction is solubilized with 1.0% (w/v) cholate. This concentration of detergent has been shown to effectively solubilize G protein $\beta\gamma$ subunits from membranes. Further centrifugation at 100,000 g for 60 min releases the soluble protein into the cholate-soluble particulate fraction and pellets the insoluble protein into the cholate-insoluble particulate fraction. The protein contents of the cytosolic, soluble particulate, and insoluble particulate fractions are resolved by SDS-PAGE followed by immunoblot analysis with γ-subunit-specific antibodies. A representative subcellular distribution is shown in Fig. 3. When cells infected with the appropriate virus are harvested at 72 hr after infection, little of the γ_2 subunit is present in the insoluble particulate fraction, whereas most of the γ_3 subunit is present in the insoluble particulate fraction. Thus, the amount of a particular γ subunit that appears in the insoluble particulate fraction varies markedly, depending on which subtype is being expressed. The reason for this is not clear, particularly in light of the high degree of homology (>70% identity) between the primary structures of the γ_2 and γ_3 subunits. Therefore, the inability to predict whether a particular γ subunit is likely to be expressed in the insoluble particulate fraction from its primary structure requires the determination of the subcellular distribution for each subtype.

Multiplicity of Infection

The importance of knowing whether a particular γ subunit is likely to be expressed in the insoluble particulate fraction will have a profound effect on the expression protocol to be used. Thus, if little of the protein is detected in the insoluble particulate fraction, as in the case of the γ_2 subunit, then increasing the multiplicity of infection is likely to have a beneficial effect by

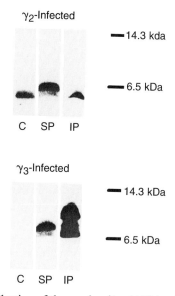

FIG. 3   Subcellular distribution of the γ subunits. At 72 hr after infection with viruses encoding the γ_2 and γ_3 subunits, Sf9 cells are fractionated to yield the cytosolic (C), cholate-soluble particulate (SP), and cholate-insoluble particulate (IP) fractions. The same percentage (15%) of each fraction is resolved by SDS-PAGE on a 15% polyacrylamide gel and transferred to nitrocellulose. The nitrocellulose blots are incubated with either the γ_2 antibody (B17) or the γ_3 antibody (B53), using a ^{125}I-labeled GAR secondary antibody for detection (25).

increasing the level of expression of functional protein in the soluble particulate fraction. On the other hand, if most of the protein is detected in the insoluble particulate fraction, as in the case of the γ_3 subunit, then raising the multiplicity of infection is likely to have detrimental effect. In this regard, raising the multiplicity of infection from 2 to 10 results in the appearance of most of the γ_3 subunit in the insoluble particulate fraction, which, in turn, proves to be toxic to cells (22). Needless to say, the multiplicity of infection resulting in the optimal level of expression for each γ subunit, either alone or in combination with β subunit, must be determined empirically.

Factors Limiting Expression and Possible Improvements

At least two factors appear to limit expression of the γ subunits in the baculovirus system. The first limitation is the tendency of a particular γ subunit, such as the γ_3 subunit, to aggregate when overexpressed. This is dependent on the characteristics of the protein being expressed, and therefore is not likely to be easily overcome. The second limitation is the inability of the Sf9 cells to process all of the γ subunit being expressed. This is not due to lack of sufficient precursor, because addition of higher concentrations of mevalonate did not result in more complete processing (22), but is more likely due to lack of sufficient enzymes for processing. This lack of processing enzymes is likely to be due to a combination of factors. First, the Sf9 cells may be compromised in their ability to process proteins by the use of the polyhedrin promoter, which directs expression of the γ subunits very late in infection, as the virus is lysing the cells. Thus, it is possible that use of a different promoter, which directs the expression of the γ subunits earlier in infection, may prove to be beneficial. Consistent with this idea, Sridhar and colleagues (26) have found that human chorionic gonadotropin is glyco-sylated to a greater extent in Sf9 cells when the basic promoter, which is expressed earlier in infection, is used in place of the polyhedrin promoter. Whether this is true for proteins that undergo processing by prenylation and carboxymethylation, such as the γ subunits, is being examined. Alterna-tively, it is possible that coexpression of a virus encoding the processing enzyme along with a virus encoding the γ subunit may prove beneficial, as discussed in greater detail in the following section on posttranslational pro-cessing.

Posttranslational Processing of γ Subunits

The membrane localization and biologic activity of the G protein βγ subunits are dependent on the posttranslational attachment of a prenyl group to a cysteine residue four removed from the carboxy terminus of the γ subunits. This attachment is catalyzed by a family of prenyltransferases (27), which recognize a tetrapeptide CAAX sequence, where C is cysteine, A is aliphatic, and X is leucine, methionine, or serine, at the carboxy terminus of all of the known γ subunits. The prenylation of the γ subunits is followed by proteolytic cleavage of the terminal -AAX residues and by carboxymethylation of the prenylated cysteine residue. This series of posttranslational reactions is re-quired for the γ subunits to localize to the plasma membrane (18, 19) and to associate with the α subunits of the G proteins (9, 14). Thus, in addition to showing expression of the γ subunits in the baculovirus system, it is important to determine whether the γ subunits undergo this series of reactions appropriately in the baculovirus system. Surprisingly, despite the widespread use of the baculovirus system for expression of G protein β and γ subunits, studies to examine whether the γ subunits undergo this series of posttransla-tional reactions in the baculovirus system have been limited to the γ$_2$ subunit

(11, 12). It is clear that more extensive studies are needed, as highlighted by our recent studies showing approximately 50% of the γ_1 subunit is modified with the wrong prenyl group in the baculovirus system (13).

Analysis of Prenylation

To study prenylation, Sf9 insect cells infected with virus encoding one of the γ subunits are incubated with [^{3}H]mevalonate (200 μCi/dish of 2×10^6 cells) (New England Nuclear, Boston, MA) for varying periods of time between 48 and 72 hr after infection. This time period represents the maximal period of prenylation, as indicated by the marked shift of the γ subunit from the cytosolic to the particulate fraction (see Fig. 2). To minimize the dilution of the [^{3}H]mevalonate, infected cells are preincubated in serum-free Grace's medium for 2 hr prior to incubation with [^{3}H]mevalonate. After being transported into cells, the [^{3}H]mevalonate is converted into one of two [^{3}H]prenyl intermediates, a portion of which is then attached to proteins in a reaction catalyzed by the appropriate prenyltransferase. To monitor the attachment of [^{3}H]prenyl groups, the proteins in the lysate, cytosolic, and particulate fractions from infected cells are resolved by SDS-PAGE on a 15% polyacrylamide gel, transferred to nitrocellulose by a high-temperature transfer procedure (28), and incubated with AMPLIFY solution (New England Nuclear) for enhancement of fluorography. If desired, following fluorography, the nitrocellulose can be blotted with the appropriate antibodies for autoradiography. A listing of antibodies specific for each of the known γ subunits and blotting procedures is described in detail elsewhere (25).

A representative [^{3}H]mevalonate labeling experiment is depicted in Fig. 4. As shown in Fig. 4A, no incorporation of [^{3}H]mevalonate-derived label into proteins is observed in the lysate fraction from cells infected with wild-type virus. Presumably, this reflects the labeling period late in the infective cycle, when synthesis of virally derived proteins has shut down synthesis of host cell proteins that might be prenylated and therefore might be expected to incorporate [^{3}H]mevalonate-derived label. In contrast, significant incorporation of [^{3}H]mevalonate-derived label into a protein of the appropriate size for the γ_2 subunit is observed in cells infected with the γ_2 virus. This indicates that at least a portion of the γ_2 subunit expressed in Sf9 cells is modified by prenylation. Based on a densitometric scan of the ^{3}H-labeled protein versus the immunodetectable protein (data not shown), we estimate that approximately 40% of the γ_2 subunit is prenylated in Sf9 cells by 72 hr after infection (12). The inclusion of higher concentrations of mevalonate in the culture medium does not increase the proportion of the γ_2 subunit that is prenylated

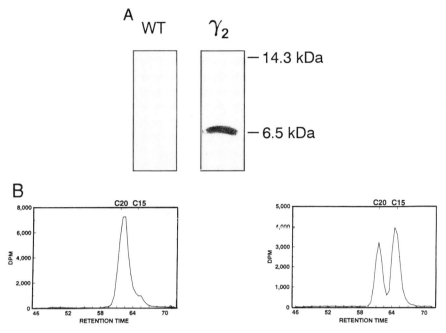

FIG. 4   Posttranslational processing of the γ subunits by prenylation. Sf9 cells (5 × 10⁶) are infected with wild-type (WT) virus, γ₂ virus, or γ₁ virus at a multiplicity of infection of 2. (A) To monitor the incorporation of [³H]mevalonate-label, WT-infected or γ₂-infected cells are labeled from 67 to 72 hr after infection with medium containing 400 μCi of [³H]mevalonate. Approximately 15% of the whole-cell lysates from WT-infected or γ₂-infected cells are resolved by SDS-PAGE on a 15% gel and transferred to a nitrocellulose blot for fluorography. (B) To determine the carbon chain length of the prenyl group(s) attached to the γ subunits, the [³H]mevalonate-derived label is released from the γ₂ subunit (left panel) or the γ₁ subunit (right panel) by Raney nickel-catalyzed cleavage and subjected to gel filtration chromatography. The retention times of the C_{15} (farnesane) and the C_{20} (phytane) standards are shown above the chromatogram.

(data not shown), indicating that the availability of this precursor is not rate limiting. Interestingly, however, when the γ₂ subunit is coexpressed with the β subunit, almost 90% of the γ₂ subunit is prenylated in Sf9 cells by 72 hr after infection (12). Although the basis for this effect of coexpression of the β subunit is not known, the β subunit has been shown to be capable of recognizing and interacting with the unprenylated γ subunit (29).

Identification of Prenyl Group

Two types of prenyl groups have been identified: a 15-carbon farnesyl and a 20-carbon geranylgeranyl group. *In vivo* studies (30, 31) and studies in the reticulolysate system (32) have identified the prenyl groups attached to the γ_1 and γ_2 subunits as farnesyl and geranylgeranyl moieties, respectively. To identify the prenyl groups attached to the γ_1 and γ_2 subunits expressed in the baculovirus system, Sf9 cells infected with the corresponding recombinant viruses are labeled with [^{3}H]mevalonate as described above. The ^{3}H-labeled γ subunits are electroeluted from SDS-polyacrylamide gels and incubated with Raney nickel (Sigma, St. Louis, MO) to disrupt the prenyl–thioether bonds (33). The ^{3}H-labeled groups released from the proteins are then extracted into pentane, reduced over platinum oxide, and subjected to high-performance gel-permeation chromatography on two tandem Phenogel columns (column size, 300 × 7.5 mm; pore size, 5 and 10 μm; Phenomenex, Torrance, CA), as described previously (33, 34). Retention times of the ^{3}H-labeled groups are determined in relation to groups of defined chain length, namely, 2,6,10-trimethyldodecane (15-carbon farnesane) and 2,6,10,14-tetramethylhexadecane (20-carbon phytane) (Wiley Organics, Coshocton, OH), which are premixed with the samples prior to chromatography. As shown in Fig. 4B, the γ_2 subunit is modified with a 20-carbon group, indicating that this protein is modified with a geranylgeranyl moiety in Sf9 insect cells in an analogous fashion to mammalian cells. In contrast, the γ_1 subunit is modified with a mixture of 15-carbon and 20-carbon groups, indicating that this protein is modified with both farnesyl and geranylgeranyl moieties in Sf9 insect cells (13). Because the γ_1 subunit is modified exclusively with a farnesyl group in mammalian cells (31), this represents a potential problem in terms of producing γ_1 subunit that is appropriately modified in the baculovirus system. Fortunately, this problem can be largely overcome by coinfection of Sf9 cells with the recombinant virus encoding a mammalian farnesyltransferase along with the virus encoding the γ_1 subunit (13). Nevertheless, these results underscore the importance of assessing both the efficiency and the accuracy of posttranslational processing in the baculovirus system, which have been frequently overlooked in the past.

Analysis of Carboxymethylation

In the case of the γ subunit, prenylation is followed by proteolytic cleavage of the last three amino acids and by methylation of the resulting α-carboxyl group of the cysteine residue (35). To study carboxymethylation, Sf9 cells are labeled with [^{3}H]methylmethionine (100 μCi/dish of 2 × 10^6 cells) (Amer-

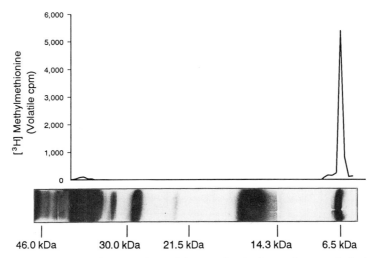

FIG. 5   Posttranslational processing of the γ subunits by carboxymethylation. Sf9 cells (2.5×10^6) are labeled with 100 μCi [³H]methylmethionine from 67 to 72 hr after infection with the γ_2 virus at a multiplicity of infection of 2. Approximately 33% of the soluble particulate fraction is resolved by SDS-PAGE on a 15% gel. To monitor incorporation of [³H]methylmethionine-derived label into proteins, the dried gel is subjected to fluorography for 2 days. Subsequently, 2-mm slices of the gel in the 6- and 30-kDa range are assayed for release of volatile radioactivity following base hydrolysis, which is indicative of carboxymethylation.

sham, Arlington Heights, IL) from 67 to 72 hr after infection with the γ virus. The soluble particulate fraction obtained from infected cells is resolved by SDS-PAGE on a 15% polyacrylamide gel. The gel is stained with Coomassie blue, incubated with AMPLIFY solution, cut into 2-mm slices, dried, and then subjected to fluorography. To determine whether the ³H-labeled proteins are carboxylmethylated, the gel slices are assayed for the release of volatile [³H]methanol in response to alkaline hydrolysis (36).

A representative [³H]methylmethionine-labeling experiment is shown in Fig. 5. Several proteins showed incorporation of the [³H]methylmethionine-derived label, including prominently labeled proteins in the 30-, 14-, and 6-kDa range. The protein in the 6-kDa range is identified as the γ_2 subunit based on its specific appearance in cells infected with γ_2 virus (12). Although no volatile radioactivity is released from the prominent ³H-labeled proteins in the 30-kDa range, a large peak of volatile radioactivity is readily released from the 6-kDa protein following alkaline hydrolysis. Because alkaline lability is characteristic of methylated α-carboxyl groups, the release of volatile radioactivity from the 6-kDa protein indicates that the majority of the γ_2

subunit is carboxyl methylated in Sf9 insect cells in an analogous fashion to mammalian cells. In this regard, it is important to determine the ratio of carboxymethylation to prenylation for each of the known γ subunits. By modifying the expression protocol (i.e., multiplicity of infection) so that the level of expression for each of the known γ subunits approximates the level of processing, it is possible to achieve virtually complete carboxymethylation of the prenylated protein. This is important if the goal is to produce a homogeneous preparation of appropriately modified γ subunits for biochemical studies.

Purification of γ Subunits

As a result of incomplete processing, a mixture of unprenylated and prenylated γ subunits is produced in the baculovirus system, requiring the development of a procedure to resolve these forms. With the development of such a procedure, this affords the opportunity to compare and contrast the properties of the unprenylated and prenylated forms of the γ subunit in a variety of assays. In this section we describe the purification of the prenylated and carboxymethylated form of the γ subunits on an α_o affinity resin (14).

Preparation of α_o Affinity Resin

The α_o subunit (α subunit of G_o) is purified from bovine brain, as described previously by Sternweis and Robishaw (37). Approximately 40 mg of the purified α_o subunit is coupled to 20 ml of ω-aminobutyl agarose (Sigma, St. Louis, MO), as described previously by Mumby *et al.* (30). The α_o affinity resin is packed into a Pharmacia XK 26/20 column fitted with flow adapters (Pharmacia, Upsala, Sweden), and then equilibrated with buffer A [20 mM HEPES (pH 8.0), 400 mM NaCl, 1 mM EDTA, 2 mM DTT, and 5 μM GDP] containing 0.1 mM phenylmethylsulfonyl fluoride (PMSF), 1 mM benzamidine, and 0.5% (v/v) polyoxyethylene 10-lauryl ether (LPX) (Sigma Corp.).

Purification of γ Subunits on α_o Affinity Resin

Approximately 68 hr after infection, the Sf9 cells are collected from 100-ml spinner flasks by centrifugation (3000 g for 10 min at 4°C), rinsed with phosphate-buffered saline, and resuspended in a solution containing 20 mM HEPES (pH 8.0), 2 mM MgCl$_2$, 1 mM EDTA, 2 mM DTT, 0.1 mM PMSF, and 1 mM benzamidine. The cells are lysed by passage through a 25-gauge needle and solubilized by addition of cholate to a final concentration of 1%. The solubilized portions from the whole-cell lysates are collected by centrifugation (100,000 g for 30 min). After diluting 10-fold with buffer A, the solubilized proteins are applied to the α_0 affinity column at a flow rate of 0.5 ml/min. The flowthrough is collected the first time and reapplied to the α_0 affinity column. The α_0 affinity column is washed successively with 2 column volumes of buffer A containing 0.5% LPX at a flow rate of 0.9 ml/min, 2 column volumes of buffer A containing 0.275% LPX at a flow rate of 2.5 ml/min, and 20 column volumes of buffer A containing 0.05% LPX at a flow rate of 2.5 ml/min. The γ subunit is eluted from the α_0 affinity column with buffer B [20 mM Na-HEPES (pH 8.0), 400 mM NaCl, 2 mM DTT, 5 μM GDP, 30 μM AlCl$_3$, 20 mM MgCl$_2$, 10 mM NaF] containing 0.05% LPX at a flow rate of 0.25 ml/min. The elution fractions containing the appropriately modified γ subunit are pooled and concentrated 10-fold, using an Amicon ultrafiltration device with a YM3 membrane (Amicon, Beverly, MA).

Additional Purification of γ Subunit on S-Sepharose Resin

To purify to homogeneity, the concentrated γ subunit pool obtained from the α_0 affinity column is loaded onto an S-Sepharose column (Sigma). After diluting sevenfold with buffer C (20 mM Na-HEPES, pH 6.8, 100 mM NaCl, 2 mM DTT, 1 mM EDTA) containing 0.05% LPX, the protein pool is loaded at a flow rate of 0.5 ml/min onto an S-Sepharose column preequilibrated with buffer C containing 0.05% LPX. The column is then washed with 5 column volumes of buffer C containing 0.05% LPX, resulting in the appearance of contaminating proteins in the flowthrough and wash fractions. The γ subunit is eluted with buffer A containing 0.05% LPX. The elution fractions containing the γ subunit are pooled and concentrated 10-fold.

Analysis of Individual Contribution of γ Subunits

The G proteins appear to function as a heterodimer, consisting of an α subunit and a $\beta\gamma$ subunit complex. Because most of the $\beta\gamma$ subunit complexes can be separated only under denaturing conditions, studies aimed at determining the individual contributions of the β and γ subunits have not been possible. To circumvent this problem, we have used the baculovirus system to direct the individual expression of the β and γ subunits. Interestingly, we have obtained evidence for a direct interaction between the γ and α subunits, which is dependent on posttranslational processing of γ (14). This interaction can be analyzed by limited tryptic proteolysis, as discussed in more detail below. This demonstration of a direct interaction between the γ and α subunits is particularly intriguing in view of the increasingly large number of structurally distinct α, β, and γ subunits, which, in turn, raises important questions regarding the assembly of these subunits into functionally distinct G proteins. We speculate that a direct interaction between the γ and α subunits, which exhibit the greatest structural diversity, may provide the basis for the selective assembly of these subunits into functionally distinct G proteins.

Analysis of α–γ Interactions by Tryptic Proteolysis

The interaction between the α and γ subunits can be demonstrated by the ability of the purified γ subunit to protect the purified α_o subunit from tryptic cleavage. As depicted in Fig. 6A, the α_o subunit is initially cleaved by trypsin at Lys-21, giving rise to a 37-kDa fragment that can be detected with an antibody made to a peptide based on amino acid residues 22–34 of this protein (A10). The 37-kDa fragment is subsequently cleaved at a second site, giving rise to 24- and 14-kDa fragments (37, 38). In contrast, the α_o subunit, when associated with the $\beta\gamma$ subunit complex, is not cleaved by trypsin at either site. Because the $\beta\gamma$ subunit complex cannot be dissociated without the use of denaturants, it has not been possible to define the individual contributions of the β and γ subunits in terms of their abilities to interact, and hence, prevent tryptic cleavage of the α_o subunit. Using the tryptic proteolysis assay, we show the γ_2 subunit alone prevents tryptic cleavage of the α_o subunit. Briefly, the α_o and γ_2 subunits are purified, as described above. Prior to starting the assay, the α_o subunit is mixed with the γ_2 subunit in a solution containing 15 mM Na-HEPES, pH 8, 250 mM NaCl, 0.6 mM EDTA, 0.6 mM DTT, 5 mM MgCl$_2$, and 0.3% LPX for 15 min at 30°C. The assay is initiated by addition of L-1-tosylamide-2-phenylethylchloromethyl ketone (TPCK)-treated trypsin (Sigma, St. Louis, MO) at a constant protein : trypsin ratio of 5 : 1 (w/w). After incubation in a 30°C water bath for 60 min, the assay is terminated by addition of SDS sample buffer containing 10 mM benzamidine and boiling for 5 min. The tryptic cleavage fragments are resolved by SDS-PAGE and are visualized by immunoblotting. As shown in Fig. 6B, in the absence of added γ_2 subunit, the α_o subunit is cleaved to a 24-kDa fragment (lane 1). In contrast, when γ_2 subunit is added to the α_o subunit at a molar ratio of 2 : 1, the α_o subunit is not cleaved to any significant extent (lane 2). When the molar ratio is reduced to 1.6 : 1 and 0.8 : 1, the α_o subunit is increasingly cleaved to a 37-kDa fragment (lanes 3 and 4, Fig. 6B), and when the molar ratio is further reduced to 0.4 : 1 and 0.2 : 1, the α_o

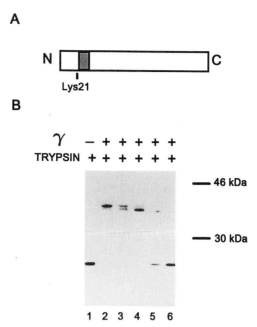

FIG. 6   Ability of processed γ subunit to prevent tryptic cleavage of the α subunit of G_o. (A) Tryptic cleavage map of $G_{o\alpha}$ showing the initial site of cleavage at Lys-21 to produce the 37-kDa fragment. The hatched area indicates the recognition site for the antibody (A10) used to detect the 37- and 24-kDa fragments. (B) Tryptic cleavage assay showing the ability of the γ subunit to prevent tryptic cleavage of $G_{o\alpha}$. Purified $G_{o\alpha}$ (5 μg) is mixed with varying amounts of the purified γ_2 subunit for 15 min at 30°C, prior to the addition of trypsin at a constant protein : trypsin ratio of 5 : 1 for 60 min at 30°C. The tryptic cleavage fragments of $G_{o\alpha}$ are resolved by SDS-PAGE on an 11% gel, and are detected following transfer to nitrocellulose by immunoblotting with antibody (A10) at a final dilution of 1 : 1000. From left to right: lane 1, $G_{o\alpha}$ alone; lanes 2–6, $G_{o\alpha}$ in combination with the γ_2 subunit at molar ratios of 1 : 2, 1 : 1.6, 1 : 0.8, 1 : 0.4, and 1 : 0.2.

subunit is increasingly cleaved to a 24-kDa fragment (lanes 5 and 6, Fig. 6B). Thus, the presence of the γ_2 subunit prevents tryptic cleavage of the α_o subunit in a stoichiometric fashion.

Acknowledgments

Work from the author's laboratory was supported by National Institutes of Health Grants GM39867 and HL49278, and an American Heart Association Established Investigatorship. The author wishes to thank Vivian Kalman for spearheading the expression of γ subunits in the baculovirus system and Eric Balcueva, Robert Erdman, Roman Ginnan, and Dr. Mohammed Rahmatullah for making important contributions to this project.

References

1. A. G. Gilman, *Annu. Rev. Biochem.* **56,** 615 (1987).
2. B. R. Conklin and H. R. Bourne, *Cell* **73,** 631 (1993).
3. L. Birnbaumer, *Cell* **71,** 1069 (1992).
4. W. J. Tang and A. G. Gilman, *Cell* **70,** 869 (1992).
5. E. Von Weizsacker, M. P. Strathmann, and M. I. Simon, *Biochem. Biophys. Res. Commun.* **183,** 350 (1992).
6. J. J. Cali, E. A. Balcueva, I. Rybalkin, and J. D. Robishaw, *J. Biol. Chem.* **267,** 24023 (1992).
7. A. N. Pronin and N. Gautam, *Proc. Natl. Acad. Sci. U.S.A.* **89,** 6220 (1992).
8. C. J. Schmidt, T. C. Thomas, M. A. Levine, and E. J. Neer, *J. Biol. Chem.* **267,** 13807 (1992).
9. J. A. Inigeuz-Lluhi, M. I. Simon, J. D. Robishaw, and A. G. Gilman, *J. Biol. Chem.* **267,** 23409 (1992).
10. N. Ueda, J. A. Iniguez-Lluhi, E. Lee, A. V. Smrcka, J. D. Robishaw, and A. G. Gilman, *J. Biol. Chem.* **269,** 4388 (1994).
11. S. Graber, R. Figler, V. Kalman-Maltese, J. D. Robishaw, and J. C. Garrison, *J. Biol. Chem.* **267,** 13123 (1992).
12. J. D. Robishaw, V. K. Kalman, and K. L. Proulx, *Biochem. J.* **286,** 677 (1992).
13. J. D. Robishaw, V. K. Kalman, R. E. Erdman, and W. A. Maltese, *J. Biol. Chem.* **270,** 14835 (1995).
14. M. Rahmatullah and J. D. Robishaw, *J. Biol. Chem.* **269,** 3574 (1994).
15. G. E. Smith, M. J. Fraser, and M. D. Summers, *J. Virol.* **46,** 584 (1983).
16. P. N. Lowe, M. Sydenham, and M. J. Page, *Oncogene* **5,** 1045 (1991).
17. J. E. Buss, L. A. Quilliam, K. Kato, P. J. Casey, P. A. Solski, G. Wong, R. Clark, F. McCormick, G. M. Bokoch, and C. J. Der, *Mol. Cell. Biol.* **11,** 1523 (1990).
18. K. H. Muntz, P. C. Sternweis, A. G. Gilman, and S. M. Mumby, *Mol. Biol. Cell* **3,** 49 (1992).
19. W. F. Simonds, J. E. Butrynski, N. Gautam, C. G. Unson, and A. M. Spiegel, *J. Biol. Chem.* **266,** 5363 (1991).
20. H. Ohguro, Y. Fukada, T. Takao, Y. Shimonishi, T. Yoshizawa, and T. Akino, *EMBO J.* **10,** 3669 (1991).
21. V. A. Luckow and M. D. Summers, *Virology* **167,** 56 (1988).
22. J. D. Robishaw, unpublished data (1995).
23. P. A. Kitts, M. D. Ayres, and R. D. Possee, *Nucleic Acids Res.* **18,** 5667 (1990).
24. M. D. Summers and G. E. Smith, *Tex. Agric. Exp. Stn. Bull.* **1555** (1987).
25. J. D. Robishaw and E. A. Balcueva, *in* "Methods in Enzymology" (R. Iyengar, ed.), p. 498. Academic Press, San Diego, 1994. Vol. 237.

26. P. Sridhar, A. K. Panda, R. Pal, G. P. Talwar, and S. E. Hasnain, *FEBS Lett.* **315,** 282 (1993).
27. W. J. Chen, J. F. Moomaw, L. Overton, T. A. Kost, and P. J. Casey, *J. Biol. Chem.* **268,** 9675 (1993).
28. J. D. Robishaw and E. A. Balcueva, *Anal. Biochem.* **208,** 283 (1993).
29. J. B. Higgins and P. J. Casey, *J. Biol. Chem.* **269,** 9067 (1994).
30. S. M. Mumby, P. J. Casey, A. G. Gilman, S. Gutowski, and P. C. Sternweis, *Proc. Natl. Acad. Sci. U.S.A.* **87,** 5873 (1990).
31. R. K. Lai, D. Perez-Sala, F. J. Canada, and R. R. Rando, *Proc. Natl. Acad. Sci. U.S.A.* **87,** 7673 (1990).
32. W. A. Maltese and J. D. Robishaw, *J. Biol. Chem.* **265,** 18071 (1990).
33. J. H. Reese and W. A. Maltese, *Mol. Cell. Biochem.* **104,** 109 (1991).
34. B. T. Kinsella and W. A. Maltese, *J. Biol. Chem.* **266,** 8540 (1991).
35. B. K. Fung, H. K. Yamane, I. M. Ota, and S. Clarke, *FEBS Lett.* **260,** 313 (1990).
36. S. Clarke, J. P. Vogel, R. J. Deschenes, and J. Stock, *Proc. Natl. Acad. Sci. U.S.A.* **85,** 4643 (1988).
37. P. C. Sternweis and J. D. Robishaw, *J. Biol. Chem.* **259,** 13806 (1984).
38. J. W. Winslow, J. R. van Amsterdam, and E. J. Neer, *J. Biol. Chem.* **261,** 7571 (1986).

[17] Expression, Purification, and Functional Reconstitution of Recombinant Phospholipase C-β Isozymes

Andrew Paterson and T. Kendall Harden*

Introduction

Phospholipase C is a phosphodiesterase that catalyzes the hydrolysis of phosphatidylinositol 4,5-bisphosphate [PtdIns(4,5)P$_2$] to the calcium-mobilizing second messenger, inositol 1,4,5-trisphosphate (InsP$_3$), and to sn-1,2-diacylglycerol, which activates protein kinase C (PKC). Receptor-mediated activation of phospholipase C (PLC) occurs in the physiological signaling pathway of an extraordinary number of hormones, neurotransmitters, chemoattractants, growth factors, and other stimuli. Multiple phospholipase C isoenzymes exist, and these can be classified into three major families based on sequence similarity, i.e., PLC-β, PLC-γ, and PLC-δ, and in two of these three classes, by signaling mechanism (1). The PLC-β and PLC-γ families consist of approximately 150-kDa proteins that are known to be intimately involved in receptor-signaling responses. The regulatory role of the 80-kDa PLC-δ class of isoenzymes has not yet been established. Growth factor receptors and a number of nonreceptor tyrosine kinases promote inositol lipid hydrolysis through a mechanism involving tyrosine phosphorylation of members of the PLC-γ class of PLC isoenzymes. However, by far the majority of extracellular stimuli promote inositol lipid signaling through G-protein-mediated activation of the PLC-β class of isoenzymes.

G proteins exist as heterotrimers consisting of an α subunit, which possesses the GTP binding site, and of a tightly associated complex of β and γ subunits. In most cases phospholipase C-β isoenzymes are activated by the receptor-promoted release of α subunits from the pertussis toxin-insensitive G$_q$ class of G proteins, i.e., α$_q$, α$_{11}$, α$_{14}$, or α$_{16}$ (1). However, pertussis toxin blocks receptor-promoted inositol lipid signaling in some cases, which implicates involvement of the G proteins, G$_i$ or G$_o$, which are ADP-ribosylated and inactivated by this bacterial toxin. Compelling evidence has been reported that indicates that it is the βγ subunit, rather than the α subunit, of G$_i$ or G$_o$ that activates phospholipase C in the pertussis toxin-sensitive signaling pathway (2).

---

* To whom correspondence should be addressed.

Methods in Neurosciences, Volume 29

Although activation of phospholipase C-β enzymes occurs through either GTP-liganded α subunits of the G_q class of G proteins or by the release of $\beta\gamma$ subunits from G_i, G_o, and possibly from other G proteins, much is yet to be learned about these pathways. For example, at least four different phospholipase C-β isoenzymes (PLC-β1, -β2, -β3, and -β4) exist in mammals, and the relative selectivities of α_q, α_{11}, α_{14}, and α_{16} for activation of each of these is yet to be established. Similarly, at least four and six genes exist for β and γ subunits, respectively, and thus a multiplicity of different $\beta\gamma$ dimers occur. These potentially could activate the individual phospholipase C-β isoenzymes with different selectivities. The actual molecular mechanism whereby any of the G protein subunits activates phospholipase C needs to be addressed in biophysical studies that to date have not been possible.

Further in-depth study of the mechanisms by which PLC-β isoenzymes are regulated requires the acquisition of substantial quantities or purified protein. Phospholipase C-β1 and C-β2 can be purified from bovine brain and from cultured polymorphonuclear neutrophils, respectively. In both cases the purification procedure is protracted, requiring access to large quantities of starting material and with only moderate or poor yield of purified protein. In addition, copurification of proteolytic cleavage products with the phospholipase C of interest can present problems. For instance, two catalytically active fragments of PLC-β1 copurify from bovine brain with this enzyme (3). The smaller 100-kDa fragment is apparently not activated by purified $G_{q\alpha}$, whereas the larger 140-kDa fragment is. These problems have been partially avoided by the use of purified recombinant PLC-β1 and -β2 purified after expression in HeLa cells (4). In both cases large cultures of cells were required for the recovery of milligram quantities of pure protein. Here we describe methodology for the production of recombinant PLC-β1 and -β2 in the *Spodoptera frugiperda* (Sf9)/baculovirus expression system. This system has the advantage of producing milligram quantities of homogeneous recombinant PLC-β1 and PLC-β2 from relatively small cultures of Sf9 cells. The smaller cultures and higher levels of recombinant protein expression facilitate their isolation by simpler purification protocols. We also describe methodology for the investigation of the mechanism by which α and $\beta\gamma$ subunits regulate the activity of the PLC-β isoforms.

Maintenance of Sf9 Cells

Culture Medium

Make Grace's medium (GGKG) containing 10% (v/v) fetal bovine serum, 3.3 g/liter yeastolate, 3.3 g/liter lactalbumin hydrolyzate, 6.1 mM glutamine, 100 μg/ml kanamycin, and 50 μg/ml gentamicin.

Establish a continuous culture of *S. frugiperda* (Sf9) cells (ATCC CRL-1711, Rockville, MD) in a spinner flask (model 1965-00250, Bellco Glass Co., Vineland NJ) at 27°C. Grow the cells in GGKG and normal atmospheric air; 120 ml of cell suspension usually is accommodated in a 250-ml spinner flask. Suspension is maintained with an impeller speed of 70 rpm. Maintain cells at a density between 1.0×10^6 and 2.5×10^6 cells/ml by dilution of the medium every 30 to 48 hr, as required.

Production of Recombinant Baculovirus Vectors

The cDNA encoding rat PLC-β1 is excised from pIBI-PLC-β1 (Dr. S. G. Rhee, National Heart, Lung, and Blood Institute, NIH, Bethesda, MD) with the restriction endonucleases *Not*I and *Bam*HI and the 3815-bp product is ligated into the multiple cloning site of the pVL1392 transfer plasmid (Invitrogen Corporation, San Diego, CA) via its *Not*I and *Bam*HI sites. Construction of a pVL1392/PLC-β2 vector requires the introduction of a *Not*I restriction site into the 5' untranslated region of the human PLC-β2 cDNA. The modified 5' sequence is amplified from pMT2-PLC-β2 (Dr. J. Knopf, Genetics Institute, Cambridge, MA) by the polymerase chain reaction (PCR) using the primers 5'-AGTCCGCGGCCGCGAGATTCTGCAA-3' and 5'-CCGGATGCTGGTGAT-3'. The 237-bp product is digested with the restriction enzymes *Not*I and *Eco*RV and ligated with *Not*I/*Eco*RI-digested pVL1392. The product of this ligation reaction is isolated and ligated with the 3' cDNA sequence of PLC-β2 excised from pMT2-PLC-β2 with *Eco*RI and *Eco*RV. Recombinant baculovirus expression vectors encoding either PLC-β1 or PLC-β2 are derived from the pVL1392/PLC-β1 and pVL1932/PLC-β2 constructs, respectively, by standard procedures (5). Recombinant baculovirus is isolated and purified through a further two rounds of plaque assay.

Infection of Sf9 Cells for Production of Recombinant Protein

Reagents

Recombinant baculovirus amplified to $(0.5–2.0) \times 10^8$ plaque-forming units (pfu)/ml. The expression studies discussed here employ purified virus recovered from the second or third passage of amplification.

In a laminar flow hood, add 5.0 ml of GGKG medium to a 150-mm tissue culture-treated dish and tilt in several directions to distribute the liquid evenly. Add 3.0×10^7 Sf9 cells to the dish, ensuring that the cells distribute evenly. Place the dish on a level surface and allow the cells to attach at 27°C for 1 hr. Greater than 95% of the cells should attach and the monolayer should appear to be 50% confluent. If required, additional cells may be added at this stage to achieve the desired level of confluency. In a sterile centrifuge tube inoculate 7.0 ml of GGKG medium with 3.0×10^7 pfu of recombinant baculovirus encoding either PLC-β1 or PLC-β2. This furnishes a multiplicity of infection (MOI) equal to 1.0, i.e., 1 pfu of recombinant baculovirus per Sf9 cell. Aspirate the medium and unattached cells from the monolayer, and add the diluted baculovirus, ensuring that the mixture distributes evenly across the cells. Return the dish to a level surface and allow the virus to infect the cells for 1 hr at 27°C, occasionally tilting the plate to redistribute the viral mixture across the cells. Finally, add 17.0 ml of GGKG medium to the viral mixture bathing the cells and seal the plate in a zip-lock bag. Humidity can be maintained in the sealed bag by inclusion of a moistened paper towel. The plate of infected cells is then incubated on a level surface at 27°C until harvest.

Recovery of Recombinant Protein from Baculovirus-Infected Sf9 Cells

After 48 hr most of the cells appear infected. They are larger, possess enlarged nuclei, and a significant proportion of the cells are detached from the monolayer. Both detached cells and those remaining adherent are collected and pooled to maximize the recovery of infected cells. Adherent cells are detached from the plate into the GGKG medium by placing the tissue culture dish on a flat surface and tapping it briskly on the side. Remove the suspended cells to a centrifuge tube, dislodge any cells remaining adherent by pipetting ice-cold PBS onto the plate at an angle of about 30°, and pool these cells with those collected previously. Pellet the cells by centrifugation (500 g, 4°C, 5 min) and aspirate the GGKG medium.

The cells are then lysed by resuspension in hypotonic buffer. Resuspend the cell pellet collected from one 150-mm dish in 20 volumes (approximately 10 ml) of ice-cold buffer containing 20 mM HEPES, pH 7.2, 1.0 mM NaHCO$_3$, 5.0 mM MgCl$_2$, 2.0 mM EGTA, 200 μM phenylmethylsulfonyl fluoride (PMSF), 200 μM benzamidine, 1.0 μM pepstatin A, and 10 μg/ml leupeptin. Incubate the suspension on ice for 10 min and then homogenize with 15 strokes of a loose-fitting Dounce homogenizer (pestle A, Wheaton, Millville, NJ). Remove unbroken cells, nuclei, and other dense organelles by centrifugation (500 g, 4°C, 5 min). Collect the supernatant and recentrifuge at 105,000 g at 4°C for 65 min. The supernatant from this ultracentrifugation step is collected and employed as a crude cytosolic fraction.

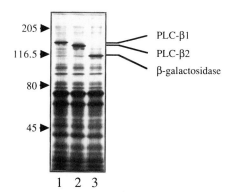

205 ▶
116.5 ▶
80 ▶
45 ▶

PLC-β1
PLC-β2
β-galactosidase

1 2 3

FIG. 1   SDS-PAGE of lysates prepared from Sf9 cells expressing PLC-β1, PLC-β2, and β-galactosidase. Sf9 cells were infected with recombinant baculovirus encoding either PLC-β1 (lane 1), PLC-β2 (lane 2), or β-galactosidase (lane 3). Cells were lysed 48 hr after infection as described in the text and 0.85 μl of each lysate (approximately 1.5 μg protein) was separated on a 7.5% SDS-polyacrylamide gel. Protein was visualized by silver staining. The migration of 205-, 116.5-, 80-, and 45-kDa markers is indicated.

Comments

The authenticity of the recombinant isoenzymes expressed in Sf9 cells can be confirmed by selective immunoreactivity with rabbit antipeptide antisera raised against the C-terminal dodecapeptides of rat PLC-β1 (Gly-Glu-Asn-Ala-Gly-Arg-Glu-Phe-Asp-Thr-Pro-Leu) and human PLC-β2 (Leu-Ile-Ala-Lys-Ala-Asp-Ala-Gln-Glu-Ser-Arg-Leu), respectively.

Ten 150-mm dishes of Sf9 cells are usually infected for production of recombinant protein. This is readily accomplished in a laboratory without any special facilities for the production of recombinant protein in the Sf9/baculovirus system. We recover 2–3 mg of PLC-β1 or PLC-β2 in the cell lysate prepared from 10 150-mm dishes of Sf9 cells infected in the above manner (Fig. 1). Fractionation of the lysate results in the recovery of 40–60 mg of total protein in the crude cytosol. Recombinant PLC-β1 is purified from the crude cytosol, where it represents greater than 5% of the total protein. In contrast, less than 10% of the recombinant PLC-β2 expressed in Sf9 cells is located in the soluble fraction, whereas the majority is recovered in a dense particulate fraction collected by low-speed centrifugation (500 *g*, 4°C, 5 min). The recombinant PLC-β2 recovered with this dense particulate fraction is both active and regulated by βγ subunits and would appear not to represent an incorrectly processed product. The reason for the appearance of this isoenzyme in this fraction is not unambiguously known. However, it is likely that it represents aggregation of the enzyme in a multimeric complex. Multimeric complexes of recombinant PLC-β2 expressed in HeLa cells have previously been reported (4). Conditions currently are being established that can be utilized to solubilize and purify the PLC-β2 from the low-speed particulate fraction. In the absence of such information, the following sections describe purification of recombinant PLC-β2 from the soluble fraction of infected cells.

Measurement of PLC Activity

The assay method detailed here is an adaptation from that described previously by Morris *et al.* (6).

Reagents

Chloroform/methanol stocks of PtdIns(4,5)P$_2$ and Ptd[^{3}H]Ins(4,5)P$_2$ are prepared as described previously (6).

4× Assay buffer: 10 mM HEPES, pH 7.2, 480 mM KCl, 40 mM NaCl, 8.0 mM EGTA, 23.2 mM MgSO$_4$, and 8.4 mM CaCl$_2$.

2.0% Cholate: sodium cholate is purified as described previously (7) and stored as a 10% (w/v) solution.

10 mM HEPES, pH 7.2.

Procedure

To assay 50 samples, mix 250 nmol of PtdIns(4,5)P$_2$ with Ptd[^{3}H]Ins(4,5)P$_2$ (750,000 cpm) in a polypropylene tube and evaporate the solvent under a stream of nitrogen. Add 1250 μl of ice-cold 10 mM HEPES to the dried lipid and sonicate the mixture (2× 15 sec, Virsonic 50-probe sonicator, Virtis Instrument Co., Gardiner, NY). Retain the sonicated lipids on ice until use. Mix 25 μl of the sonicated lipid with 25 μl of 4× assay buffer and 25 μl of 2.0% sodium cholate in 10 mM HEPES, pH 7.2. Initiate the reaction by adding PLC in 25 μl of 10 mM HEPES, vortex, and transfer to a 30°C water bath. The final reaction mixture contains 10 mM HEPES, pH 7.2, 50 μM PtdIns(4,5)P$_2$ [15,000 cpm Ptd[^{3}H]Ins(4,5)P$_2$], 0.5% sodium cholate, 120 mM KCl, 10 mM NaCl, 2.0 mM EGTA, 5.8 mM MgSO$_4$, 2.1 mM CaCl$_2$ ($\sim$100 μM free Ca^{2+}). Terminate the reaction after 15 min with 375 μl of CHCl$_3$: CH$_3$OH : concentrated HCl (40 : 80 : 1, v/v). Add 125 μl of CHCl$_3$ and

125 μl of 0.1 M HCl, vortex, and split phases by centrifugation (2000 g, 5 min, room temperature). Remove 400 μl of the upper phase and quantitate the tritium by liquid scintillation counting. The velocity of the PLC-catalyzed reaction is determined from the release of water-soluble [^{3}H]InsP$_3$ from Ptd[^{3}H]Ins(4,5)P$_2$.

Comments

This assay is routinely employed to quantitate the PLC activity eluting from columns during chromatography. Under the conditions described, the PLC-catalyzed reaction velocity is linear over the range 0–25 pmol/min with respect both to time (0–20 min) and increasing PLC concentration. This assay also has proved suitable for estimating apparent K_m and V_{max} values for the purified PLC isozymes under conditions of pseudo-first-order rate kinetics. For this purpose the reaction is incubated at 30°C for 5 min only and the PtdIns(4,5)P$_2$ concentration varied through the range of 10 μM to 1.0 mM. It is also necessary to limit the amount of PLC added such that less than 10% of the PtdIns(4,5)P$_2$ is hydrolyzed. The concentration of PtdIns(4,5)P$_2$ employed to determine the PLC activity present in column fractions is below the apparent K_m calculated with the above assay conditions. Therefore, it must be appreciated that any measurement of enzyme activity with only 50 μM PtdIns(4,5)P$_2$ does not represent the maximal values that would be obtained at saturating concentrations of substrate.

Purification of Recombinant Phospholipase C-β1

All procedures are executed at 4°C.

Step 1: Q-Sepharose FF Chromatography of Recombinant PLC-β1

Prepare crude cytosol from 10 150-mm dishes of cells infected with PLC-β1-encoding recombinant baculovirus as described above and apply at 3 ml/min to a 10 ml (1.6 × 5 cm) column of Q-Sepharose FF (Pharmacia, Piscataway, NJ) equilibrated in buffer A [25 mM HEPES, pH 7.2, 2.0 mM dithiothreitol (DTT), 2.0 mM EDTA, 2.0 mM EGTA, 200 μM PMSF, 200 μM benzamidine, 1.0 μM pepstatin A] containing 10 mM NaCl. Wash the column with 200 ml of buffer A containing 110 mM NaCl and elute recombinant PLC-β1 with a 200-ml linear gradient of 110–410 mM NaCl in buffer A, collecting the column eluate in 6-ml fractions. Assay 5 μl of each fraction for PLC activity. The recombinant PLC-β1 elutes between 150 and 250 mM NaCl (Fig. 2A). Pool the fractions containing PLC activity (fractions 8–20).

Step 2: Heparin–Sepharose CL-6B Chromatography of Recombinant PLC-β1

Dilute the pooled enzyme with 0.5 volumes of buffer A and apply at 0.4 ml/min to a 4.0-ml (1.0 × 5.1 cm) column of heparin–Sepharose CL-6B (Pharmacia) equilibrated in buffer A. Wash the column with 60 ml of 100 mM NaCl in buffer A and elute the recombinant PLC-β1 with a 60-ml linear gradient of 0.1–1.0 M NaCl in buffer A, followed by 20 ml of 1.0 M NaCl in buffer A. Collect the column eluate in 2.0-ml fractions and assay 2 μl of each fraction for PLC activity. Recombinant PLC-β1 elutes between 0.4 and 0.7 M NaCl (Fig. 2B). Pool the fractions containing PLC activity (fractions 13–20).

Step 3: Mono Q Chromatography of Recombinant PLC-β1

Dilute the PLC-β1 pooled from the heparin–Sepharose CL-6B column in 2 volumes of buffer A containing 10 mM NaCl and apply at 0.5 ml/min to an FPLC Mono Q HR 5/5 column (Pharmacia) equilibrated in buffer A containing 10 mM NaCl. Wash the column with 5 ml of 10 mM NaCl in buffer A and elute the column with a 20-ml linear gradient of 10–460 mM NaCl in buffer A followed by 5 ml of 460 mM NaCl in buffer A. Collect the column eluate in 0.5-ml fractions. Assay 0.5 μl of each fraction for PLC activity and pool those containing PLC activity (fractions 32–36). Recombinant PLC-β1 elutes between 200 and 300 mM NaCl (Fig. 2C). The pooled enzyme is diluted in an equal volume of 40% (v/v) glycerol in buffer A and stored at −80°C.

Comments

The recombinant PLC-β1 expressed and purified by this method was found to be homogeneous when analyzed by SDS-PAGE and visualized by silver staining (Fig. 3). The final yield of purified enzyme was 1.9 mg.

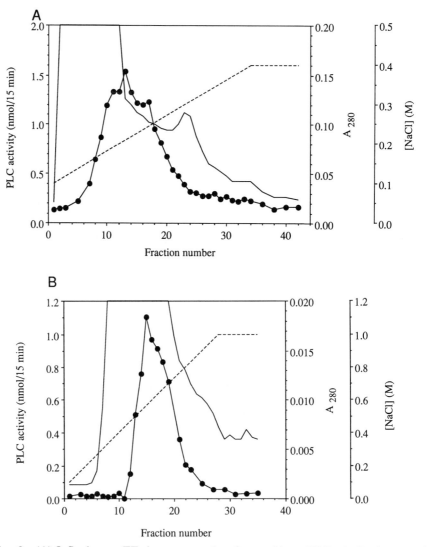

FIG. 2 (A) Q-Sepharose FF chromatography of recombinant PLC-β1. Crude cytosol prepared from Sf9 cells infected with recombinant baculovirus encoding PLC-β1 was prepared and applied to a column of Q-Sepharose FF as described in the text. The column was eluted and 5.0 μl of each fraction assayed for PLC activity (●). The elution of protein was determined by absorbance at 280 nm (—) and NaCl concentration (---) is indicated. (B) Heparin-Sepharose CL-6B chromatography of recombinant PLC-β1. Fractions collected from the Q-Sepharose FF column containing PLC activity were pooled, diluted, and applied to a column of heparin-Sepharose CL-6B as described in the text. The column was eluted and 2.0 μl of each fraction assayed for PLC activity (●). The elution of protein was determined by absorbance at 280 nm (—) and NaCl concentration (---) is indicated. (C) Mono Q chromatography of

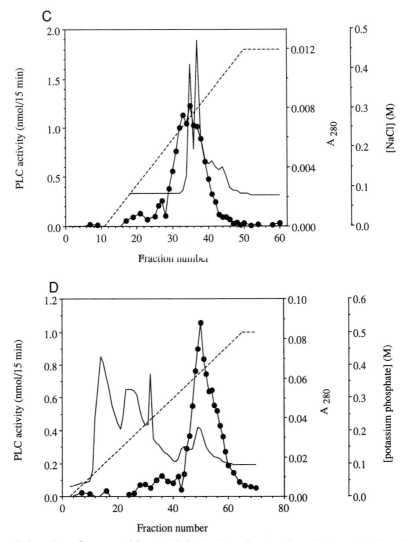

FIG. 2 (*continued*)   recombinant PLC-β1. Fractions collected from the heparin-Sepharose CL-6B column containing PLC activity were pooled, diluted, and applied to an FPLC Mono Q HR 5/5 column as described in the text. The column was eluted and 0.5 μl of each fraction assayed for PLC activity (●). The elution of protein was determined by absorbance at 280 nm (—) and NaCl concentration (- - -) is indicated. (D) Bio-Gel HPHT chromatography of recombinant PLC-β2. Fractions collected from the heparin-Sepharose CL-6B column containing PLC activity were pooled, diluted, and applied to a Bio-Gel HPHT column as described in the text. The column was eluted and 1.0 μl of each fraction assayed for PLC activity (●). The elution of protein was determined by absorbance at 280 nm (—) and potassium phosphate concentration (- - -) is indicated.

Purification of Recombinant Phospholipase C-β2

All procedures are executed at 4°C.

Step 1: Q-Sepharose FF Chromatography of Recombinant PLC-β2

Prepare crude cytosol from 10 150-mm dishes of Sf9 cells infected with recombinant baculovirus encoding PLC-β2 as detailed previously and fractionate the recombinant PLC-β2 by Q-Sepharose FF chromatography as described for recombinant PLC-β1 in step 1 above. Recombinant PLC-β2 elutes at a position in the gradient equivalent to between 250 and 350 mM NaCl.

Step 2: Heparin–Sepharose CL-6B Chromatography of Recombinant PLC-β2

Pool the fractions eluting from the Q-Sepharose FF column with PLC activity, dilute in an equal volume of buffer A, and load onto a 4.0-ml column of heparin–Sepharose CL-6B as described in step 2 above for PLC-β1. Elute the column with an 80-ml gradient of 0.1–1.0 M NaCl in buffer A, collecting 2-ml fractions. Assay 2.0 μl of each fraction for PLC activity and pool those containing recombinant PLC-β2. Recombinant PLC-β2 elutes at a position in the gradient equivalent to between 0.3 and 0.6 M NaCl.

Step 3: Bio-Gel HPHT Chromatography of Recombinant PLC-β2

Dilute the pooled recombinant PLC-β2 in an equal volume of buffer B (25 mM HEPES, pH 7.2, 10 mM KCl, 2.0 mM DTT, 200 μM PMSF, 200 μM benzamidine, 1.0 μM pepstatin A) and apply at 0.3 ml/min to a Bio-Gel HPHT (7.8 × 100 mm) hydroxylapatite column/guard-column configuration (Bio-Rad Laboratories, Hercules, CA). The Bio-Gel HPHT column configuration is equilibrated in buffer B and operated in conjunction with a fast protein liquid chromatography (FPLC) apparatus. Wash the column with 20 ml of buffer B and elute recombinant PLC-β2 with a 62.5-ml linear gradient of 0–500 mM potassium phosphate in buffer B, collecting the eluate in 1.0-ml fractions. Assay 1.0 μl of each fraction for PLC activity and pool those containing rPLC-β2 (fractions 45–59). The recombinant PLC-β2 elutes between 325 and 450 mM potassium phosphate (Fig. 2D).

Step 4: Mono Q Chromatography of Recombinant PLC-β2

Dilute the recombinant PLC-β2 pooled from the Bio-Gel HPHT column with 4 volumes of buffer A containing 10 mM NaCl. Load the diluted enzyme at 0.5 ml/min onto an FPLC Mono Q HR 5/5 column equilibrated in buffer A containing 10 mM NaCl. Unbound protein is removed with 5 ml of 10 mM NaCl in buffer A, and recombinant PLC-β2 is eluted with a 10-ml linear gradient of 0.01–1.01 M NaCl in buffer A. Collect the column eluate in 0.5-ml fractions and assay 0.5 μl of each for PLC activity. The recombinant PLC-β2 elutes from the column between 400 and 600 mM NaCl. Pool the fractions containing PLC activity, dilute in an equal volume of 40% (v/v) glycerol in buffer A, and store at $-80°$C.

Comments

The Bio-Gel HPHT column described in step 3 above can be replaced by 10 ml (1.6 × 5.0 cm) of Bio-Gel HTP (Bio-Rad Laboratories). The Bio-Gel HTP column is equilibrated in buffer B and operated at a flow rate of 0.33 ml/min. Elute the recombinant PLC-β2 from this column with a 200-ml gradient of 0–500 mM potassium phosphate in buffer B, collecting 3.0-ml fractions. Although use of a Bio-Gel HTP column eliminates the backpressure problem often experienced with an FPLC/Bio-Gel HPHT column configuration, it results in elution of recombinant PLC-β2 from the hydroxylapatite column in a larger volume (30 ml compared to 11 ml). However, the recombinant PLC can be concentrated adequately by Mono Q chromatography (step 4).

This method has allowed a 390-fold purification of rPLC-β2 from the soluble fraction of recombinant baculovirus-infected Sf9 cells. The final yield of rPLC-β2 after purification was 80 μg, and this protein preparation, although not homogeneous, contained only three minor contaminants after SDS-PAGE analysis and visualization by silver staining (Fig. 3).

Functional Reconstitution of Phospholipase C-β with G$_{\alpha 11}$

The reconstitution method described here is an adaptation of that previously reported by Waldo et al. (8).

Reagents

Chloroform/methanol stocks of PtdIns(4,5)P$_2$ and Ptd[^{3}H]Ins(4,5)P$_2$ are prepared as described previously (6).

G$_{\alpha 11}$ is purified from turkey erythrocyte membranes as described previously (8).

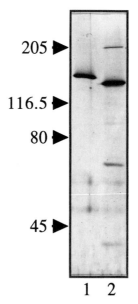

FIG. 3   SDS-PAGE of purified recombinant PLC-β1 and PLC-β2; 100 ng of purified recombinant PLC-β1 (lane 1) and PLC-β2 (lane 2) were separated on a 7.5% SDS-polyacrylamide gel. Protein was visualized by silver staining. The migration of 205-, 116.5-, 80-, and 45-kDa markers is indicated.

Dialysis buffer: 20 mM HEPES, pH 7.4, 1.0 mM MgCl$_2$, 100 mM NaCl, 2.0 mM DTT, 100 μM PMSF, 100 μM benzamidine.

0.8% Cholate solution: 20 mM HEPES, pH 7.4, 0.8% (v/v) sodium cholate, 1.0 mM MgCl$_2$, 100 mM NaCl, 2.0 mM DTT, 100 μM PMSF, 100 μM benzamidine.

4$\times$ Assay buffer: 150 mM HEPES, pH 7.0, 300 mM NaCl, 16 mM MgCl$_2$, 8.0 mM EGTA, 6.4 mM CaCl$_2$, 2.0 mg/ml fatty acid-free BSA (cat. no. A 3803, Sigma Chemical Co., St. Louis, MO).

10$\times$ AlF$_4$: 200 μM AlCl$_3$, 100 mM NaF, prepared by a 10-fold dilution of a 2.0 mM AlCl$_3$, 1.0 M NaF stock solution in 20 mM HEPES, pH 7.0, immediately prior to use.

20 mM HEPES, pH 7.0.

Procedure

For 15 assays, mix 30 nmol PtdIns(4,5)P$_2$, Ptd[^{3}H]Ins(4,5)P$_2$ (225,000 cpm), 120 nmol phosphatidylethanolamine (Avanti Polar Lipids, Alabaster, AL), and 30 nmol phosphatidylserine (Avanti Polar Lipids) in a polypropylene tube and evaporate the solvent under a stream of nitrogen. Add 150 μl of 0.8% cholate solution to the dried lipids and resuspend by sonication (2× 15 sec, Virsonic 50-probe sonicator). Combine 150 ng of $G_{\alpha 11}$ (diluted to 25 μl with 0.8% cholate solution) with the resuspended lipids, adjust the volume to 300 μl with cholate solution, mix, and dialyze for 14–20 hr against 2 liters of dialysis buffer. Collect the dialyzed vesicles, dilute to 750 μl with dialysis buffer, and store on ice until use. For each assay tube mix 25 μl of 4× assay buffer with 10 μl 10× AlF_4^-, 5 μl of 20 mM HEPES, and 50 μl of the diluted vesicles. The enzyme reaction is initiated by adding 10 μl of 20 mM HEPES containing 1–10 ng of PLC-β1 or PLC-β2 and warming to 30°C in a water bath. Each assay tube contains a final reaction mixture consisting of 52.5 mM HEPES/NaOH, pH 7.4, 20 μM PtdIns(4,5)P$_2$ [15,000 cpm Ptd[^{3}H]Ins(4,5)P$_2$], 80 μM phosphatidylethanolamine, 20 μM phosphatidyl-serine, 125 mM NaCl, 4.5 mM MgCl$_2$, 2.0 mM EGTA, 1.6 mM CaCl$_2$ (approximately 300 nM free Ca^{2+}), 1.0 mM DTT, 20 μM AlCl$_3$, 10 mM NaF, and 0.5 mg/ml fatty acid-free BSA. Incubate at 30°C for 5–15 min and terminate the reaction with 375 μl of CHCl$_3$: CH$_3$OH : concentrated HCl (40 : 80 : 1). Vortex the mixture briefly and incubate at room temperature for 5 min. Add 125 μl of CHCl$_3$ and 125 μl of 0.1 M HCl, vortex, and split phases by centrifugation (2000 g, 5 min, room temperature). Remove 400 μl of the upper aqueous phase and quantitate the release of [^{3}H]InsP$_3$ from Ptd[^{3}H]Ins(4,5)P$_2$ by liquid scintillation counting.

Comments

Dialysis of the $G_{\alpha 11}$/cholate/phospholipid mixture is performed in a 15-well microdialyzing chamber (model D 1451, MRA, Naples, FL). Fresh dialysis buffer is introduced to the reservoir of the chamber at a rate of approximately 100 ml/hr.

Under these assay conditions PLC-β1 and PLC-β2 also will catalyze the hydrolysis of PtdIns(4,5)P$_2$ in an AlF_4^--independent mechanism. The velocity of this reaction can be measured by omitting AlF_4^- from the final reaction mixture. The $G_{\alpha 11}$-stimulated PLC reaction rate can be determined by comparing the rates measured in the presence and absence of AlF_4^-. The method described above also is suitable for examining the concentration dependence of PLC on $G_{\alpha 11}$ for activation. For this purpose the final $G_{\alpha 11}$ concentration in the reaction mixture may be varied over the range 0.01–3.0 nM (Fig. 4). Although this chapter concerns itself with reconstitution of $G_{\alpha 11}$ with recombinant PLC-β1 and PLC-β2, this method is equally applicable to the reconstitution of other G protein α subunits and other PLC-β isoforms, i.e., recombinant $G_{q\alpha}$ and turkey erythrocyte PLC, respectively.

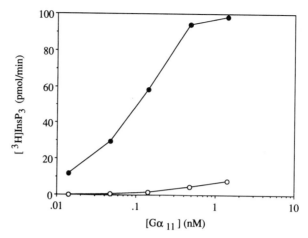

FIG. 4   Activation of purified recombinant PLC-β1 and PLC-β2 by $G_{\alpha11}$. Phospho-lipid vesicles containing the indicated concentrations of $G_{\alpha11}$ were prepared as de-scribed in the text. The reaction was initiated by the addition of 5 ng of either recombinant PLC-β1 (●) or recombinant PLC-β2 (○) and incubated for 10 min at 30°C. Reactions were terminated and PLC activity determined as described in the text. $G_{\alpha11}$-regulated PLC activity was determined in the presence of AlF_4^- and cor-rected for the activity measured in the presence of vehicle alone.

Functional Reconstitution of Phospholipase C-β with βγ Subunits

The G protein βγ subunits are reconstituted with either recombinant PLC-β1 or PLC-β2 by the method originally described by Boyer *et al.* (9).

Reagents

The βγ subunits are purified from bovine brain membranes or turkey erythrocyte plasma membranes as described previously (10).
All other reagents are identical to those described for measurement of PLC activation by $G_{\alpha11}$.

Procedure

Activation of PLC-β1 and PLC-β2 by G protein βγ subunits is assessed by methodology that, except for the substitution of βγ subunits for $G_{\alpha11}$ and omission of AlF_4^-, is identical to that described above for $G_{\alpha11}$-mediated activation. In brief, 30 nmol PtdIns(4,5)P$_2$, Ptd[^{3}H]Ins(4,5)P$_2$ (225,000 cpm), 120 nmol phosphatidylethanolamine, and 30 nmol phosphatidylserine are dried under a stream of nitrogen in a polypropylene tube and resuspended by sonication in 150 μl of 0.8% cholate solution. The $\beta\gamma$ subunits (1.5 μg) are diluted to 25 μl in 0.8% cholate solution and added to the sonicated lipids. The volume is adjusted to 300 μl with 0.8% cholate solution and the mixture dialyzed for 14–20 hr against 2 liters of dialysis buffer. The dialyzed vesicles are collected, adjusted to 750 μl with dialysis buffer, and stored on ice until use. On ice, mix 25 μl of 4× assay buffer, 15 μl of 20 mM HEPES, and 50 μl of phospholipid/$\beta\gamma$ subunit vesicles. Initiate the enzyme reaction by adding 10 μl of 20 mM HEPES containing 1–10 ng PLC-β1 or PLC-β2, vortex, and incubate for 15 min at 30°C. Terminate the reaction by adding 375 μl of CHCl$_3$: CH$_3$OH : concentrated HCl (40 : 80 : 1, v/v) and isolate and quantitate [^{3}H]Ins(1,4,5)P$_3$ as described for the G$_{\alpha 11}$ reconstitution assay.

Comments

The assay conditions described above are suitable for examining the concentration dependence of $\beta\gamma$ subunits for activation of PLC. For this purpose the final concentration of $\beta\gamma$ subunits in the reaction mixture may be varied through the range 0.2–60 nM (Fig. 5).

Recombinant Phospholipase C-β1 and C-β2 as Models of Native Enzymes

The recombinant PLC-β1 and PLC-β2 expressed and purified by the methods described here do not differ markedly from the native isoenzymes purified from bovine brain and HL60 polymorphonuclear neutrophils, respectively. The reaction velocities catalyzed by the recombinant isoenzymes have been compared to those of the native isoenzymes under varying assay conditions and found not to differ significantly (11). This is the case for the effect of PtdIns(4,5)P$_2$ concentration, Ca^{2+} concentration, and pH on the reaction velocities catalyzed by the recombinant PLC-β1 and PLC-β2. Similarly, the activities of recombinant PLC-β1 and PLC-β2 are regulated by G$_{\alpha 11}$ and free $\beta\gamma$ subunits in a manner that is identical to the modulation of the native isoenzymes by these G protein subunits (11).

In summary, utilization of the recombinant baculovirus/Sf9 expression system has allowed production of recombinant PLC-β1 and PLC-β2, both with properties that are indistinct from those of the same isoenzymes purified

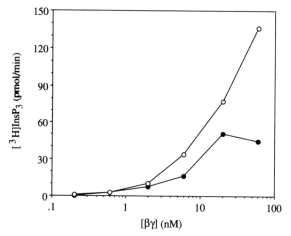

FIG. 5   Activation of purified recombinant PLC-β1 and PLC-β2 by G protein $\beta\gamma$ subunits. Phospholipid vesicles containing the indicated concentration of $\beta\gamma$ subunits were prepared as described in the text. The reaction was initiated by the addition of 30 ng of either recombinant PLC-β1 (●) or recombinant PLC-β2 (○) and incubated for 5 min at 30°C. Reactions were terminated and PLC activity determined as described in the text.

from native sources. The relatively large quantities of recombinant protein that can be obtained in a state of high purity from comparatively small cultures of Sf9 cells allow the investigator to follow many paths of investigation not feasible with the previously limited access to PLC-β1 and PLC-β2.

References

1. S. G. Rhee and K. D. Choi, *J. Biol. Chem.* **267,** 12393 (1992).
2. J. L. Boyer, A. Paterson, and T. K. Harden, *Trends Cardiov. Med.* **4,** 88 (1994).
3. D. Park, D.-Y. Jhon, C.-W. Lee, S. H. Ryu, and S. G. Rhee, *J. Biol. Chem.* **268,** 3710 (1993).
4. D. Park, D.-Y. Jhon, R. Kriz, J. Knopf, and S. G. Rhee, *J. Biol. Chem.* **267,** 16048 (1992).
5. D. R. O'Rielly, L. K. Miller, and V. A. Luckow, "Baculovirus Expression Vectors: A Laboratory Manual." W. H. Freeman, New York, 1992.
6. A. J. Morris, G. L. Waldo, C. P. Downes, and T. K. Harden, *J. Biol. Chem.* **265,** 13501 (1990).
7. G. L. Waldo, T. Evans, E. D. Fraser, J. K. Northup, M. W. Martin, and T. K. Harden, *Biochem. J.* **246,** 431 (1987).

8. G. L. Waldo, J. L. Boyer, A. J. Morris, and T. K. Harden, *J. Biol. Chem.* **266,** 14217 (1991).
9. J. L. Boyer, G. L. Waldo, and T. K. Harden, *J. Biol. Chem.* **267,** 25451 (1992).
10. J. L. Boyer, G. L. Waldo, T. Evans, J. K. Northup, C. P. Downes, and T. K. Harden, *J. Biol. Chem.* **164,** 13917 (1989).
11. A. Paterson, J. L. Boyer, V. J. Watts, A. J. Morris, E. M. Price, and T. K. Harden, *Cell. Signalling* **7,** 709 (1995).

[18] Measuring Cooperative Binding between Retinal Rod G Protein, Transducin, and Activated Rhodopsin

Barry M. Willardson, Tatsuro Yoshida, and Mark W. Bitensky

Introduction

Many signal transduction processes use G-protein-linked transmembrane receptors to communicate extracellular chemical and physical information to the interior of the cell. The ability of the activated receptor to catalyze nucleotide exchange with, and subunit dissociation of, its corresponding G protein (G) permits transmission of the signal via G protein subunits. Subsequent activation of effector enzymes by the $G_\alpha \cdot GTP$ subunit or the $G_{\beta\gamma}$ subunits serves to modulate the intracellular concentration of second messenger molecules such as cyclic nucleotides, inositol phosphates, and Ca^{2+}. The concentration of second messengers determines the ultimate cellular response to the signal that was initially received by the receptor. Because of their central role in G-protein-mediated signaling, interactions between G proteins and their receptors have been the focus of a considerable amount of research.

In the retinal rod, the photon receptor 11-*cis*-retinal is embedded in the binding pocket of rhodopsin (Rho). Its photoisomerization to all-*trans*-retinal causes a change in the structure of Rho, which initiates a well-documented G-protein-dependent cascade that modulates neurotransmitter release at the synaptic terminus of the photoreceptor cell (1, 2). Because of the striking abundance of the cascade components, the visual system has served well as a general model for the study of the complex protein–protein interactions of G-protein-mediated signaling cascades. Research into the nature of the interaction between activated rhodopsin (Rho*) and the retinal G protein transducin (G_t) has yielded a considerable amount of information about the interactions between an activated receptor and its G protein. A number of methods have been developed to measure the catalytic capacity of Rho to induce the dissociation of GDP and the binding of GTP to G_t. These methods include filtration assays for the binding of $[^{35}S]GTP\gamma S$ to G_t (3) and light-scattering measurements that report changes in G_t binding to rod outer segment membranes on activation by Rho (4). A method that has proved useful

Methods in Neurosciences, Volume 29

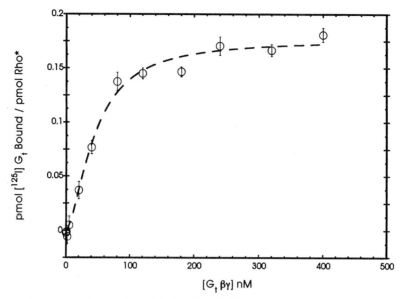

FIG. 1 $G_{t\beta\gamma}$ dependence of ^{125}I-labeled $G_{t\alpha}$ binding to Rho*. Data points are average values ± standard error from three separate experiments. Reproduced with permission from Ref. 7.

for direct assessment of the binding of activated Rho to G_t is the spectroscopic measurement of the extra *meta*-II-Rho formed as a result of the stabilization of the active photoconformer of Rho (*meta*-II) when G_t is bound at high pH (8.0) and low temperature (4°C) (5). These conditions favor the inactive *meta*-I photoconformer of Rho in the absence of G_t. However, this method cannot be readily transferred to other G protein systems, because the conformational changes brought about by activation of these receptors cannot be readily measured spectroscopically. Measurement of radiolabeled G_t binding to rod outer segment membranes has also been used to directly assess the binding of Rho* to G_t (6). Kinetic measurements using [^{35}S]GTPγS binding to $G_{t\alpha}$ revealed positive cooperativity in G_t activation by Rho* with respect to G_t concentration (3). The Hill coefficient for this phenomenon was approximately 2. Although such kinetic measurements suggested that the binding of G_t to Rho* was cooperative, they did not rule out a Rho* kinetic hysteresis model for G_t activation.

We have developed a method to measure cooperative binding of G_t to Rho* directly. In this procedure, $G_{t\alpha}$ is labeled with ^{125}I, and the light-induced association of $G_{t\alpha}$ to rod outer segment membrane is measured in the presence of $G_{t\beta\gamma}$ (see Fig. 1) (7). We found that Rho* · G_t binding was cooperative with respect to G_t concentration, with a Hill coefficient of approximately 2. These experiments demonstrate that cooperativity in the $[^{35}S]GTP\gamma S$ binding experiments was due to cooperative binding of G_t to Rho*. Here, we describe in detail the method for labeling G_t and measuring light-induced association of G_t with Rho* in rod disk membranes. As molecular cloning techniques make a variety of G proteins and their receptors available in larger quantities, this approach may become more widely applicable to the study of receptor–G protein interactions. Among the many questions that will be addressed is whether cooperative interactions are a general feature of receptor–G protein binding or whether they are specific to Rho* and G_t. The unique ability of rods to amplify and detect very small signals constitutes one of the most remarkable examples of G-protein-dependent signal transduction. The possible significance of the positively cooperative binding of G_t to Rho* for rod signal amplification is now being studied.

Purification of $G_{t\alpha}$ and $G_{t\beta\gamma}$

A number of purification schemes for $G_{t\alpha}$ and $G_{t\beta\gamma}$ have been published (8, 9). The procedures described below are a modification of those purification schemes, which we have found to give consistently high yields of very pure $G_{t\alpha}$ and $G_{t\beta\gamma}$ without detectable loss in activity.

Buffers

HEPES/Ringer buffer: 10 mM HEPES, pH 7.5, 120 mM NaCl, 3.5 mM KCl, 0.2 mM CaCl$_2$, 0.2 mM MgCl$_2$, 0.1 mM EDTA, 10 mM glucose, 1 mM dithiothreitol (DTT) (2.5 liter).

Sucrose flotation buffers: 45% (w/w) sucrose in HEPES/Ringer buffer (600 ml); 33% (w/w) sucrose in HEPES/Ringer buffer (100 ml).

Hypotonic buffer A: 10 mM HEPES, pH 7.5, 1 mM EDTA, 1 mM DTT, 0.2 mM phenylmethylsulfonyl fluoride (PMSF), 1 μM leupeptin, 1 μM pepstatin A (500 ml).

Hypotonic buffer B: hypotonic buffer A plus 200 μM GTP (500 ml).

Blue Sepharose column buffer A: 10 mM HEPES, pH 7.5, 6 mM MgSO$_4$, 1 mM EDTA, 1 mM DTT, 0.2 mM PMSF (300 ml).

Blue Sepharose column buffer B: Blue Sepharose column buffer A plus 1 M KCl (100 ml).

Size-exclusion column buffer: 10 mM HEPES, pH 7.5, 500 mM NaCl, 5 mM MgCl$_2$, 1 mM EDTA, 1 mM DTT, 0.1 mM PMSF (1 liter).

Materials

> 300 fresh bovine retinas (J. A. and W. L. Lawson, Lincoln, NE).
> Dark room equipped with infrared image convertors (Night Vision Equipment Co. Inc., Emmaus, PA) and infrared diodes as light source.
> 60-ml syringe drilled out to give a 0.8-cm-diameter orifice, 3-ml syringe, and needles (23 and 25 gauge).
> 6 × 250-ml fixed angle rotor (Sorvall GSA, Newtown, CT, or Beckman JA-14, Fullerton, CA).
> 6 × 30-ml swinging bucket rotor (Beckman SW-28).
> 8 × 20-ml high-speed fixed-angle rotor (Beckman Ti 70).
> 62-mm ultrafiltration concentrator with 30- and 10-kDa filter cutoffs (Amicon, Danvers, MA, concentrator with YM30 and YM10 filters).
> Centrifuge microconcentrator with 10-kDa cutoff (Amicon Centricon 10).
> 1.5 × 30-cm Blue Sepharose CL-6B column (Pharmacia, Piscataway, NJ).
> Column monitor absorbance detector (ISCO V$_4$ Absorbance Detector) and fraction collector (ISCO Cygnet, Lincoln, NE).
> 0.78 × 30-cm Bio-Sil SEC 250-5 HPLC size-exclusion column (Bio-Rad, Richmond, CA) and biocompatible HPLC system (Millipore/Waters 625 LC, Milford, MA).

Procedure

Fresh bovine retinas, shipped on ice overnight by express courier, are immediately processed on arrival. Initial procedures are carried out in the dark using infrared illumination and infrared image converters. All procedures are done on ice unless otherwise indicated, and all buffers are precooled to 4°C. Rod outer segments (ROS) are released from the retinas by adding approximately two volumes of HEPES/Ringer buffer plus 45% (w/w) sucrose for each volume of retinal material and passing the suspension three times through a 60-ml syringe with the orifice modified to 0.8 cm diameter. The disrupted retinal suspension is then passed through two layers of cheesecloth. The cheesecloth is rinsed with ~100 ml of HEPES/Ringer buffer plus 45% sucrose, and the retinal material retained on the cheesecloth is disrupted again in the same manner. This suspension is passed through two layers of fresh cheesecloth and washed with buffer as before. The material that does not pass through the cheesecloth is discarded. The filtrate from the cheesecloth is centrifuged in four 250-ml centrifuge bottles for 5 min at 4000 g. All centrifugations are done at 4°C. The supernatant containing the ROS is decanted and diluted with two volumes of HEPES/Ringer buffer without sucrose. The pellet is discarded. The diluted supernatant is centrifuged in 250-ml centrifuge bottles for 10 min at 15,000 g. The supernatant is discarded and the pellet containing the ROS is resuspended with a plastic transfer pipette in ~60 ml of HEPES/Ringer buffer without sucrose. This suspension is layered over six 15-ml cushions of 33% sucrose in 20-ml centrifuge tubes and is centrifuged for 15 min at 30,000 g in a Beckman SW28 swinging bucket rotor. The ROS are collected at the 33% sucrose interface and diluted twofold with HEPES/Ringer buffer without sucrose. The ROS are pelleted by centrifuging in 50-ml centrifuge tubes for 10 min at 12,000 g; 20 ml of hypotonic buffer is added to each pellet. The pellets are brought out into room light and illuminated with a 300-W tungsten lamp at 60 cm distance until the color of the ROS changes from deep red to orange. During the illumination, the pellets are resuspended and disrupted with a 3-ml syringe attached to a 23-gauge needle. Illumination causes G_t to bind tightly to ROS disk membranes under hypotonic buffer conditions via its interaction with Rho* (10). After thoroughly disrupting the ROS, the suspension is centrifuged in six to eight 20-ml ultracentrifuge tubes for 15 min at 140,000 g. The pellets are washed in this manner three times. The supernatants are used to purify phosphodiesterase or are discarded. After the three washes, the pellet is resuspended with a 3-ml syringe and 25-gauge needle in hypotonic buffer B, which contains 200 μM GTP. GTP breaks up the complex between Rho* and G_t, thus eluting G_t from the disk membrane (10). The suspension is centrifuged for 15 min at 140,000 g. The supernatant, containing the G_t, is saved. The pellets are washed in hypotonic buffer B a total of three times. The supernatants are pooled and concentrated to ~10 ml in an Amicon ultrafiltration concentrator using a YM30 filter (molecular weight cutoff of 30K). The pellets are discarded.

The concentrated supernatant is loaded at 1.0 ml/min on a 1.5 × 30-cm Blue Sepharose column that has been previously equilibrated in Blue Sepharose column buffer A. The effluent from the column is monitored at 280 nm using an ISCO V_4 absorbance detector; 4.0-ml fractions are collected. When material that does not bind has passed through the column and the 280-nm absorbance returns to baseline, a 200-ml linear gradient from column buffer A to column buffer B is applied to the column at 1.0 ml/min. $G_{t\beta\gamma}$ elutes from the column very early in the gradient at ~150 mM KCl. $G_{t\alpha}$ elutes much later in the gradient at ~650 mM KCl and is the last peak to elute. Other unidentified proteins elute between the $G_{t\beta\gamma}$ and $G_{t\alpha}$ peaks. The $G_{t\beta\gamma}$ peak and the $G_{t\alpha}$ peak are pooled separately and concentrated by ultrafiltration to ~5 ml and then to ~0.5 ml by centrifuge microconcentration using filters with molecular weight cutoffs of 10K (Amicon Centricon 10).

The $G_{t\alpha}$ is generally >90% pure at this point and the $G_{t\beta\gamma}$ is ~80% pure. They are further purified by size-exclusion HPLC using a Waters 625 LC system controlled by Millennium 2010 Chromatography Manager software (Millipore/Waters, Milford, MA). The concentrated $G_{t\alpha}$ or $G_{t\beta\gamma}$ (0.5 ml, 2–3 mg/ml) is injected onto a Bio-Sil SEC-250 size-exclusion column that has been equilibrated in size-exclusion column buffer. For $G_{t\alpha}$, DTT is omitted from the column buffer because it can interfere with the subsequent iodination step (see below). The column is run isocratically at 23°C at a flow rate of 1.0 ml/min. The absorbance of the column effluent is monitored for protein at 254 nm using a Waters 486 tunable absorbance detector. Under these conditions $G_{t\alpha}$ elutes at 10.4 min, which is later than expected for its 40-kDa size, corresponding to a M_r of ~20,000. Thus, some interaction between $G_{t\alpha}$ and the column is occurring. $G_{t\beta\gamma}$ binds to the column even more than does $G_{t\alpha}$. $G_{t\beta\gamma}$ elutes at 12.7 min, which is later than the elution time of the total column volume (12.0 min). It appears that the long retention times of $G_{t\beta\gamma}$ and $G_{t\alpha}$ are a result of hydrophobic interactions with the column matrix because increasing the NaCl concentration in the column buffer increases the retention times. These hydrophobic interactions may be caused by the long-chain hydrocarbon moieties of the subunits, a farnesyl group at the C terminus of $G_{t\gamma}$ and a myristyl group at the N terminus of $G_{t\alpha}$. The unusual behavior of $G_{t\beta\gamma}$ on the size-exclusion column yields very pure $G_{t\beta\gamma}$. It is >98% pure as determined by SDS-PAGE. The $G_{t\alpha}$ is >95% pure as determined by SDS-PAGE.

Fractions containing $G_{t\beta\gamma}$ are pooled and concentrated in Amicon Centricon microconcentrators to ~100 µl. Glycerol is added to give 50% glycerol and the $G_{t\beta\gamma}$ is stored at −20°C. No loss of $G_{t\beta\gamma}$ activity, as measured by its ability to enhance ^{125}I-labeled $G_{t\alpha}$ binding to Rho*, is observed for 3–4 weeks when stored in this manner. Fractions containing $G_{t\alpha}$ are pooled and concentrated to ~300 µl. This $G_{t\alpha}$ is then iodinated. The final yield of $G_{t\beta\gamma}$ and $G_{t\alpha}$ is approximately 1.0 mg for each species when starting with 300 retinas.

Iodination of $G_{t\alpha}$

Buffer

Isotonic buffer: 10 mM HEPES, pH 7.5, 100 mM KCl, 20 mM NaCl, 2 mM MgCl$_2$, 1 mM EDTA, 1 mM DTT, 0.2 mM PMSF.

Materials

250 µCi ^{125}I-labeled Bolton–Hunter reagent with charcoal trap (NEN-Dupont, Boston, MA).

Two 1.5-inch, 21-gauge needles, one 1.0-inch, 21-gauge needle, and one 1.0-ml syringe.

Dry N_2 gas cylinder with regulator.

1.2×30-cm Sephadex G-25 column (Pharmacia).

Gamma counter (LKB/Wallac 1272, Clinigamma, Gaithersburg, MD).

Procedure

$G_{t\alpha}$ is radiolabeled with ^{125}I-labeled Bolton–Hunter reagent (11). Labeling just the α subunit of G_t results in much greater binding to Rho*, in the presence of unlabeled $G_{t\beta\gamma}$, than when the entire G_t heterotrimer is labeled. This may be due to inhibition of binding to Rho* when sites on $G_{t\beta\gamma}$ are labeled. The N-hydrosuccinimide ester of Bolton–Hunter reagent reacts preferentially with primary amines, thus lysine and arginine residues and the N terminus of proteins are the primary targets of the reagent. The reaction requires a free electron pair on the nitrogen atom of the primary amine, so the reaction is favored at higher pH values (pH 7.5–8.5), at which the amino group is less protonated. Reaction buffers with primary amino groups should not be used, to avoid competing reactions with the buffer. Also, sulfhydryl reagents such as DTT or 2-mercaptoethanol must be excluded from the reaction buffer, because the Bolton–Hunter reagent will react to a lesser extent with sulfhydryl groups. For this reason, DTT is omitted from the size-exclusion column buffer for $G_{t\alpha}$ purification, allowing the radiolabeling to be done directly after size-exclusion chromatography without exchanging buffers. Because the N-hydrosuccinimide ester of ^{125}I-labeled Bolton–Hunter reagent is slowly hydrolyzed by H_2O, precautions should be taken to avoid introducing moisture to the reagent prior to use. For this reason, the reagent is never removed from the vial supplied by the manufacturer, and the entire reaction is carried out in this vial.

The ^{125}I-labeled Bolton–Hunter reagent (250 μCi) is used to label 0.5–1.0 mg of $G_{t\alpha}$ in 100–300 ml of size-exclusion column buffer without DTT. The reagent is shipped in 100 μl of benzene, which must be removed before adding $G_{t\alpha}$. This is accomplished by evaporating the benzene with a gentle stream of N_2 gas in a vented fume hood. The N_2 is introduced into the vial supplied by the manufacturer with a 1.5-inch, 21-gauge needle through the septum on the vial. The vial is vented with a 1-inch, 21-gauge needle attached to a charcoal trap that is also supplied by the manufacturer. The N_2 stream is adjusted so that the surface of the benzene solution is gently agitated without splashing up the walls of the vial. When the benzene is thoroughly evaporated, $G_{t\alpha}$ is injected into the vial. The iodination is allowed to proceed on ice for 2 hr, at which time the reaction is essentially complete. By then, any ^{125}I-labeled Bolton–Hunter reagent that does not react with $G_{t\alpha}$ has been hydrolyzed by the aqueous buffer. ^{125}I-Labeled $G_{t\alpha}$ is separated from hydrolyzed Bolton–Hunter reagent on a 1.2×30-cm Sephadex G-25 column using the Pharmacia FPLC system. After preequilibration in isotonic buffer, the column is run at 1.0 ml/min, and the effluent is monitored for protein at 280 nm. Fractions (2 ml) are collected and monitored for radioactivity by counting 5-μl samples from each fraction in a gamma counter. ^{125}I-Labeled $G_{t\alpha}$ elutes in the void volume, whereas the hydrolyzed reagent elutes in the total column volume. The ^{125}I-labeled $G_{t\alpha}$ is concentrated in Centricon 10 microconcentrators to ~100 μl, and 50% glycerol is added. The protein concentration and the specific activity of the ^{125}I-labeled $G_{t\alpha}$ are determined by counting 1–5 μl in the gamma counter and then measuring the amount of protein in these same samples using a Pierce (Rockford, IL) Coomassie blue protein assay reagent with bovine serum albumin (BSA) as a standard. Typical specific activities range from 5000 to 10,000 cpm/pmol. Higher specific activities are not recommended because the loss of binding activity of the ^{125}I-labeled $G_{t\alpha}$ is accelerated. When stored in 50% glycerol, ^{125}I-labeled $G_{t\alpha}$ shows no loss of Rho* binding activity for 7–10 days. Beyond 10 days, a gradual decay in ^{125}I-labeled $G_{t\alpha}$ activity is observed. This decline in activity could result from autoradiation of the $G_{t\alpha}$ by ^{125}I radioactive decay products.

Addition of BSA at 1–10 mg/ml to the ^{125}I-labeled $G_{t\alpha}$ alleviates the loss of activity to some degree. It should be noted that BSA blocks dark binding of $G_{t\alpha}$ to rod outer segment membranes, the half-maximal effect occurring at ~60 μM BSA (12). However, BSA does not effect light-induced binding of G_t to Rho* up to 50 μM. Thus, ^{125}I-labeled $G_{t\alpha}$ can be stored in BSA if in Rho* $\cdot$ $G_{t\alpha}$ binding experiments, the BSA concentration is below 50 μM.

Preparation of Urea-Stripped ROS Membranes

Buffers

> Hypotonic buffer B (100 ml).
> Hypotonic buffer B plus 5 M urea, pH 6.1 (20 ml).
> Hypotonic buffer A (20 ml).
> Isotonic buffer (20 ml).
> Isotonic binding buffer: isotonic buffer plus 1.0 mg/ml BSA (2.0 ml).
> Isotonic binding buffer plus 20% (w/v) sucrose (2.0 ml).

Materials

> ROS membranes (1–2 ml of 100 μM Rho) prepared as described above.
> 3-ml syringe with a 25-gauge needle.
> 8 × 20-ml high-speed fixed angle rotor (Beckman Ti 70).

8 × 50-ml fixed angle rotor (Sorvall SS-34 or Beckman JA-20) equipped with microcentrifuge tube adaptors.

Procedure

In order to measure the binding of ^{125}I-labeled $G_{t\alpha}$ to Rho*, ROS membranes must first be depleted of endogenous $G_{t\alpha}$. If they are not, endogenous $G_{t\alpha}$ competes with ^{125}I-labeled $G_{t\alpha}$ for binding to Rho*, and the resulting measurement of ^{125}I-labeled $G_{t\alpha}$ binding to Rho* is an underestimate. ROS membranes are depleted of endogenous $G_{t\alpha}$ and other peripheral proteins by treatment with 5 M urea (13). All steps are carried out in the dark under infrared illumination. A portion of the ROS membranes is washed three times in 20 ml of hypotonic buffer B. All centrifugations are for 15 min at 140,000 g at 4°C. The supernatants are discarded and the pellets are resuspended in 5 ml of hypotonic buffer B by drawing 10 times through a 25-gauge needle and syringe and then are diluted to 20 ml with this same buffer. The GTP-washed ROS are resuspended as just described in 5 ml of 5 M urea in hypotonic buffer B at pH 6.1 and diluted to 20 ml of this buffer. A pH of 6.1 is chosen because Rho has been shown to be most stable at pH 6.1 (14). The suspension is incubated for 1 hr on ice. The suspension is then centrifuged and the supernatant is discarded. The pellet is washed once each in 20 ml of hypotonic buffer A and isotonic buffer. Finally, the pellet is resuspended in 1.0 ml of isotonic binding buffer and layered over two 1.0-ml cushions of 20% sucrose in the same buffer. The samples are centrifuged for 20 min at 43,000 g, and the pellets are resuspended in ~500 μl of isotonic binding buffer. Thus, a subpopulation of larger membrane fragments that readily pellet through 20% sucrose is selected. Without this precaution small, sedimentation-resistant vesicles could complicate the ^{125}I-labeled $G_{t\alpha}$ binding assay. The Rho concentration is determined by difference spectroscopy in the presence of 1% Triton X-100 and 50 mM hydroxylamine (15). These urea-stripped ROS membranes (UROS) contain no residual endogenous $G_{t\alpha}$ as determined by measuring light-induced [^{35}S]GTPγS binding using nitrocellulose filter assays (7). They do, however, retain 5% of the endogenous $G_{t\beta\gamma}$ (16). This residual $G_{t\beta\gamma}$ does not interfere with the measurement of ^{125}I-labeled $G_{t\alpha}$ binding to Rho*, but simply decreases the amount of exogenous $G_{t\beta\gamma}$ that has to be added back to the urea-stripped ROS membranes to obtain maximal binding of ^{125}I-labeled $G_{t\alpha}$. The 5 M urea treatment appears not to denature Rho, because 100% of the photolyzable Rho can be recovered after urea stripping if losses of membrane from the washing steps are taken into account. UROS are stored at ~50 μM Rho at 4°C for at least 3 weeks with no loss in ^{125}I-$G_{t\alpha}$ binding activity.

Binding of ^{125}I-Labeled G_t to ROS Membranes

Buffer

Isotonic binding buffer (50 ml).

Materials

Two 8 × 50-ml fixed angle rotors (Sorvall SS-34 or Beckman JA-20) equipped with microcentrifuge tube adapters.

Gamma counter (LKB/Wallac 1272 Clinigamma)

Graphics software with curve-fitting capabilities (KaleidaGraph data analysis/graphics application, Synergy Software, Reading, PA).

The binding of ^{125}I-labeled $G_{t\alpha}$ to Rho* in the presence of excess $G_{t\beta\gamma}$ is measured by sedimentation analysis. The amount of ^{125}I-labeled $G_{t\alpha}$ that sediments with illuminated UROS membranes is ~10-fold greater than with unilluminated ROS membranes (Fig. 2A). This difference is a result of the high-affinity binding of ^{125}I-labeled $G_{t\alpha}/G_{t\beta\gamma}$ to Rho* and is taken as a measure of the binding of ^{125}I-labeled $G_{t\alpha}/G_{t\beta\gamma}$ to Rho*. Portions of ^{125}I-labeled $G_{t\alpha}$ and $G_{t\beta\gamma}$ stock solutions (20–50 μM for each subunit species) are diluted in isotonic binding buffer to give the desired concentration in 500-μl microcentrifuge tubes. In the dark, a 2 μM Rho suspension is prepared from stock UROS suspensions by diluting in isotonic binding buffer. Of this UROS suspension, 32 μl is added to the ^{125}I-labeled $G_{t\alpha}$ and $G_{t\beta\gamma}$ solution to give a final Rho concentration of 200 nM and a total volume of 320 μl. Samples are incubated at 30°C for 20 min in the dark, during which time they are divided into three 90-μl aliquots in 500-μl microcentrifuge tubes. One aliquot is illuminated for 1 min by two 100-W tungsten lamps at a distance of 20 cm, which bleaches 50% of the Rho pool. This aliquot and one unilluminated aliquot are immediately centrifuged for 20 min at 43,000 g in a Sorvall SS-34 rotor equipped with microcentrifuge tube adaptors to pellet the UROS. The third aliquot is reserved for total ^{125}I-labeled $G_{t\alpha}$ determination. The free concentration of $G_{t\alpha}$ is determined by removing three 20-μl aliquots from the supernatants of the centrifuged samples and measuring ^{125}I-labeled $G_{t\alpha}$ by counting in the gamma counter. The total concentration of ^{125}I-labeled $G_{t\alpha}$ is measured by removing similar aliquots from the uncentrifuged sample and counting. The amount of ^{125}I-labeled $G_{t\alpha}$ bound to the UROS pellets is then determined by subtracting the average of the three 20-μl total count aliquots from a similar average of the supernatant counts. The volume of the UROS pellets is negligible, so no correction for pellet volume is made.

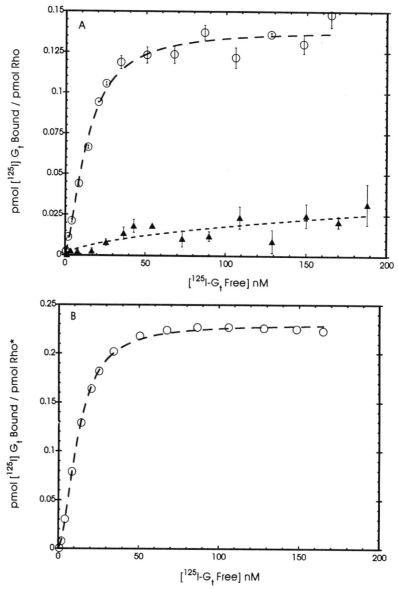

FIG. 2 Binding of ^{125}I-labeled $G_{t\alpha}$ to ROS membranes. (A) The light (○) and dark (▲) binding of ^{125}I-labeled $G_{t\alpha}$ to UROS membranes was measured in the presence of $G_{t\beta\gamma}$. Data points are average values ± standard error. (B) Binding caused by the interaction of ^{125}I-labeled $G_{t\alpha}/G_{t\beta\gamma}$ with Rho* was assessed by subtracting the curve fit of the light data from the curve fit of the dark data in A at the concentrations indicated. (C) Hill plot of the ^{125}I-labeled $G_{t\alpha}/G_{t\beta\gamma} \cdot$ Rho* binding data from B. Data points in the linear range of the Hill plot (between 10 and 90% saturation) are shown. Reproduced with permission from Ref. 7.

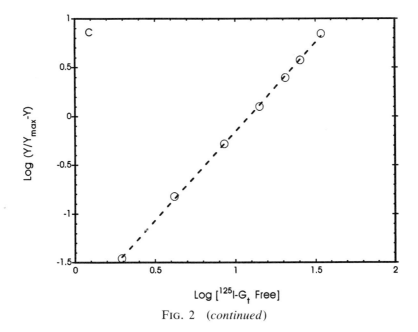

FIG. 2   (*continued*)

Pipetting errors between aliquots are between 0.5 and 2%. No loss of counts from the supernatant is observed in the absence of UROS.

The amount of $G_{t\beta\gamma}$ required to obtain maximal ^{125}I-labeled $G_{t\alpha}$ binding to Rho* is determined by varying the $G_{t\beta\gamma}$ concentration from 0 to 400 nM $G_{t\beta\gamma}$ at fixed ^{125}I-labeled $G_{t\alpha}$ and Rho concentrations (50 and 200 nM, respectively). The results of these $G_{t\beta\gamma}$ titration experiments are shown in Fig. 1. The data represent the difference between the binding in the light and dark at each $G_{t\beta\gamma}$ concentration. The curve represents a nonlinear least-squares fit of the data to the Hill equation using KaleidaGraph graphics software: $B = B_{max}(X)^n / [K_H + (X)^n]$, where B is the amount bound, B_{max} is the total binding site concentration, K_H is the Hill constant, and n is the apparent Hill coefficient. The n_{app} is 1.65 ± 0.22, and half-maximal binding occurred at ~50 nM $G_{t\beta\gamma}$. The binding is saturated by ~400 nM $G_{t\beta\gamma}$. This concentration of $G_{t\beta\gamma}$ is used in all subsequent ^{125}I-labeled $G_{t\alpha}$ binding experiments.

The concentration dependence of ^{125}I-labeled $G_{t\alpha}$ binding is assessed by measuring the binding of ^{125}I-labeled $G_{t\alpha}$ at concentrations ranging from 1 to 200 nM with 400 nM $G_{t\beta\gamma}$ and 200 nM Rho in UROS membranes. The results of ^{125}I-labeled $G_{t\alpha}$ binding experiments in the light and dark are shown in Fig. 2A. The dashed lines represent nonlinear least-squares fits of the data to the Hill equation. These fits reveal an n_{app} of 1.6 ± 0.2 for the light curve, suggesting positive cooperative binding of ^{125}I-labeled $G_{t\alpha}/G_{t\beta\gamma}$ to Rho*. No positive cooperativity is observed with the dark curve ($n_{app} = 0.64 \pm 0.6$).

The most accurate measurement of light-induced binding to Rho* is obtained by subtracting the light curve from the dark curve. This method avoids errors that can be introduced by individual points on the light or dark curves. In addition, at a given total ^{125}I-labeled $G_{t\alpha}$ concentration, the free concentration of ^{125}I-labeled $G_{t\alpha}$ in the light is always less than the free concentration of ^{125}I-labeled $G_{t\alpha}$ in the dark. This is because much more ^{125}I-labeled $G_{t\alpha}$ is found in the bound fraction in the light. Thus, individual light and dark paired data points never have the same concentration of free ^{125}I-labeled $G_{t\alpha}$. Subtracting the dark curve from the light curve avoids this problem. This subtraction to obtain the Rho*-specific binding is valid if illumination adds Rho* sites but does not significantly decrease the number of dark binding sites available. This appears to be the case, for when UROS are heat denatured at 95°C for 5 min, light binding is abolished, whereas no change in dark binding occurs. This result can be justified if the dark binding of ^{125}I-labeled $G_{t\alpha}$ to the membrane occurs via interactions with lipid and not dark Rho or another membrane protein that would be denatured on heating.

The binding curve generated by subtracting the light curve from the dark curve is also sigmoidal in nature, and fitting the data to the Hill equation yields an n_{app} of 1.84 ± 0.05 (Fig. 2B). A Hill plot of the light-induced binding data confirms the cooperativity, giving an n_{app} of 1.8 in the linear range of the data (between 10 and 90% saturation; Fig. 2C). In the Hill plot, Y represents the amount of ^{125}I-labeled $G_{t\alpha}$ bound at a particular concentration of ^{125}I-labeled $G_{t\alpha}$, and Y_{max} represents the ^{125}I-labeled $G_{t\alpha}$ bound at saturation, as determined by the curve fit in Fig. 2B. The n_{app} values are consistently within this range, showing little variation between G_t preparations (1.85 ± 0.12, $n = 4$). However, n_{app} values consistently decrease after 7–10 days of storage of the ^{125}I-labeled $G_{t\alpha}$, until at ~21 days no cooperative binding is observed. The source of this decay of cooperativity is not known, but it can be avoided if ^{125}I-labeled $G_{t\alpha}$ binding is measured within 10 days of iodination.

Because the Hill coefficient provides a lower limit for the order of cooperativity, an n_{app} of 1.8 suggests at least two sites of cooperative interaction but does not rule out a greater number of interacting sites. To address the number of allosteric sites to which G_t binds, the data are compared with what would be predicted by a simple two-interaction site model. In this model, the binding of the first molecule of G_t to either of the two sites facilitates the binding of a second molecule of G_t to the remaining free site.

$$\text{Rho*} + G_t \overset{K_{d1}}{\rightleftharpoons} \text{Rho*} \cdot G_t$$

$$\text{Rho*} \cdot G_t + G_t \overset{K_{d2}}{\rightleftharpoons} \text{Rho*} \cdot (G_t)_2$$

The amount of ^{125}I-labeled $G_{t\alpha}$ bound is described by the following equation:

$$B = \frac{\frac{1}{2}B_{max}\{[G]/K_{d1} + 2[G]^2/(K_{d1}K_{d2})\}}{1 + [G]/K_{d1} + [G]^2/(K_{d1}K_{d2})}$$

where [G] is the concentration of free G_t, B and B_{max} are as described above, and K_{d1} and K_{d2} are the dissociation constants for the binding to the first and second G_t sites, respectively. The fit of the data to this equation is represented as the dashed line in Fig. 2B. The resulting K_d values from the curve fit are $K_{d1} = 80 \pm 30$ nM and $K_{d2} = 1.9 \pm 0.7$ nM. The excellent fit of the data to this model suggests that a two-site binding model for G_t binding to Rho* is sufficient to describe the binding. The binding of the first G_t to Rho* enhances the binding of the second G_t approximately 40-fold. The possible impact of this 40-fold increase in binding affinity on the rapid amplification of signal that occurs when one Rho* activates many G_t molecules is a topic of current research. The molecular structure of the cooperatively interacting sites is also being investigated.

The iodination of $G_{t\alpha}$ may have reduced the ability of ^{125}I-labeled $G_{t\alpha}$ to bind Rho* or $G_{t\beta\gamma}$ by reacting with residues involved in the interaction. This would result in apparent K_{d1} and K_{d2} values that are higher than the true dissociation constants. Such is the case with sulfhydryl reagents that have been used to label G_t in the past (17). Sulfhydryl reagents modify Cys-346 in $G_{t\alpha}$, which is part of the C-terminal Rho* binding domain, and decrease the binding of labeled G_t to Rho*. The ^{125}I-labeled Bolton–Hunter labeling method used here reacts with primary amino groups on $G_{t\alpha}$ and may not interfere with binding to Rho*. To test this possibility, increasing concentrations of unlabeled $G_{t\alpha}$ (0–400 nM) are mixed with 45 nM ^{125}I-labeled $G_{t\alpha}$, 400 nM $G_{t\beta\gamma}$, and 100 nM Rho, and light-induced binding to UROS is measured as described above. Unlabeled $G_{t\alpha}$ is found to inhibit the binding of ^{125}I-labeled $G_{t\alpha}$ over a concentration range that would be expected if the iodination did not alter the Rho* $\cdot G_t$ interaction (Fig. 3). The light-induced binding data fit closely the dashed line, which represents an ideal competitive inhibition in which K_{d1} and K_{d2} are the same for ^{125}I-labeled $G_{t\alpha}$ and unlabeled $G_{t\alpha}$. In this case,

$$B = \frac{\frac{1}{2}B_{max}\{[G]/K_{d1} + 2[G]^2/(K_{d1}K_{d2}) + [G][I]/K_{d1}K_{d2})\}}{\{1 + [G]/K_{d1} + [G]^2/(K_{d1}K_{d2}) \\ \quad + [I]/K_{d1} + 2[G][I]/(K_{d1}K_{d2}) + [I]^2/(K_{d1}K_{d2})\}},$$

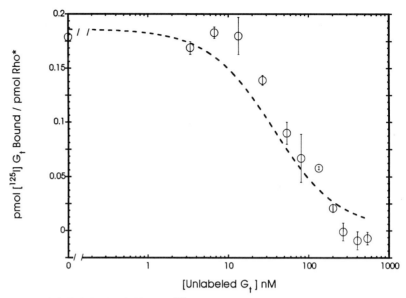

FIG. 3   Unlabeled $G_{t\alpha}$ inhibition of ^{125}I-labeled $G_{t\alpha}$ binding to Rho*. Data points are average values $\pm$ standard error from four separate experiments. Reproduced with permission from Ref. 7.

where [I] is the free concentration of unlabeled G_t, and the other parameters are as described above. Thus, iodination with Bolton–Hunter reagent does not perturb the binding properties of $G_{t\alpha}$.

References

1. L. Stryer, *Annu. Rev. Neurosci.* **9**, 87 (1986).
2. M. Chabre and P. Deterre, *Eur. J. Biochem.* **179**, 255 (1989).
3. M. Wessling-Resnick and G. Johnson, *J. Biol. Chem.* **262**, 3697 (1987).
4. H. Kuhn, N. Bennett, M. Michel-Villaz, and M. Chabre, *Proc. Natl. Acad. Sci. U.S.A.* **78**, 6873 (1981).
5. D. Emeis, H. Kuhn, J. Reichert, and K. P. Hofmann, *FEBS Lett.* **143**, 29 (1982).
6. B. K.-K. Fung, *J. Biol. Chem.* **258**, 10495 (1983).
7. B. M. Willardson, B. Pou, T. Yoshida, and M. W. Bitensky, *J. Biol. Chem.* **268**, 6371 (1993).
8. K. Kleuss, M. Pallast, S. Brendel, W. Roshethal, and G. Schultz, *J. Chromatog.* **407**, 281 (1987).
9. A. Yamazaki, M. Tatsumi, and M. W. Bitensky, *in* "Methods in Enzymology," Vol. 159, p. 702. Academic Press, San Diego, 1988.

10. H. Kuhn, *Nature (London)* **283,** 587 (1980).
11. A. E. Bolton and W. M. Hunter, *Biochem. J.* **133,** 529 (1973).
12. B. E. Buzdygon and P. A. Liebman, *J. Biol. Chem.* **259,** 14567 (1984).
13. A. Yamazaki, F. Bartucca, A. Ting, and M. W. Bitensky, *Proc. Natl. Acad. Sci. U.S.A.* **79,** 3702 (1982).
14. S. M. A. Khan, W. Bolen, P. A. Hargrave, M. M. Santoro, and J. H. McDowell, *Eur. J. Biochem.* **200,** 53 (1991).
15. M. D. Bownds, A. Gordon-Walker, A. C. Gaide-Hugnenin, and W. Robinson, *J. Gen. Physiol.* **58,** 225 (1971).
16. T. Yoshida, B. M. Willardson, J. F. Wilkins, G. J. Jensen, B. D. Thornton, and M. W. Bitensky, *J. Biol. Chem.* **269,** 24050 (1994).
17. K. P. Hofmann and J. Reichert, *J. Biol. Chem.* **260,** 7990 (1985).

[19] Crosstalk between Tyrosine Kinase and G-Protein-Linked Signal Transduction Systems

Ross D. Feldman and Michel Bouvier

Introduction

Increasing interest has focused on the effects of tyrosine kinase-mediated phosphorylation on G-protein-linked receptor–effector systems. The approaches have been variable, including assessment of the effects of hormones that activate tyrosine kinase activity (insulin and tyrosine kinase-linked growth factor receptor hormones), purified enzymes, and the use of tyrosine phosphatase inhibitors (e.g., oxygen-derived free radicals; see below). Additionally, the effects of tyrosine kinase activation have been inferred based on the characterization of the effects of insulin deficiency on G-protein-linked systems (as in diabetic models).

As might be expected, these activators of tyrosine kinase-mediated phosphorylation have resulted in variable effects on G-protein-linked transmembrane signaling mechanisms (i.e., receptor, G protein, or G-protein-linked effector). Variability in the effect of tyrosine kinase activation has also been noted with different cell types and may be dependent on whether cells have been freshly isolated or cultured (1).

In this chapter we present a description of techniques that we have found useful in dissecting out tyrosine kinase-mediated effects on adenylylcyclase-linked G protein receptor systems (primarily β-adrenergic receptor systems). We have studied these effects in several cell lines that demonstrate a common effect—i.e., increased responsiveness/function associated with perturbations that increase tyrosine kinase activity.

Functional/Cellular Approaches to Study of Tyrosine Kinase-Mediated Alterations in G-Protein-Linked Receptor Systems

Techniques for Stimulation of Tyrosine Kinase-Mediated Effects in Intact Cell Models

The techniques outlined below have been utilized in studies wherein either (i) tyrosine kinase has been stimulated physiologically by hormones that activate receptors linked to activation of the enzyme (insulin), (ii) phospha-

Methods in Neurosciences, Volume 29

tase inhibitors have been used to inhibit tyrosine dephosphorylation, or (iii) hydroxyl radical-generating systems have been used. In utilizing insulin as a stimulator of tyrosine kinase activity, it should be appreciated that at higher concentrations (greater than 1 nM) insulin-mediated effects may not necessarily be due to insulin receptor activation, but could reflect cross-reactivity with other growth factor receptor systems. This is especially important because these pathways, although also tyrosine kinase-dependent, diverge significantly in their "downstream" mechanisms. Thus, many of the tyrosine kinase-linked growth factor receptors directly stimulate phospholipase C activity whereas insulin receptor activation does not. This divergence is of functional importance, e.g., in the vasculature insulin mediates vasodilation whereas other growth factors [e.g., epidermal growth factor (EGF)] mediate vasoconstriction.

Hydrogen peroxide and hydroxyl radical-generating systems (e.g., xanthine oxidase and xanthine in the presence of superoxide dismutase) have been used extensively as insulinomimetic agents. Their effects have been linked to both tyrosine kinase activation and to tyrosine phosphatase inhibition. Oxygen-derived free radicals have proved to be potent modulators of adenylylcyclase activation *in vitro*. However, it should be recalled that oxidants may have widespread and nontyrosine kinase-specific effects. Thus, the specificity of any effects of oxidants acting via a tyrosine kinase-dependent pathway should be confirmed with the use of tyrosine kinase-selective inhibitors (typhostins, genistein, etc.) to attenuate the effect and/or the use of phosphatase inhibitors to mimic the effect.

The protocols we have used have examined the persistent effects of these agents i.e., cells are pretreated with insulin, hydrogen peroxide, and/or phosphatase inhibitors then are well-washed and subsequently assayed. This approach has allowed us to differentiate those effects that were dependent on an intact cell and generation of second messengers vs. direct, membrane-delimited effects.

For the assessment of the effects of insulin on peripheral blood lymphocytes, the cells are incubated with human insulin for 10 min at 37°C. Increasing the incubation time up to 60 min does not significantly alter the effect.

For the assessment of the effects of hydrogen peroxide on A10 cells, cells in multiwell plates were incubated for 30 min at 37°C. The effect of hydrogen peroxide increases with incubation time up to approximately 20 min. The response is maximal at a hydrogen peroxide concentration of approximately 300 μM.

An alternative approach has been to utilize phosphatase inhibitors to enhance the signal of tyrosine kinase-mediated phosphorylation. In studies of the effects of insulin on β_2-adrenergic receptor (β_2AR) tyrosine phosphorylation, CHW cells (see below) are pretreated with the nonselective phosphatase inhibitor Na$_3$VO$_4$ (2 hr, 100 μM) before the addition of insulin.

Utilization of Permeabilized Cell Preparations in Study of Tyrosine Kinase-Mediated Regulation of G-Protein-Linked Transmembrane Signaling Systems

The use of permeabilized cell systems have been useful in assessing alterations in G-protein-linked transmembrane signaling systems. These techniques have been especially useful in the study of persistent effects of tyrosine kinase activation on G-protein-linked receptor systems. Their utility, in part, relates to the appreciation that they demonstrate some of the efficiencies of broken cell preparation in regard to *in vitro* enzyme activity and radioligand binding assays but maintain characteristics of intact cell systems. They have been especially useful in assays of receptor-stimulated activation of cAMP-dependent protein kinase activity (see below). In the study of membrane-bound enzymes (adenylylcyclase), permeabilized preparations tend to demonstrate the lower basal activity characteristic of intact cell systems. Additionally, permeabilized cell preparations have proved useful as systems in which to study alterations in radioligand binding characteristics. In β-adrenergic receptor radioligand binding assays, the permeabilized cell preparations are still useful in identifying agonist-induced sequestration (as demonstrated by alterations in CGP-12177-sensitive binding) and also demonstrate guanine nucleotide-sensitive agonist competition for the binding of β-adrenergic selective radioligands (a characteristic of membrane preparations) (2).

General Considerations in Use of Permeabilized Cell Preparations

Several techniques have been developed for assessing receptor and second-messenger systems using permeabilized preparations, including electroporation, bacterial toxins, and low concentrations of detergents. We have utilized low concentrations of digitonin in our permeabilization preparations. Although the details of permeabilization protocols vary among cell lines, and especially differ for adherent cells and cells in suspension, certain general principles apply. Optimal permeabilization is generally achieved within a range from 10 to 100 μg/ml for digitonin using an exposure time from 15 to 30 min. The optimal detergent concentration and duration of exposure to detergent are dependent not only on the cell type but on the assay to be performed. Optimal permeabilization represents the balance between the increasing accessibility of substrates for enzyme assays (or radioligand binding assays) seen with increasing digitonin concentration versus the adverse effects of increasing detergent concentrations on enzyme function. This is most notable in the assessment of adenylylcyclase activity. In our hands, optimal detergent concentrations for enzyme assays are generally in the range of 10–30 μg/ml. Higher concentrations are used in the assessment of radioligand binding properties (i.e., up to a digitonin concentration of 75 μg/ml for guanine nucleotide-sensitive agonist competition studies). Again, detailed dose–response curves should be performed during the development of these assays because the determination of the optimal digitonin concentrations for each assay is strictly empirical. Additionally, in cell suspension permeabilization protocols, the cell density at the time of digitonin exposure affects the extent of permeabilization (as assessed by trypan blue exclusion). We have found optimal permeabilization occurs at a concentration of approximately 2×10^6 cells/ml.

Permeabilized Preparation in Suspended Cells

Our protocol for lymphocyte permeabilization (3) was adapted from the method of Brooker and Pedone (4). We have utilized these approaches in the study of peripheral blood mononuclear leukocytes (as described below), cultured lymphocyte lines (S_{49} murine lymphoma cells and Jurkat T cells), as well as freshly isolated cardiocytes. The technique for permeabilization of peripheral blood mononuclear leukocytes is described below. EDTA-anticoagulated blood is drawn. Mononuclear leukocytes are isolated according to the method of Boyum (4a). Following separation on Ficoll–Hypaque gradients, mononuclear cells are harvested, resuspended in Hanks' balanced salt solution (without Mg^{2+} or Ca^{2+}), centrifuged for 10 min at 400 g at room temperature, and resuspended in Hanks' balanced salt solution (HBSS), pH 7.4, 4°C, with 33 mM HEPES, 0.5 mM EDTA (buffer A). Cells are permeabilized with digitonin (Sigma, St. Louis, MO) that is prepared by dissolving in boiling water (6 g/60 ml), cooling overnight at room temperature, filtering to remove insoluble material, and lyophilizing. A final digitonin concentration of 10 μg/ml is used for studies of adenylylcyclase activity, cAMP-dependent protein kinase activity, and radioligand binding studies involving determination of the proportion of CGP-12177-accessible sites. A final digitonin concentration of 75 μg/ml is used for radioligand binding studies relating to isoproterenol competition binding and guanine nucleotide sensitivity. Adenylylcyclase activity is markedly decreased at digitonin concentrations greater than 30 μg/ml. Treatment with 10 μg/ml digitonin results in optimal detection of adenylylcyclase activity. However, the ability of guanine nucleotides to shift agonist competition curves is significantly diminished at digitonin concentrations of 10 μg/ml as compared to guanine nucleotide sensitivity achieved at a concentration of 75 μg/ml digitonin. In contrast, agonist-induced alterations in CGP-12177-accessible labeling are comparable at 10 or 75 μg/ml digitonin. Cells are incubated in digitonin for 15 min at 4°C, washed at 400 g for 10 min at 4°C, resuspended in buffer A, washed again, and resuspended for radioligand binding studies or assays of adenylyl-cyclase activity.

Assays of Adenylycyclase Activity in Permeabilized Lymphocytes

Assays of adenylylcyclase activity are performed on permeabilized cell preparations according to our previously published methods, modified after Salomon *et al.* (5). Permeabilized cells that have been resuspended in buffer A are added in an aliquot of 40 μl to give a final incubation volume of 100 μl with 1 μCi [α-^{32}P]ATP, 0.3 mM ATP, 2 mM MgSO$_4$, 0.1 mM cAMP (used in lieu of a phosphodiesterase inhibitor), 5 mM phosphoenol pyruvate, 40 μg/ml pyruvate kinase, and 20 μg/ml myokinase. Incubations are carried out at 37°C for 10 min and terminated by addition of 1 ml of a solution containing 100 μg ATP, 50 μg cAMP, and 15,000 cpm [^{3}H]cAMP. Cells are then pelleted by centrifugation at 300 g for 5 min and cAMP is isolated from the supernatant by sequential Dowex and alumina chromatography corrected for recovery with [^{3}H]cAMP as the internal standard. Adenylylcyclase activity should be linear with time and cell number over the ranges used in the studies. Adenylylcyclase activity can be stimulated by the use of receptor-specific, G-protein-selective, and catalytic subunit-selective stimulators. Hormonal stimulation of adenylylcyclase activity (as with isoproterenol) should be performed in the presence of either GTP (up to 100 μM) or a GTP analog, such as GPP(NH)p. Prostaglandin E$_1$ (PGE$_1$) is a more effective receptor-specific stimulator of adenylylcyclase activity of lymphocytes. G-Protein-selective stimulation can be achieved with either nonhydrolyzable analogs of GTP [e.g., GPP(NH)p) or NaF/AlCl$_2$ combinations]. Catalytic subunit activity can be assessed with manganese chloride or manganese chloride/forskolin combinations.

As noted above, proportional stimulation of adenylylcyclase activity (over basal) by activators is significantly higher in permeabilized cells than is seen with crude broken cell preparations, probably due to lower basal activities with this method.

Assays of cAMP-Dependent Protein Kinase Activation in Permeabilized Lymphocytes

Assays of cAMP-dependent protein kinase activity are performed in permeabilized cells according to modifications of the methods of Roskowski (6). Permeabilized cells prepared as above and resuspended in buffer A with 2 mM magnesium sulfate are incubated for 12 min at 30°C with 1 mM kemptide (Sigma Chemical Co.), 0.5 mM isobutylmethylxanthine, 1 μg bovine serum albumin (BSA), 0.5 mM ascorbic acid, 0.8 mM ATP, 1–2 μCi [α-^{32}P]ATP, 100 μM GTP, in a final volume of 100 μl. Reactions are terminated by spotting aliquots (80 μl) on 2 $\times$ 3-cm phosphocellulose strips (Whatman P81) and immersing them in 75 mM phosphoric acid. The strips are swirled gently for 2 min, the phosphoric acid decanted, and the strips washed twice more, as above. Radioactivity is measured by liquid scintillation spectroscopy. Background is determined by blanks incubated in the absence of kemptide, cells, or [^{32}P]ATP alone and should account for less than 15% of basal activity. Protein kinase activity in permeabilized lymphocytes is linear with time up to at least 30 min. Maximal levels of stimulated activity are determined in the presence of PGE$_1$ (100 μM), isoproterenol (100 μM), or cAMP (100 μM). We express hormonal activity as the ratio of hormonal-stimulated activity to cAMP-stimulated activity.

β-Adrenergic Receptor Radioligand Binding Studies in Permeabilized Lymphocytes

Radioligand binding studies can be performed in permeabilized lymphocytes resuspended in buffer A (as above) with the use of [^{125}I]iodopindolol (IPIN; 2.2 Ci/μmol, New England Nuclear Corporation, Boston, MA). IPIN and other drugs are prepared in 1.25 mM ascorbic acid as an antioxidant and with 10 μg/ml bovine serum albumin. For saturation binding curves, eight concentrations of IPIN (15–150 pM) are used in each assay. For competition curves, one concentration of IPIN (44–67 pM) is used in each assay. Competition binding studies are performed at 11 concentrations of isoproterenol from 1×10^{-3} to 1×10^{-8} M. Studies are performed at 25°C, with steady-state conditions being reached at 90 min at all IPIN concentrations used. Binding assays started by adding 100 μl of the cell preparation containing $(4-6) \times 10^5$ cells in buffer A to the incubation mixture containing the radioligand (added in a 50-μl aliquot in buffer A). Guanine nucleotides (for saturation binding studies and for "low-affinity" agonist competition curves) are added as a 50-μl addition in buffer A. Competing adrenergic agonists and antagonists are added as a 50-μl addition in the BSA/ascorbic acid solution as above. The final volume of the incubation is 250 μl. The assays are performed in disposable polystyrene test tubes (Fisher Scientific). After steady-state conditions have been achieved, reactions are diluted to 10 ml with a 1:10 dilution of buffer A at 25°C. The assay tubes are incubated for an additional 3 min. This step substantially lowers nonspecific binding. Incubation is terminated by rapid vacuum filtration (Whatman GF/C filters). Each filter is washed with an additional 10 ml of diluted buffer (at 25°C) and radioactivity is determined.

Saturation binding curves are used to determine receptor density ($B_{\max}$) and receptor affinity for the radioligand (K_D). IPIN labeling in the presence of propranolol (1 μM) is defined as nonspecific binding. The proportion of specific (IPIN) binding blocked by the hydrophilic ligand CGP-12177 (30 nM) is used as an index of "nonsequestered" β-receptor sites.

Receptor affinity for agonists is determined from radioligand–agonist competition curves using permeabilized cells in the presence and absence of guanylimidodiphosphate [Gpp(NH)p]. As compared to the previous broken cell preparation we have used previously (7), β-adrenoceptor affinity for isoproterenol in the absence of Gpp(NH)p is increased up to 10-fold in the permeabilized cell preparation.

Data from competition curves and saturation binding curves should be analyzed by nonlinear curve-fitting algorithms. We have used Scatfit software (by Andre De Lean) and more recently Inplot software (GraphPad, San Diego, CA). For competition curves performed in the absence of guanine nucleotides, fitted estimates for two-affinity states are obtained in the setting where the fit of the data are significantly better than that obtained in a one-affinity state model.

Permeabilized Preparation in Adherent Cells

We have successfully utilized permeabilized adherent cell lines (A10 and CHW cells). The approaches used in permeabilization and in enzyme activity assays are analogous to those used in suspended cell preparations. Adherent cells are cultured and permeabilized, and assays are performed in multiwell culture plates. Cell adherence following digitonin exposure should be confirmed microscopically. To start adenylylcyclase or cAMP-dependent protein kinase assays, 75 μl of buffer A is added to each well. The rest of the assay constituents (ATP-regenerating cocktail, [^{32}P]ATP, etc.) are premixed. A premix "soup" is prepared for each assay condition (i.e., basal, GTP-stimulated, isoproterenol) and 125 μl of "soup" is added as a single aliquot to each well at 5-sec intervals to start the assays for adenylylcyclase activity and 10-sec intervals for cAMP-dependent protein kinase activity.

A10 cells (obtained from American Type Culture Collection, Rockville, MD) are cultured in 48-well plates in Dulbecco's modified Eagle's medium (DMEM) with 5% (v/v) fetal calf serum. Assays are performed on cells in stationary growth-phase cultures and at least 4 days following addition of fresh media.

The medium is aspirated from the multiwell culture plates and cells are washed in Hanks' balanced salt solution (pH 7.4 at 4°C) with 33 mM HEPES and 0.5 mM EDTA (buffer A), with 2 mM magnesium sulfate. Cells are permeabilized with the addition of digitonin (10 μg/ml) in buffer A and are incubated for 25 min at 4°C. Cells are then washed in buffer A without digitonin.

Adenylylcyclase activity is assessed by the conversion of [^{32}P]ATP to [^{32}P]cAMP as described above. Permeabilized cells are incubated in a final volume of 200 μl with 1 μCi [α-^{32}P]ATP, 0.3 mM ATP, 2 mM magnesium sulfate, 0.1 mmol cAMP, 5 mmol phosphoenolpyruvate, 40 mg/ml pyruvate kinase, 20 mg/ml myokinase, and 0.5 mM EDTA. Assays are begun with the addition of premixed "soups" as above. Cells are incubated for 20 min at 37°C. Incubations are terminated by the addition of 1 ml of a solution containing 100 μg ATP, 50 μg cAMP, and 15,000 cpm [^{3}H]cAMP (New England Nuclear). cAMP is isolated in the supernatant by sequential Dowex and alumina chromatography and corrected for recovery with [^{3}H]cAMP as the internal standard. Protein concentration in permeabilized cells is determined by the method of Bradford (Bio-Rad Protein Assay, Bio-Rad Laboratory).

Assays of cAMP-dependent protein kinase activity are performed in permeabilized cells according to modifications of our previous methods utilizing suspended cell preparations, as above. Permeabilized cells in buffer A are incubated for 20 min at 30°C with 1 mM kemptide (Sigma), 0.5 mM isobutyl-methylxanthine, 1 μg/ml bovine serum albumin, 0.5 mM ascorbic acid, 0.8 mM ATP, 1 μCi [α-^{32}P]ATP in a final volume of 125 μl. Assays are begun with addition of premixed "soups" and terminated following immersion of the multiwell plate in an ice slurry and sequential spotting of 100-μl aliquots from each well onto phosphocellulose paper. Papers are washed and counted as described above.

Assessment of Tyrosine Phosphorylation of G-Protein-Coupled Receptors

Numerous studies have demonstrated the importance of phosphorylation in the regulation of G-protein-coupled receptors (GPCRs) (reviewed in Ref. 3). In particular, phosphorylation of serine and threonine residues in cytosolic domains of many GPCR leads to functional uncoupling—the first step of agonist-promoted desensitization. In many respects, the β_2-adrenergic receptor may be considered to be prototypical of GPCRs and has been used as a model for the study of this receptor family. Two distinct protein kinases, the cAMP-dependent protein kinase (PKA) and the β-adrenergic receptor kinase (βARK), have been implicated in the phosphorylation events associated with functional uncoupling and leading to rapid desensitization. Whereas PKA has broad substrate specificity, βARK demonstrates a much more restricted specificity and phosphorylates only agonist-occupied receptor (8). Using site-directed mutagenesis we and others identified several phosphorylation sites involved in desensitization (9–11). βARK-mediated phosphorylation of serine and threonine residues in the distal portion of the carboxyl terminus of the receptor has been shown to promote the association of the protein β-arrestin with the β_2AR, thus inhibiting functional coupling of the receptor to G$_s$ (12, 13). In contrast, PKA mediates phosphorylation of the

β_2AR on sites located in the third cytoplasmic loop and in the carboxyl tail. This results in functional uncoupling but not via β-arrestin association.

In contrast to the abundant literature concerning the phosphorylation of serine and threonine residues in GPCRs, very little is known about phosphorylation of these receptors on tyrosine residues. Several studies have documented that activation of tyrosine kinase receptors modulates GPCR functions, but, until recently, neither the molecular processes involved nor the existence of a direct link between tyrosine phosphorylation and GPCR-mediated signaling have been clearly established. However, recent studies reported that stimulation of cells with insulin leads to the phosphorylation of the β_2AR on tyrosine residues (14, 15). Also, tyrosine phosphorylation of the B_2-bradykinin receptor has recently been documented (16). The lability of phosphotyrosines and the low expression level of GPCRs in most tissues make tyrosine phosphorylation of these receptors particularly difficult to study. In the following sections, we present various approaches used to study tyrosine phosphorylation of the human β_2AR.

Use of Heterologous Expression Systems

As mentioned above, one of the major difficulties in directly studying GPCR phosphorylation is the low amount of receptor that can be isolated from tissues or cell lines naturally expressing these receptors. One approach that we and others have selected to circumvent this problem has been to use heterologous expression allowing various level of overexpression. In addition, these systems permit the study of selected receptor subtypes in a chosen genetic background.

Transient vs. Stable Expression Systems

Heterologous expression systems can be divided into two classes: transient and stable. In both systems, the objective is to express a sufficient quantity of receptor in a surrogate cell line that normally expresses very little (or preferably none) of this receptor. For that purpose, the gene or cDNA of the receptor under study is cloned in an expression vector so that its expression is under the control of a strong promoter. The expression vector encoding the receptor is then transfected into the selected cell line. In transient expression systems, cells can be harvested several days later (usually 48 hr) and the receptor studied. In such expression systems, the expression vector is not inserted in the genome of the cell and it functions as an independent transcription unit. However, the proportion of cells effectively harboring the expression vector is relatively small (5–20%). It follows that, on cell division, the population of cells possessing the expression vector will become diluted and that the level of expression of the receptor will rapidly decline. Thus, the expression is transient and new transfections must be performed for each experiment. Stable expression takes advantage of relatively rare recombination events that lead to random insertion of the expression vector into the genome of the cells. Such recombination events occur in a very small proportion of the cells that have internalized the expression vector. The idea is then to select those cells into which one or more copies of this vector have been stably incorporated into the genome. These cells will then constitutively express the receptor. Selective expression of these cells is achieved by virtue of selectable markers, such as the antibiotic resistance gene, which are part of the expression vectors or are cotransfected as part of independent vectors. Selection of the expressing cells is then realized by treatments with high concentrations of the antibiotic. Once obtained, stable cell lines heterologously expressing the receptor represent valuable biological reagents to study phosphorylation.

The major advantage of transient over stable expression systems is that one does not need to go through the tedious process of selection (4–6 weeks) and can readily use the cells 48 hr after the transfection. Moreover, higher levels of expression are generally obtained using transient systems. However, stable expression systems provide a homogeneous population of cells that all express the receptor and allow for repeated experiments under identical conditions. In contrast, in transient systems, only a small proportion of the total cells expresses large amounts of receptor. This leads to a heterogeneous cell population that may vary from experiment to experiment. These characteristics should be kept in mind when designing and interpreting experiments using these systems.

Choice of Cell Lines

African green monkey kidney fibroblasts (Cos 7 cells; ATCC CRL 1651) are certainly the mammalian cell line most widely used for transient expression. These cells are transformed by an origin-defective mutant of simian virus 40 (SV40) that codes for wild-type T antigen. The line thus contains the T antigen and can support the replication of SV40 virus. Expression vectors encoding an SV40 origin of replication will therefore be replicated to high copy number in these cells. This accounts for the very high level of expression obtained in these cells. Using the expression vector pBC12BI (17) encoding the human β_2AR, one can achieve levels of receptor expression of up to 15 pmol/mg membrane proteins. It should be remembered that only a small portion of the total cell population will incorporate the vector and this subpopulation will be responsible for the massive expression. It follows that although these cells may be used to determine if tyrosine phosphorylation does occur, they might not represent the best system with which to estimate the ambient level of receptor tyrosine phosphorylation nor to study the regulation of this phosphorylation.

A number of cell lines can be used to generate stable expression systems. Considerations in selecting a cell line should include the ease in cultivating large quantities of these cells, levels of endogenous expression of the receptor to be transfected, the presence of the signaling pathway(s) to which the receptor should be functionally coupled, and the presence of other signaling molecules that could be considered in the course of the study. Chinese hamster fibroblasts (CHW-1102; NIGMS GM0459) and mouse fibroblasts [L(TK$^-$);, ATCC CCL 1.3] have been used commonly and successfully to express stably various GPCR. Both cell lines are devoid of endogenous β_2AR and have been used extensively to study heterologous expression of the human β_2AR. For example, we wished to assess the potential crosstalk between insulin and β_2AR. Therefore, the endogenous expression of insulin receptors in CHW cells made them a particularly attractive system. In the following sections, we describe methodologies used to study tyrosine phosphorylation of the human β_2AR in CHW cells. It should be emphasized that using heterologous expression systems is always a compromise between having access to a system allowing purification of sufficient quantity of receptors for biochemical studies and studying a system that is physiologically relevant. In particular, problems of relative stoichiometries between signaling molecules is always a concern when using overexpression systems.

Stable Expression of β_2AR in CHW Cells

The human β_2AR cDNA clone pTF (18) containing 190 bp of 5' untranslated sequence was cloned into the *Bgl*II site of the eukaryotic expression vector pKSV10 containing the SV40 early promoter. The resulting plasmid was cotransfected with the neomycin-resistance plasmid pSV2-Neo into CHW-1102 fibroblasts (CHW) by calcium phosphate precipitation (19). Neomycin-resistant cells were selected in Dulbecco's modified Eagle's medium (DMEM) supplemented with 10% fetal bovine serum and 150 μg/ml of the neomycin analog G418. Selected clones were then screened for β_2AR expression by radioligand assay using the β_2AR antagonist [^{125}I]iodocyanopindolol (CYP). Individual cell lines expressing between 0.2 and 8 pmol of β_2AR/mg of membrane protein were identified and maintained as clonal stable cell lines.

Insulin Treatment of CHW Cells Expressing Human β_2AR and Sample Preparation

Cells are grown as monolayers in 600-cm^2 dishes in DMEM supplemented with 10% fetal bovine serum, penicillin (100 U/ml), streptomycin (100 μg/ml), Fungizone (0.25 μg/ml), and glutamine (1 mM) in an atmosphere of 95%

air/5% CO_2 at 37°C. The day before the experiment, nearly confluent cells are incubated in serum-free DMEM for 18 hr. Following this period, cells are incubated with the nonspecific tyrosine phosphatase inhibitor Na_3VO_4 (100 μM) for 2 hr. Insulin (1 μM) or the vehicle alone is added for the last 60 min. Thereafter, all the buffers contain 100 μM Na_3VO_4. The cells are then rinsed twice with ice-cold PBS, mechanically detached, and disrupted by sonication in ice-cold buffer containing 5 mM Tris-HCl, pH 7.4, 2 mM EDTA, 5 mg/ml leupeptin, 10 mg/ml benzamidine, and 5 mg/ml soybean trypsin inhibitor. Lysates are centrifuged at 500 g for 5 min at 4°C, the pellets are sonicated as before, spun again, and the supernatants are pooled. The pooled supernatant is then centrifuged at 45,000 g and the pelleted membranes are washed thrice in the same buffer.

Detection by Antiphosphotyrosine Antibodies

The availability of high-affinity antibodies specifically directed against phosphotyrosine residues has been an important advance in the study of protein tyrosine phosphorylation. In many instances, immunoprecipitation of a protein using antiphosphotyrosine antibody has been interpreted as evidence that this protein is indeed phosphorylated on tyrosine residues. Similarly, specific immunoblotting of an immunoaffinity-purified protein using antiphosphotyrosine antibodies has been routinely used to demonstrate and quantitate tyrosine phosphorylation. For the GPCR, the use of these tools has been hampered by the difficulty of raising receptor-specific antibodies. Therefore, the ability of antiphosphotyrosine antibody to immunoprecipitate GPCR can only be assessed based on the binding activity that can be immunoprecipitated. Unfortunately, most conditions used for efficacious immunoprecipitation do not preserve the binding activity of the receptors. Indeed only few detergents (e.g., digitonin and n-dodecylmaltoside) preserve binding activity. Further, it has been very difficult to demonstrate stoichiometric immunoprecipitation of GPCRs using antiphosphotyrosine antibodies in these detergents. Selective immunoblotting of tyrosine-phosphorylated GPCRs with antiphosphotyrosine antibodies is not feasible without initial purification of the receptor. Indeed, even in overexpression systems, these receptors represent only a very small proportion of the membrane proteins and they cannot be readily identified in immunoblots of total membrane extracts. For the β_2AR a purification procedure based on the binding properties of the receptor exists.

Affinity Purification of β_2AR
The β_2AR is solubilized from membrane preparations (see above) in 12 ml of a buffer containing 100 mM NaCl, 10 mM Tris-HCl, 5 mM EDTA, pH 7.4, 2.0% digitonin, and protease and phosphatase inhibitors with mild agitation at

4°C for 90 min. Nonsolubilized material is removed by centrifugation at 45,000 g for 20 min at 4°C. The solubilized β_2AR can then be purified by affinity chromatography using a Sepharose matrix coupled to the β-adrenergic antagonist alprenolol (20). Alprenolol-Sepharose columns (20 ml of gel/column) are equilibrated with 5 bed volumes of buffer B (100 mM NaCl, 10 mM Tris-HCl, 2 mM EDTA, pH 7.4, 0.05% (w/v) digitonin, protease, and phosphatase inhibitors). Solubilized receptor (5 ml) is loaded on the column and shaken gently for 2 hr at room temperature to allow binding of the receptor to the matrix. The supernatant is then allowed to flow through and the columns are washed with 15 ml of a buffer containing 500 mM NaCl, 50 mM Tris-HCl, 2 mM EDTA, pH 7.4, 0.05% digitonin, and protease inhibitors maintained at 4°C. The original ionic strength is restored by washing with 30 ml of buffer B. The columns are returned to room temperature and receptors are eluted with buffer B containing 60 μM alprenolol. This eluate is concentrated with Centriprep and Centricon cartridges (Amicon, Danvers, MA) down to 50 μl. β_2AR recovery after affinity purification is measured by soluble binding after desalting on a G-50 gel filtration column to remove alprenolol using [^{125}I]CYP as the radioligand. Nonspecific binding is assessed in the presence of (−)-alprenolol (10 μM). Receptor-bound [^{125}I]CYP is separated from free ligand through gel filtration over Sephadex G-50 columns. The purified sample can then be prepared for electrophoretic analysis according to Laemmli (21) with 10% (w/v) acrylamide slab gels. Sample buffer consist of 8% SDS, 10% (v/v) glycerol, 5% (v/v) 2-mercaptoethanol, 25 mM Tris-HCl, pH (v/v) 6.5, and 0.003% (v/v) bromphenol blue.

Immunoblotting

Following SDS-PAGE, proteins are electrophoretically transferred to a nitrocellulose membrane (Schleicher & Schuell, Keene, NH). Membranes are incubated for 3 hr at 25°C in a blocking buffer consisting of 4% BSA, 10 mM Tris, pH 7.4, 150 mM NaCl, 0.05% (v/v) Nonidet P-40, 0.05% (v/v) Tween-20. The polyclonal phosphotyrosine antibody (Upstate Biotechnology Incorporated) is then added at a final concentration of 2 μg/ml and incubated for 3 hr. Selectivity of the immunoreactivity is assessed by preincubating the antibody with 80 mM pTyr or pSer/pThr at room temperature for 1 hr. Blocked antibody is then diluted (1 : 1000) in blocking buffer (as above). The concentration of pTyr or pSer/pThre is kept at 5 mM throughout the incubation. After five washes of 5 min each in rinse buffer (blocking buffer without BSA), the membranes are incubated for 1 hr with 1 μCi/ml ^{125}I-labeled protein A, washed, dried, and exposed for autoradiography.

Using this protocol, we have shown that human β_2AR is recognized by the antiphosphotyrosine antibody, suggesting that phosphorylation of this receptor on tyrosine residues does occur (15). Moreover, insulin treatment increased the immunoreactivity of the receptor, suggesting increased levels of tyrosine phosphorylation of the β_2AR.

Detection of Phosphotyrosine Residues by Metabolic Labeling

The presence of phosphotyrosines can also be assessed by [^{32}P]P$_i$ metabolic labeling followed by phosphoamino acid analysis. This approach allowed us to confirm directly that the β_2AR is indeed phosphorylated on tyrosine residues.

Cell Labeling

At 18 hr before the [^{32}P]P$_i$ labeling, the culture medium is replaced by phosphate- and serum-free minimal essential media (MEM). Following this period, Na$_3$VO$_4$ (100 μM) is added for 2 hr. Cells are then incubated with 5 mCi of ortho[^{32}P]phosphoric acid in MEM containing 50 μM NaH$_2$PO$_4$ and 100 μM Na$_3$VO$_4$ for 2 hr at 37°C. These labeling conditions allow for the isotopic equilibrium of the ATP pools. Insulin (1 μM) or the vehicle alone is added for the last 30 min. Membranes are prepared and receptor is solubilized and affinity purified as described above. Known amounts of purified receptor are prepared for SDS-PAGE. Following electrophoresis, proteins are transferred to Immobilon Membranes (Millipore Continental Water Systems) and exposed to Kodak XAR-5 film at −70°C. Laser densitometric scanning of the autoradiograms can be used to assessed the incorporation of ^{32}P into β_2AR.

Phosphoamino Acid Analysis

Phosphoamino acid analysis procedures are based on the ability to separate labeled phosphoamino acids following complete hydrolysis of the protein of interest. Various high-voltage electrophoresis and chromatographic procedures have been described. Here we describe a two-dimensional separation procedure based on sequential high-voltage electrophoresis and ascending chromatography. This technique allows an excellent resolution of the phosphotyrosines from the other phosphorylated species and does not require any sophisticated electrophoresis apparatus. Acid hydrolysis of the β_2AR with 5.7 M HCl for 1 hr at 110°C can be performed directly on excised pieces of Immobilon membranes. Nonradioactive pSer, pThr, and pTyr standards (1 μg) are then added to the hydrolyzate. The mixture is subjected to separation on cellulose-coated thin-layer sheets (Sigma) by electrophoresis at 500 V for 1 hr at 4°C in solution consisting of formic acid (88%) : glacial acetic acid : H_2O (50 : 156 : 1794), pH 1.9, followed by ascending chromatography in isobutyric acid : 0.5 M ammonium hydroxide (5 : 3) (22, 23). The plates are then dried and phosphoamino acid standards are visualized by colorimetric reaction with sprayed ninhydrin (following drying with a hot air blower). The radiolabeled phosphoamino acids are then revealed by autoradiography of the chromatographic plates. Semiquantitative assessment of the phospho-amino acid species can be obtained by laser scanning densitometry of the autoradiograms. However, it should be noted that this procedure is not absolutely quantitative, in part due to differences in the stability and recovery of the individual phosphoamino acids. For example, they show distinct patterns of sensitivity to alkali treatment. Phosphotyrosines are relatively stable to alkali whereas phosphoserine and phosphothreonine are virtually destroyed by such treatment. This property of the phosphoamino acids can be exploited to reveal proteins containing few phosphotyrosines. Indeed, alkali treatment of the samples will facilitate detection of a weak phosphotyrosine signal for proteins heavily phosphorylated on serine and threonine residues. However, the resistance to alkali treatment cannot be used as an absolute criterion for the presence of phosphotyrosine because alkaline-mediated serine/threonine phosphoamino acid elimination is rarely complete under most conditions. Furthermore, some phosphoserine and phosphothreonine residues were found to be especially resistant to alkali.

Stoichiometry of Phosphorylation and Determination of Phosphorylation Sites

An important aspect concerning the assessment of the potential functional importance of tyrosine phosphorylation of GPCRs is the stoichiometry of such phosphorylation. Conventional wisdom predicts that a very low (substoichiometric) phosphorylation would have little effect on the function of the receptor. Phosphorylation levels approximating 1 mol of phosphate per mole of receptor would predict (generally) that the modification could have significant impact on receptor activity. Lower levels of phosphorylation would be less likely. Determining the stoichiometry of phosphorylation on tyrosine residues is not a trivial problem for GPCRs. The overall phosphorylation stoichiometry can be assessed by determining the specific activity of the ATP pools at equilibrium in metabolic labeling experiments. The total amount of phosphate incorporated is then determined by direct scintillation counting of the purified receptor following SDS-PAGE electrophoresis. However, the presence of multiple phosphorylated serines and threonines in most GPCRs makes it impossible to determine the proportion of the phosphate molecules specifically transferred to tyrosine. Moreover, as discussed above, phosphoamino acid analysis is not sufficiently quantitative to determine accurate stoichiometries. A more direct approach, such as phosphopeptide mapping or direct sequencing, would be more appropriate. However, this approach is confounded by the small quantities of receptor that can be purified (even from heterologous expression systems). Indirect approaches using antiphosphotyrosine could be considered to estimate the proportion of receptor-bearing phosphotyrosines. Indeed, the proportion of receptor being selectively immunoprecipitated by these antibodies would provide a conservative approximation. Unfortunately, and as described above, technical difficulties have made it difficult to assess immunoprecipitation of GPCRs with antiphosphotyrosine antibodies. Moreover, this approach would not provide quantitative information concerning the number of tyrosine phosphorylation sites present in a given receptor.

In order to determine both the tyrosine phosphorylation site(s) and the potential functional implications of this phosphorylation, we chose the approach of site-directed mutagenesis. The use of heterologous expression systems allows one to express, in the same genetic background, mutant forms of the receptor that lack potential phosphorylation sites. The proposed secondary structure of the β_2AR predicts that five tyrosine residues are present in the cytoplasmic domains. Using a polymerase chain reaction (PCR)-based mutagenesis technique (24), four of these tyrosines were individually changed to alanine or phenylalanine. For Y141F-β_2AR, a first PCR was done on the wild-type pBCβ_2AR cDNA using the (+) nonmutant oligonucleotide (5'-GTCTCTCATCGTCCTGGC-3') and the (−) mutant oligonucleotide in which the tyrosine codon is mutated to a phenylalanine (3'-GTGGAAAGT-**TTAAG**GTCTCGGACGACTGG-5'). The generated fragment (318 bp) was purified by agarose gel electrophoresis and used as a matrix in a second PCR reaction. The reaction mixture contained the wild-type β_2AR cDNA restriction fragment *Bst*EII/*Eco*RV (898 bp), the (+) nonmutant oligonucleotide (5'-AAAAGGCAGCTCCAGAAGATTGACAAA-3'), and the (−) nonmutant oligonucleotide (3'-CCTCGAAGACCCGGACGCGTC-5'). The resulting 911-bp PCR product encoding a fragment of the β_2AR from Val-67 to Arg-259 and containing a phenylalanine residue instead of a tyrosine at position 141 was digested with *Bst*EII/*Bgl*II and subcloned into the *Bgl*II/*Bst*EII sites of the wild-type β_2AR cDNA already cloned in the mammalian expression vector pBC12BI. The other mutants were constructed using similar strategies. Mutations are confirmed by dideoxynucleotide sequencing. The resulting plasmids are cotransfected with the neomycin-resistance plasmid pSV2-Neo in CHW cells, and clonal cell lines expressing the mutant receptors are selected as described above.

Mutation of Tyr-141 completely abolishes tyrosine phosphorylation of the β_2AR, suggesting that it represents the major site of tyrosine phosphorylation and that the maximal stoichiometry of tyrosine phosphorylation to be expected would be 1 mol/mol (15). Also, this mutation completely abolished the supersensitization of the receptor that accompanied the insulin-promoted tyrosine phosphorylation in the wild-type receptor. These data argue that tyrosine phosphorylation of GPCRs may be stoichiometric and can have important functional effects.

References

1. R. D. Feldman, *Br. J. Pharmacol.* **110,** 1640–1644 (1993).
2. R. D. Feldman, *Mol. Pharmacol.* **35,** 304–310 (1989).
3. C. M. Tan, S. Xenoyannis, R. D. Feldman, *Circ. Res.* **77,** 710–717 (1995).
4. G. Brooker and C. Pedone, *J. Cyclic Nucleotide Protein Phosphorylation Res.* **11,** 113–121 (1986).
4a. A. Boyum, *Scand. J. Clin. Lab. Invest.* **21,** 77–89 (1968).
5. Y. Salomon, C. Londos, and M. Rodbell, *Anal. Biochem.* **58,** 541–548 (1974).
6. R. Roskowski, *in* ''Methods in Enzymology,'' Vol. 99, pp. 3–6. Academic Press, San Diego, 1983.
7. R. D. Feldman, G. D. Park, and C. Y. C. Lai, *Circulation* **72,** 547–554 (1985).
8. J. L. Benovic, R. H. Strasser, M. G. Caron, and R. J. Lefkowitz, *Proc. Natl. Acad. Sci. U.S.A.* **83,** 2797–2801 (1986).
9. M. Bouvier, W. P. Hausdorff, A. De Blasi, B. F. O'Dowd, B. K. Kobilka, M. G. Caron, and R. J. Lefkowitz, *Nature (London)* **333,** 370–373 (1988).
10. W. P. Hausdorff, M. Bouvier, B. F. O'Dowd, G. P. Irons, M. G. Caron, and R. J. Lefkowitz, *J. Biol. Chem.* **264,** 12657–12665 (1989).
11. R. B. Clark, J. Friedman, R. A. F. Dixon, and C. D. Strader, *Mol. Pharmacol.* **36,** 343–348 (1989).
12. M. J. Lohse, J. L. Benovic, J. Codina, M. G. Caron, and R. J. Lefkowitz, *Science* **248,** 1547–1550 (1990).
13. J. Pitcher, M. J. Lohse, J. Codina, M. G. Caron, and R. J. Lefkowitz, *Biochemistry* **31,** 3193–3197 (1992).
14. J. R. Hadcock, J. D. Port, M. S. Gelman, and C. C. Malbon, *J. Biol. Chem.* **267,** 26017–26022 (1992).
15. M. Valiquette, S. Parent, T. P. Loisel, and M. Bouvier, *EMBO J.* **14,** 5542–5549 (1995).
16. Y. J. I. Jong, L. R. Dalemar, B. Wilhelm, and N. L. Baenziger, *Proc. Natl. Acad. Sci. U.S.A.* **90,** 10994–10998 (1993).
17. B. R. Cullen, *in* ''Methods in Enzymology,'' Vol. 152, pp. 684–704. Academic Press, San Diego, 1987.
18. J. Vanecek, D. Sugden, J. Weller, and D. C. Klein, *Endocrinology* **116** 2167–2173 (1985).
19. P. L. Mellon, V. Parker, Y. Gluzman, and T. Maniatis, *Cell* **27,** 279–288 (1981).
20. J. L. Benovic, R. G. L. Shorr, M. G. Caron, and R. J. Lefkowitz, *Biochemistry* **23,** 4510–4518 (1984).
21. U. K. Laemmli, *Nature (London)* **227,** 680–686 (1970).

22. B. Duclos, S. Marcandier, and A. J. Cozzone, *in* ''Methods in Enzymology,'' Vol. 201, pp. 10–27. Academic Press, San Diego, 1991.
23. J. A. Cooper, B. M. Sefton, and T. Hunter, *in* ''Methods in Enzymology,'' Vol. 99, pp. 387–405. Academic Press, San Diego, 1983.
24. S. Herlitze and M. Koenen, *Gene* **91,** 143–147 (1990).

[20] Subcellular Distribution and Posttranslational Modifications of GTP-Binding Proteins in Insulin-Secreting Cells

Anjaneyulu Kowluru* and Stewart A. Metz

Introduction

Using selective blockers of the guanosine triphosphate (GTP) biosynthesis pathway (e.g., mycophenolic acid), we have documented a permissive role for guanine nucleotides in insulin secretion (1, 2). Although the exact molecular and cellular mechanisms underlying the regulatory role(s) of GTP remain uncertain, available evidence indicates that it might involve activation of one (or more) GTP-binding proteins (GBPs). Both of the two major groups of GBPs have been identified in β cells. The first group consists of trimeric GBPs composed of α (39–43 kDa), β (35–37 kDa), and γ (6–8 kDa) subunits. These GBPs are involved in the coupling of various receptors to their intracellular effectors, such as adenylate cyclase, phosphodiesterase, or several phospholipases (3, 4). The second group of GBPs is composed of the low molecular mass (~20–25 kDa) monomeric GBPs, which are involved in sorting of proteins as well as trafficking of secretory vesicles (5). A growing body of evidence indicates that both low molecular mass GBPs as well as the γ subunits of trimeric GBPs undergo posttranslational modifications, such as isoprenylation and carboxyl methylation, at their C-terminal cysteinyl residue (5, 6).

The first of a four-step modification sequence includes incorporation of a 15-carbon (farnesyl) or 20-carbon (geranylgeranyl) isoprenoid moiety [which is derived from mevalonic acid (MVA)] onto a cysteine residue on the carboxyl terminus of the GBP. This is followed by the proteolytic cleavage of several amino acids (up to a maximum of three). Often a carboxyl methylation step then modifies the newly exposed carboxylate anion of the cysteine. In some cases, the covalent addition of a long-chain fatty acid, typically palmitate, completes the cascade. This is felt to render the GBPs more hydrophobic and more able to associate tightly with membranes (their putative site of

---

* To whom correspondence should be addressed.

Methods in Neurosciences, Volume 29

action). Because the isoprenylation of GBPs occurs shortly after their synthesis, and because "half-lives" of prenylated proteins are rather long (5), this is not likely to be an acute regulatory step; however, in many cases, prenylation is necessary to allow GBPs to intercalate into the relevant membrane compartment. In contrast, the methylation and acylation steps are subject to acute regulation at the level of the "on" step (addition of methyl or acyl groups) and/or the "off" step (demethylation and deacylation).

Recent experimental evidence from our laboratory (7) and from that of others (8) has indicated that at least some low molecular mass GBPs undergo posttranslational modifications in insulin-secreting cells. Furthermore, these studies have linked such posttranslational modifications of GBPs to the phenomenon of insulin secretion (7, 8). In this chapter, we primarily focus on the methodological aspects underlying the identification and quantitation of posttranslational modifications of low molecular mass GBPs in normal rat islets and pure β cells, and their possible implications in the phenomenon of nutrient-induced insulin secretion. This chapter is divided into three major sections: (1) isolation of subcellular fractions from islets and pure β cells, (2) identification of low molecular mass GBPs in normal islets and pure β cells, and (3) assessment of posttranslational modifications of low molecular mass GBPs in normal rat islets and pure β cells.

Isolation of Subcellular Fractions from Islets and Pure β Cells

Sources of β Cells

Despite severe limitations in the amount of islet tissue available (i.e., typically 200–250 islets per rat represents 100–125 μg protein), we routinely carry out all of our studies using normal rat islets. This is because we recently observed significant (quantitative and qualitative) differences in the abundance of both trimeric and low molecular mass GBPs between normal rat islets and transformed β cells (see Ref. 9 for a recent review). Pancreatic islets are isolated from male Sprague–Dawley rats (250–300 g) by collagenase digestion followed by a Ficoll density gradient centrifugation method (1, 2, 7, 10). In order to prevent contamination by acinar tissue, islets are picked twice under stereomicroscopic control. For comparison purposes, we repeat key studies using normal human islets (kindly provided by Dr. David Scharp, University of Washington Medical School, St. Louis, MO), and clonal β (HIT) cells (generously provided by Drs. Paul Robertson and Hui-Jian Zhang; University of Minnesota School of Medicine, Minneapolis, MN).

Subcellular Fractionation

Differential Centrifugation Method

Subcellular fractions of rat islets are isolated using a Beckman Optima TL-100 ultracentrifuge (11–13). All procedures are carried out at 4°C, unless stated otherwise. Briefly, 1000–1500 islets (corresponding to 500–750 μg protein) are washed once with 230 mM mannitol/70 mM sucrose/5 mM HEPES buffer, pH 7.4, containing 1 mM EGTA, and twice more with the same buffer without EGTA. They are then homogenized manually (8–10 strokes) in the same buffer containing 1 mM dithiothreitol (DTT) and 2.5 μg each of leupeptin and pepstatin/ml, in a 2-ml glass homogenizer. The homogenate is spun at 600 g for 5 min to remove the nuclear and cell debris pellet. The resulting supernatant fraction is centrifuged at 5500 g for 10 min to yield a pellet enriched in mitochondria. The resulting supernatant fraction is centrifuged at 25,000 g for 20 min to yield a secretory granule-enriched pellet. Each of these pellets is washed twice with homogenization medium and resuspended in a suitable volume (typically 500–600 μl) of the homogenization buffer.

Sucrose Density Gradient Method

In some studies, the secretory granule fraction obtained after the differential centrifugation procedure is reconstituted in 750 μl of 0.3 M sucrose/5 mM sodium phosphate buffer, pH 6.8, and subjected to a further density gradient centrifugation. The secretory granule-enriched fraction is layered over 800-μl gradients of 1.5 and 2.0 M sucrose in polyallomer (11 × 34 mm) tubes and centrifuged at 105,000 g for 2 hr using a TL-55 rotor. After the centrifugation, the interfaces between the 0.3, 1.5, and 2.0 M sucrose layers, as well as pellets at the bottom of the tubes, are carefully removed by aspiration using siliconized pipettes. Insulin content (Table I) as well as content of marker enzymes (11–13) are determined for each fraction.

Purity of Subcellular Fractions

Current formulations about the role of GBPs in vesicle fusion and protein traffic would suggest that one or more GBPs might be located in the insulin-containing dense-core secretory granules. Testing this hypothesis requires the isolation of a fraction largely, if not totally, devoid of contamination by

TABLE I   Relative Distribution of Insulin in Islet Subcellular Fractions[a]

Fraction	Insulin (mU/mg protein)	Relative specific activity
Differential centrifugation		
Homogenate	1544 ± 127 (4)	1
Nuclear and cell debris	1918 ± 121 (4)	1.24
Mitochondria	4116 ± 408 (4)	2.66
Secretory granules	6460 ± 510 (4)	4.18
Postsecretory granule supernatant	178 ± 32 (4)	0.12
Sucrose gradients of secretory granule fraction		
0.3/1.5 M interface	1255 ± 487 (2)[b]	0.81
1.5/2.0 M interface	10338 ± 1401 (4)	6.69
2.0 M pellet	Undetectable	0

[a] Subcellular fractions were isolated from normal rat islets as described in the text. Insulin content in each fraction was quantitated by radioimmunoassay. Relative specific activity is defined as the ratio of final to initial specific activity (homogenates) of insulin in each fraction. Data are mean ± SEM of four individual preparations.

[b] We failed to detect any insulin in these fractions in two other preparations.

other subcellular organelles. We assess the possible contamination of the secretory granule fraction by plasma membrane (using ouabain-sensitive Na^+,K^+-ATPase as a marker), Golgi complex (using UDP-galactosyltransferase and thiamin pyrophosphatase as markers), mitochondria [using succinate-iodonitrotetrazolium (INT) reductase as a marker], and cytosol (using lactate dehydrogenase as a marker). The enzyme activities are quantified in homogenates as well as in the secretory granule fraction in order to assess the degree of enrichment of the latter. Insulin is used as a marker for intact secretory granules. The relative specific activities of various marker enzymes (defined as the ratio of the final specific activity in the secretory granule fraction to the initial specific activity in the homogenate) were total ATPase, 0.09; acid phosphatase, 0.39; succinate-INT reductase, 0.39; UDP-galactosyl transferase, 1.50; lactate dehydrogenase, 0.26; insulin, 4.18. We failed to detect any Na^+,K^+-ATPase activity in this fraction. These data indicate that the fraction isolated by this method is highly enriched in intact secretory granules (11–13). Furthermore, the purity of the secretory granule fraction isolated by the differential centrifugation was examined by electron microscopy (13), which indicated that this fraction consisted of >90% β cell granules, with a modest (<10%) contamination by heavy mitochondria. Further fractionation of this fraction on a sucrose density gradient (Table I) yields a greater degree of enrichment of insulin-containing secretory granules (6.7-fold) compared to that obtained by differential centrifugation (4.2-fold; Refs. 11 and 12).

Identification of Low Molecular Mass GBPs in Subcellular Fractions of Insulin-Secreting Cells

Unlike trimeric GBPs, the majority of low molecular mass GBPs (with the conspicuous exception of those belonging to the rho family, e.g., CDC42) bind GTP even after SDS-PAGE and renaturation after transfer to nitrocellulose (13, 14). This property can be used to localize low molecular mass GBPs in β cell subcellular fractions; furthermore, immunoblotting can be used to specifically identify some of these GBPs in those fractions.

Specific Methods

The β cell proteins (6–24 μg protein) are separated by SDS-PAGE (12.5% acrylamide) and are then transferred (at 60 V for 60 min at 4°C) to nitrocellulose membranes (Trans-Blot transfer medium; pure nitrocellulose membranes; 0.45 μm, Bio-Rad, Hercules, CA) using a Bio-Rad Trans-Blot apparatus. The blots are washed in a medium consisting of 50 mM Tris-HCl, 5 mM MgCl$_2$, 1 mM EGTA and 0.3% Tween-20, pH 7.4, for 10 min. The membranes are then soaked in the same medium containing [α-^{32}P]GTP (3000 Ci/mol; 1 μCi/ml) for 1 hr at 25°C followed by extensive washing (at least three times) in the same buffer to remove unbound label from the membranes. Radioactivity bound to the proteins is detected by autoradiography using Kodak (Rochester, NY) X-OMAT AR film (typically exposed for 1–3 days at −70°C). Apparent molecular weights of the labeled proteins are determined using prestained standards (Bio-Rad, Richmond, CA).

The presence of rab3A, rac2, rhoA, ras, and CDC42 in insulin-secreting cells is studied by Western blotting. For this purpose, proteins are separated by SDS-PAGE (12% acrylamide) and transferred onto nitrocellulose membranes as described above. The membranes are incubated with monoclonal antiserum directed against rab3A (kindly provided by Dr. Reinhard Jahn, Yale University Medical School, New Haven, CT), polyclonal antisera against CDC42 (generously provided by Dr. Tony Evans, Onyx Pharmaceuticals, Richmond, CA), or polyclonal antisera raised against rac2 and rhoA (purchased from Santa Cruz Biotechnology, Santa Cruz, CA), in 1:200 dilution for 90 min at 37°C (for rab3A) or for 15 hr at 25°C for rac2, rhoA, or CDC42. Immune complexes are detected using ^{125}I-labeled protein A or by color development using alkaline phosphatase coupled to antirabbit IgG (as described in Ref. 13).

Specificity of the antisera used in present studies is evaluated by using adrenal homogenates and purified CDC42 (provided by Dr. Evans) as positive

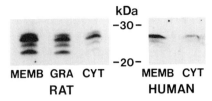

FIG. 1   Identification of low molecular weight GBPs in normal rat and human islets by the GTP-overlay method. Normal rat islet membranes (MEMB; 24 μg), secretory granules (GRA; 6 μg), and cytosol (CYT; 10 μg) and human islet membranes (20 μg) and cytosol (20 μg) proteins were separated by SDS-PAGE. Following this, proteins were transferred onto nitrocellulose membranes, which were soaked in 50 mM Tris-HCl, pH 7.4, 5 mM MgCl$_2$, 1 mM EGTA, 0.3% Tween-20 containing [α-^{32}P]GTP (1 μCi/ml) for 45 min at 25°C. Labeled proteins were identified by autoradiography. Data are representative of three (using rat islets) and two (using islets from two donors) individual experiments. Molecular weights of labeled proteins were determined using prestained molecular weight standards (Bio-Rad). Reproduced with permission from Ref. 13.

controls for rab3A and CDC42, respectively. In both cases, we observed a single band by Western blotting. Information provided by the supplier (Santa Cruz Biotechnology, Santa Cruz, CA) indicates a lack of cross-reactivity of rac2 antisera with either rac1 or CDC42.

GTP-Overlay Studies

Data from these studies indicated that at least four low molecular weight GBPs, in the molecular mass region of 21–27 kDa were labeled in the membrane and secretory granule fraction of rat islets (Fig. 1). However, only two proteins were clearly labeled in the cytosolic fraction of normal rat islets. All labeling was specific for GTP because it was completely abolished by coprovision of 10 μM GTPγS, but not ATP or ATPγS. The specific activity (defined as the degree of labeling per milligram of protein) of these proteins in the secretory granule fraction was 3.7 times that of the crude membrane fraction, which in turn was 2.5 times that of cytosol (Fig. 1). At least three proteins (two major bands and one minor band) in the molecular mass region of 21–27 kDa were detected in the membrane fraction derived from normal human islets (Fig. 1), with only one protein apparent in the cytosolic fraction.

To determine whether β cells (which comprise only 70–75% of islet mass) contribute to the labeling, GTP binding (as assessed by the overlay method)

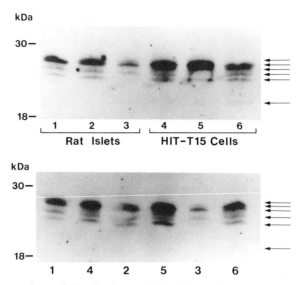

FIG. 2   A comparison of distribution and abundance of low molecular weight GBPs in fractions from normal rat islet and transformed β cells. Normal rat islet homogenates (30 μg; lane 1), rat islet membranes (20 μg; lane 2), rat islet cytosol (20 μg; lane 3), HIT cell homogenates (30 μg; lane 4), HIT cell membranes (20 μg; lane 5), and HIT cell cytosol (20 μg; lane 6) were separated by SDS-PAGE and low molecular weight GBPs were identified by GTP-overlay method. To facilitate comparison, in a second experiment (bottom) comparable fractions of rat islet or HIT cell homogenates (lanes 1 and 4), membranes (lanes 2 and 5), or cytosol (lanes 3 and 6) were separated by SDS-PAGE (next to each other); data are presented in side-by-side comparisons. Arrows indicate positions of labeled bands seen in one or more preparations. These data suggest that significant qualitative and quantitative differences exist in GBPs between normal rat islets and transformed β cells, although the GTP-overlay technique has limitations as a tool to quantify GBP abundances. Reproduced with permission from Ref. 13.

in normal rat islets was compared to that observed using clonal, β (HIT) cells. In HIT cells, at least six proteins (indicated by arrows, Fig. 2) were labeled with the same apparent molecular weight distribution as in rat islets except that (at least) one additional band was visible in the 26- to 27-kDa region (Fig. 2). Densitometric quantitation of the labeling in the cytosolic fraction indicated that at least five proteins were labeled in HIT cells. Intensity of labeling of a sixth protein (26–27 kDa) in HIT cells could not be quantitated because of lack of separation of this protein from other GBPs (13). It is evident from Fig. 2 that the relative labeling of proteins in the soluble compartment was much higher in HIT cells compared to normal rat

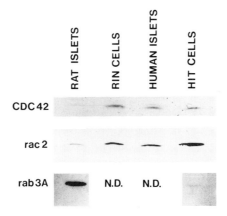

FIG. 3   Immunologic detection of CDC42, rac2, and rab3A in insulin secreting cells. Homogenates of normal rat islets, human islets, HIT cells or rat insulinoma (RIN) cells (50 μg each, protein) were separated by SDS-PAGE and transferred onto nitrocellulose membranes. After transfer, membranes were incubated with antisera (1 : 200 dilution) directed against rab3A (monoclonal), CDC42 (monoclonal), or rac2 (polyclonal); immune complexes were identified using ^{125}I-labeled protein A or alkaline phosphatase coupled to antirabbit IgG. ND, Not determined. Reproduced with permission from Ref. 13.

islets or human islets (Figs. 1 and 2). These data indicate some discordance (both qualitative and quantitative) between islets and HIT cells, which may in part be due to changes in GBP metabolism associated with cellular transformation (see Refs. 5 and 9 for reviews) or cell culture. However, the major finding is that no additional bands were seen in islets, which were not observed in pure β cell preparations, indicating that most of the findings using islets could be explicable by their β cell population.

Immunoblotting Studies

Immunoblotting experiments using homogenates of normal rat islets and HIT cells revealed that three of these proteins reacted with antisera directed against rab3A, rac2, and CDC42. Data in Fig. 3 indicate that both CDC42 and rac2 are present in all the four insulin-secreting cells, namely, normal rat islets, human islets, HIT cells, and rat insulinoma (RIN) cells (provided by Dr. Chris Rhodes, Joslin Diabetes Center, Boston, MA). The rhoA protein was also localized in normal rat islets, human islets, and HIT cells (13). The abundance of these proteins, as judged by Western blotting, was higher in transformed clonal β (HIT or RIN) cells compared to normal rat and human islets (Fig. 3).

Additional studies of the subcellular distribution of rac2 have shown that most (>90%) of rac2 was associated with the membrane fraction of an unstimulated islet. The data in Fig. 3 also indicate that rab3A is enriched in rat islets compared to HIT cells. However, none of the proteins in islet or HIT cell homogenates cross-reacted detectably with antisera directed against pan-ras (Oncogene Science, Uniondale, NY).

Posttranslational Modification of Low Molecular Mass GBPs in β Cells

Analysis of Acetate and Mevalonate Metabolism in Islets

Mevalonte (MVA) is the precursor for isoprenoid biosynthesis, which in turn is needed for the first modification step (i.e., isoprenylation) at the C-terminal cysteine residue of GBPs (5). The ability of pancreatic islets to synthesize MVA endogenously or to metabolize MVA appropriately had not been verified previously. In order to address this issue, normal rat islets are labeled with metabolic precursors (e.g., [^{14}C]acetate or [^{14}C]MVA) and the incorporation of these precursors into the sterol and nonsterol limbs of MVA metabolism in pancreatic islets is assessed.

Labeling with [^{14}C]Acetate

Groups of 400 islets are cultured in RPMI medium (containing 10% fetal bovine serum, 100 U/ml penicillin, 100 μg/ml streptomycin, and 2.8 mM glucose) overnight in the presence or absence of 30 μM lovastatin.* The next morning, the medium is carefully removed and cells are incubated for an additional 4 hr in a fresh medium containing [^{14}C]acetate (59 mCi/mmol) in the continued presence or absence of 30 μM lovastatin. Lipids are extracted with chloroform : methanol (2 : 1, v/v) and separated by one-dimensional thin-layer chromatography for sterols and nonsterol compounds, using a hexane : diethylether : acetic acid (70 : 30 : 1.5, v/v) solvent system. A mixture of authentic standards is added to islet lipid extracts, spotted onto silica gel 60 plates (0.2 mm thickness; EM Science, Gibbstown, NJ), and lipid

---

* Prior to use, lovastatin (in its lactone form) is converted to the free acid form by incubation in 0.1 N NaOH for 2 hr at 50°C. Following this, the pH of the medium is adjusted to 7.2 using 0.1 N HCl.

TABLE II   [^{14}C]Acetate Incorporation into Islet Sterols and
Nonsterol Compounds; Effect of Lovastatin[a]

Lipid	Incorporation (dpm/400 islets)		Inhibition (%)
	− Lovastatin	+ Lovastatin	
Nonsterols			
Dolichol	365	169	54
Ubiquinones	79	41	48
			Mean: 51
Sterols			
Cholesterol	403	74	82
Lanosterol	315	41	87
Squalene	1816	121	93
			Mean: 87

[a] Islets (400 per group) were cultured overnight in the presence or absence of
lovastatin (30 μM) and then incubated with [^{14}C]acetate (30 μCi/ml) for an
additional 4 hr at 37°C. Lipids were extracted and separated by thin-layer
chromatography as described in the text. Incorporation of radioactivity into
individual lipids was quantitated by scintillation spectrometry.

spots are identified in an iodine chamber. Relative migration values are
determined for authentic standards on at least three occasions to verify the
reproducibility of separation using this method. Radioactivity in individual
lipid spots is quantitated by scintillation spectrometry.

Labeling with [^{14}C]Mevalonate

Groups of 300 islets are cultured for 16 hr in the presence or absence of 30
μM lovastatin. Incubations with lovastatin are necessary to prevent the
production of endogenous MVA in islets. A culture period of (at least) this
length was used because the "half-life" of previously isoprenylated GBPs
is quite long (5) and because the prenyl group is not subject to rapid turnover.
Following this, the cells are labeled with RS-[2-^{14}C]mevalonolactone (50 mCi/
mmol) for an additional 4 hr at 37°C in the continued presence or absence
of lovastatin. Lipids are extracted and separated by thin-layer chromatography as described above.

Data from these studies indicated that islets converted labeled [^{14}C]acetate
into compounds comigrating with authentic standards for cholesterol, lano-
sterol, dolichols, and ubiquinones (Table II). Lovastatin (30 μM) markedly

TABLE III   Incorporation of [^{14}C]MVA into Sterols
and Nonsterol Lipids in Islets[a]

Compound	Incorporation (dpm)	
	− Lovastatin	+ Lovastatin
Sterols		
Cholesterol	2178 (66)	7268 (92)
Squalene	744 (23)	90 (1)
Nonsterols		
Dolichols	232 (7)	323 (4)
Ubiquinones	116 (4)	233 (3)
	3270	7914

[a] Values in the parentheses represent percent of total incorporation, taking the sum of incorporation of the four measured species as 100%. Islets (300 per group) were labeled in the presence of 50 μCi/ml of [^{3}H]MVA for 4 hr after preincubation in the absence or presence of 30 μM lovastatin for 16 hr. Islets preincubated with lovastatin also had lovastatin present during the labeling period. Lipids were extracted with chloroform–methanol and separated by thin-layer chromatography using hexane : diethyl ether : acetic acid (70 : 30 : 1.5, v/v) as the solvent system. Individual lipids were identified using authentic standards and incorporation of radioactivity was quantitated by scintillation spectrometry.

reduced the synthesis of all compounds. This provides the basis for using lovastatin to impede synthesis of MVA and, therefore, of isoprenoid moieties. In addition, islets converted [^{14}C]MVA to similar compounds (Table III). In the presence of lovatatin, exogenous MVA was converted more efficiently (242% of control) to labeled products, especially cholesterol, presumably indicating a reduction in isotopic dilution and thus providing indirect evidence of depletion of endogenous MVA by lovastatin. Thus, these data indicate that pancreatic islets have the enzymatic machinery required to synthesize isoprenoid precursors endogenously.

Isoprenylation of Islet GBPs

To investigate whether some islet GBPs undergo isoprenylation, normal rat islets are labeled in the presence of [^{14}C]MVA (in the continued presence of lovastatin) and the incorporation of radioactivity into proteins is monitored by autofluorography following SDS-PAGE as described above. During such studies, one can determine whether prevention of isoprenylation of these proteins by coprovision of lovastatin results in alterations in the subcellular distribution (i.e., membrane vs. cytosolic) of these proteins. Furthermore, one can examine the effects of lovastatin and other inhibitors of isoprenylation (e.g., perillic acid; see below) on nutrient-induced insulin secretion.

Specific Methods

Groups of 300 islets are incubated in the presence or absence of lovastatin (15–30 μM, although effects of lovastatin on insulin secretion are essentially already maximal at 5 μM; see below). This was essential to deplete endogenous MVA and facilitate labeling by exogenous MVA. Incubations are carried out in RPMI 1640 medium, containing 2.8 mM glucose and 10% (v/v) fetal calf serum, for 8 hr in a metabolic chamber (VWR Scientific; model 2300) at 37°C. This is followed by an additional 8- to 12-hr incubation in the continued presence of 15–30 μM lovastatin plus [^{14}C]MVA (40 μCi/ml). After labeling, cells are washed quickly with an isotonic medium consisting of 230 mM mannitol, 70 mM sucrose, and 5 mM HEPES buffer, pH 7.4. Islets are homogenized (manually, as described above) in the same medium and proteins are separated by SDS-PAGE (12% acrylamide); labeled proteins are identified by fluorography (exposed at −70°C for 4 weeks). For studies involving the subcellular distribution of low molecular mass GBPs, homogenates are centrifuged at 105,000 g for 90 min to separate total particulate (membrane) and soluble (cytosolic) fractions (13). Abundance of GBPs in the membrane and cytosolic fractions is quantitated by the GTP-overlay method.

Data from these studies (Fig. 4) indicated that when islets were incubated with [^{14}C]MVA, at least five proteins in the molecular mass region of 20–27 kDa were labeled. These proteins seem to be similar to the low molecular mass GBPs identified by the GTP-overlay method (Figs. 1 and 2). However, it should be reiterated that GBPs of the *rho* family may be labeled using [^{14}C]MVA, but are not detected by the GTP-overlay technique. In islets exposed to lovastatin, a shift in the subcellular localization of these GBPs was observed as assessed by the GTP-overlay method (Fig. 4B)—that is, a relative increase in the labeling of proteins in the cytosolic fraction was observed. These findings were confirmed by cutting the nitrocellulose membranes and directly counting the radioactivity (9). The cytosolic : membrane ratios of these labeled proteins were significantly higher in lovastatin-treated islets compared to control islets (i.e., 0.66 ± 0.06 in lovastatin-treated islets vs. 0.43 ± 0.07 in control islets; n = 5 experiments). These data suggest that prevention of isoprenylation of proteins results in their accumulation in the cytosolic fraction (7, 9). Similar lovastatin-induced alterations in the

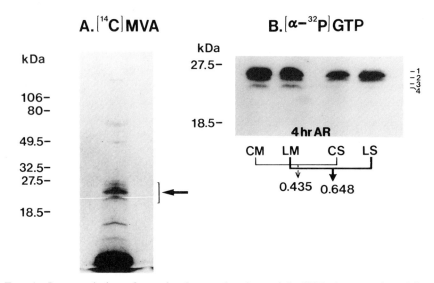

FIG. 4   Isoprenylation of putative low molecular weight GBPs in normal rat islets.
(A) Homogenates of islets (300 per group) labeled with [^{14}C]MVA and labeled proteins
were separated by SDS-PAGE and identified by autoradiography. These data indicate
that at least four to five proteins in the molecular mass region of 18.5–27.5 kDa were
labeled, presumably via isoprenylation. These proteins seem to be similar to GBPs
identified by GTP-overlay method (B) Pretreatment of islets with lovastatin (15 μM
for 18 hr) to block MVA production (and thereby to reduce protein isoprenylation)
resulted in accumulation of these proteins in the cytosolic fraction. The cytosolic :
membrane ratios of these proteins increased from 0.435 in control islets to 0.648 in
lovastatin-treated islets. These data suggest a redistribution of these proteins as a
consequence of inhibition of their isoprenylation. The autoradiograph in A was exposed at −70°C for 4 weeks; exposure in B was at −70°C for 4 hr. Numbers to the
right in B are used to identify individual labeled bands. Reproduced with permission
from Ref. 7. CM, Membrane from control islets; CS, soluble fraction from control
islets; LM, membranes from lovastatin-treated islets; LS, soluble fraction from lovastatin-treated islets.

subcellular distribution of low molecular mass GBPs in β (HIT) cells were
reported by Li *et al.* (8).

Effects of Lovastatin on Glucose-Induced Insulin Secretion

We observed that incubation of normal rat islets with lovastatin (5–30 μM)
significantly reduced (by 46–57%) glucose-induced insulin secretion (7). The
inhibitory effect of lovastatin was visible at a drug concentration of 1 μM

or less, and was maximal at 5 μM; at higher concentrations (i.e., 75 μM) the inhibitory effect was somewhat less (i.e., 28%). Furthermore, the inhibitory effect of lovastatin was totally prevented by coprovision of either mevalonolactone or MVA–sodium salt (100 or 200 μM), which bypasses the lovastatin-induced block in the synthesis of endogenous MVA. These data are compatible with recent reports of the inhibition by lovastatin of bombesin- and vasopressin-potentiated insulin secretion in HIT cells (8). Taken together, these data are congruous with the formulation that isoprenylation of certain GBPs may be required for nutrient-induced insulin secretion in β cells.

In addition to lovastatin, which is an inhibitor of HMG-CoA reductase inhibitor, several inhibitors of ras farnesyltransferase (an enzyme that catalyzes the incorporation of isoprenoids into C-terminal cysteine of GBPs) have been reported recently (15–17). These include agents such as manumycin, gliotoxin, limonene, and perillic acid (see Ref. 16 for a review). We have recently reported a marked (>80%) reduction in nutrient-induced insulin secretion from normal rat islets by high concentrations of perillic acid (7). Although limonene (the precursor of perillic acid) is said to share its ability to inhibit isoprenylation of ras-like GBPs (17), we did not pursue the use of limonene because of difficulties with its solubilization, and recent reports of its cytotoxicity in pure β cells (8). In fact, limonene, as well as perillic acid methyl ester, directly augmented basal rats of insulin release, probably reflecting a toxic effect. It is suggested that such drugs must be used with caution as "specific" inhibitors of prenylation.

Carboxyl Methylation of Islet GBPs

This modification step, which is catalyzed by prenylcysteine methyltransferase (18), involves the incorporation of a methyl group into the carboxylate anion of prenylated cysteine via an ester linkage. S-Adenosylmethionine (SAM) serves as the methyl donor. N-Acetyl-S-trans,trans-farnesyl-L-cysteine (AFC), which is a competitive substrate for prenylcysteine methyltransferases (type III methyltransferase), has been widely used as an inhibitor of carboxyl methylation at the C-terminal cysteine (7, 18). The methylated status of a protein is controlled by the balance of methyltransferase and methylesterase activities, which induce the addition and removal of methyl groups, respectively (18, 19). Using [^{3}H]methionine (in intact cells) or S-adenosyl[^{3}H]methionine (in cell-free preparations), at least four major proteins (~36, 23, 21, and <8 kDa) were methylated in five insulin-secreting cells [e.g., normal rat islets, human islets, and clonal β (HIT, RIN, or INS-1) cells]. Based on molecular size and immunoblotting, the 36-kDa protein was identified as the catalytic subunit of protein phosphatase 2A* (20). Unlike the 23- and <8-kDa proteins, the methylation of the 36- and 21-kDa proteins was resistant to AFC, suggesting that they may not be carboxylmethylated at the C-terminal cysteine residue (20). Immunoblotting and immunoprecipitation experiments suggested that the 23-kDa protein is CDC42, a low molecular mass GBP, which is a calcium-binding protein involved in cytoskeletal organization (5). Preliminary evidence suggested that the <8-kDa protein(s) represents the γ subunit of trimeric GBPs, which has been shown to undergo isoprenylation and carboxyl methylation in several cell types (22, 23). It has been demonstrated that these modifications of γ subunits are required for optimal interaction of $\beta\gamma$ complex with the α subunit to attain a trimeric conformation (22, 23). Putative involvement of a regulatory role for the carboxyl methylation of one or more GBPs in insulin secretion came from our recent studies demonstrating a marked reduction in glucose- and ketoisocaproate-induced insulin secretion by AFC (7).

Labeling of Proteins

Protein carboxyl methylation assays are carried out (in a total volume of 100 μl) in homogenates or subcellular fractions (25–30 μg protein) using 50 mM sodium phosphate buffer, pH 6.8, consisting of 1 mM EGTA at 37°C for different time intervals using S-adenosyl[^{3}H]methionine (100 μCi/ml) as a methyl donor. GTPγS and other reagents were present in their respective concentrations, as indicated in the text. Usually, the reaction is started by the addition of [^{3}H]SAM, and terminated by the addition of SDS-PAGE sample buffer. Labeled proteins are separated by SDS-PAGE (usually 12% gels for the study of CDC42 and 17% gels for the study of γ subunits). The degree of labeling is quantitated either by fluorography of dried gels or by vapor-phase equilibration assay (see below).

For intact islets, groups of islets (100–150 islets per group) are incubated in an isotonic Krebs–Ringer bicarbonate medium in a metabolic chamber (at 37°C) for 2–4 hr in the presence of *methyl*-[^{3}H]methionine (40–100 μCi/ml). Cycloheximide (5 μM) and actinomycin D (2 μM) are included 1 hr

---

* This protein was predominantly cytosolic. Its carboxyl methylation was resistant to inhibition by AFC, which is a competitive inhibitor of prenylcysteine methyltransferases. These data suggest that its methylated amino acid was not cysteine, and therefore that the protein is not a classical GBP. Okadaic acid, a specific inhibitor of protein phosphatase 2A, but not its inactive analog 1-norokadaone, completely inhibited the carboxyl methylation of this protein ($K_i < 20$ nM). These data may indicate that okadaic acid-induced inhibition of protein phosphatase 2A activity may in part be due to its ability to inhibit the carboxyl methylation of PP2A; the latter has been shown to occur on a C-terminal leucine and to increase the catalytic activity of PP2A (21).

prior to the addition of the labeled methionine. Preincubation with methionine is necessary to prelabel the endogenous SAM pools. After prelabeling, cells are washed once with an isotonic medium to remove label, and various modulators are added and islets are incubated for an additional 30 min. Labeled proteins are separated by SDS-PAGE as described above.

Vapor-Phase Equilibration Assay

The α-carboxyl methyl groups on prenylcysteine residues of modified GBPs are base labile, being released as volatile [^{3}H]methanol (18). To assess this specific modification in insulin-secreting cells, methyl esters are quantified by vapor-phase equilibration assay (7, 8, 24). After separation by SDS-PAGE (as described above), individual lanes of dried gels are cut into 3- or 5-mm slices and are placed in 1.5-ml Eppendorf centrifuge tubes (without caps) containig 500–750 μl of 1 N NaOH.[†] Tubes are then placed in 20-ml scintillation vials containing 5 ml of scintillation fluid (Ultima Gold; Packard Instrument Co., Meriden, CT). The vials are then capped and left at 37°C overnight to maximize the base-catalyzed release of [^{3}H]methanol due to hydrolysis of methyl esters. Tubes are then gently removed from the vials and the sides of the tubes are rinsed (into the vials) with an additional 2 ml of scintillant and the radioactivity is determined by scintillation spectrometry.

Studies of Carboxyl Methylation of CDC42

Considerable experimental evidence was obtained that linked the carboxyl methylation of CDC42 to nutrient-induced insulin secretion. These findings are summarized below.

CDC42 is present in rat islets, human islets, and insulin-secreting, pure β cells (Fig. 3) based on immunoblotting method. Each of these cells has endogenous methyltransferase activity[‡] that can mediate the guanine nucleotide-dependent carboxyl methylation of endogenous (26, 27) as well as puri-

---

[†] Some of the other bases employed to quantitate base-labile methanol include sodium bicarbonate (pH 11), sodium borate (pH 11), and 90% hyamine hydroxide. Instead of incubating at 37° C for overnight, Floer and Stock (25) have used 90% hyamine hydroxide for 1 hr at 37°C.

[‡] Using AFC as the substrate, we have recently observed that prenylcystine methyltransferase activity is predominantly associated with the plasma membrane fraction of pure β (INS-1) cells (G. Li, A. Kowluru, and S. A. Metz, unpublished, 1995).

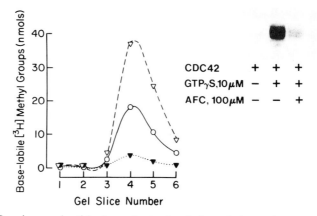

FIG. 5 Guanine nucleotide-dependent stimulation of the carboxyl methylation of authentic CDC42 by pancreatic islet homogenates. A homogenate of normal rat islets (50 μg protein) was incubated with authentic CDC42 (1 μg) in the absence or presence of GTPγS (10 μM) and AFC (100 μM), as indicated in the inset. S-Adenosyl[^{3}H]methionine was used as methyl donor. Labeled proteins were separated by SDS-PAGE (12%) and the degree of methylation was quantitated in gel slices (3 mm) by vapor-phase equilibration assay. ($\blacktriangledown$) Homogenate incubated with CDC42 alone; ($\triangledown$) homogenate plus CDC42 plus 10 μM GTPγS; ($\bigcirc$) homogenate plus CDC42 plus 10 μM GTPγS plus 100 μM AFC. Upper right: An autoradiogram representing the stimulation of the carboxyl methylation of CDC42 and its inhibition by AFC. Exposure period was 4 weeks at $-70°$C. These data indicate that carboxyl methylation of CDC42 is stimulated in a guanine nucleotide-dependent manner and is sensitive to AFC.

fied authentic CDC42 (Fig. 5). Incubation of homogenates of insulin-secreting cells with GTPγS (10 μM) results in the translocation of CDC42 from the cytosol to the membrane fraction as assessed by vapor-phase equilibration assay and by immunoblotting method (26, 27). This effect may partially explain the stimulatory role of GTP in nutrient-induced insulin release (1, 2). Inhibition of the carboxyl methylation of CDC42 by AFC correlated with a concomitant reduction of nutrient-induced insulin secretion from norml rat islets (7, 26, 27); this effect was confirmed by using other methylation inhibitors, such as 3-deazaadenosine and S-adenosylhomocysteine (7). N-Acetyl-S-trans-geranyl-L-cysteine, an analog of AFC, which did not inhibit carboxyl methylation of CDC42 and the γ subunit of trimeric GBPs in islets, also failed to alter secretion (7, 26, 27). In contrast, AFC does not inhibit methylation of the 21-kDa protein in islets; although AFC did reduce the carboxyl methylation of another major protein in islets (putatively, the γ subunit of trimeric GBPs; see below), its methylation was not stimulated by GTPγS (7). Furthermore, in transformed β cells, lovastatin reduces the prenylation of CDC42,

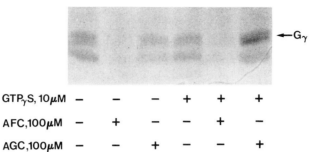

GTP$_\gamma$S, 10 μM   −   −   −   +   +   +

AFC,100 μM   −   +   −   −   +   −

AGC,100 μM   −   −   +   −   −   +

FIG. 6   Preliminary evidence indicating the carboxyl methylation of two proteins in the molecular mass region of 6–8 kDa in normal rat islet homogenates. Normal rat islet homogenate proteins were carboxyl methylated using [³H]SAM as a methyl donor. GTPγS, AFC, or AGC was present as indicated in the figure. Labeled proteins were separated on a 17% SDS-PAGE and identified by autofluorography. At least two proteins (one comigrating with the authentic γ subunit of transducin, and the other with an apparent molecular weight lower than the γ subunit) were carboxyl methylated. However, unlike the carboxyl methylation of CDC42 (Fig. 5), the carboxyl methylation of these two proteins was not stimulated by GTPγS. AFC (but not its inactive analog, AGC) completely inhibited the carboxyl methylation of these proteins. These data were also confirmed by base-labile methanol assay. The lower molecular weight may represent an isoform of the G_γ subunit.

which, in turn, leads to its redistribution from membrane to cytosol; similar effects were not seen with certain other GBPs (e.g., rho or ADP-ribosylation factor; Ref. 28). Concomitantly, lovastatin, like AFC, impedes nutrient-induced insulin release (7). In contrast to nutrients, the effects of agonists that induce secretion by directly activating distal components in signal transduction (such as a phorbol ester or 40 mM K$^+$) or activating or mimicking events induced by GBP-coupled receptors (e.g., carbachol or mastoparan) were either unaffected or were enhanced by lovastatin or AFC (7). These data are compatible with the hypothesis that prenylation and subsequent carboxyl methylation are required for the efficient functioning of one (or more) stimulatory GBP(s) to promote proximal steps in fuel-induced insulin secretion, whereas one (or more) inhibitory GBP(s) reduces secretion at a more distal locus.

Carboxyl Methylation of γ Subunits

In addition to the carboxyl methylation of CDC42, the carboxyl methylation of <8-kDa protein(s) was also inhibited by AFC, but not its inactive analog AGC. Data in Fig. 6 indicate that this protein comigrated with the purified γ subunit of transducin, a retinal trimeric GBP. However, the methylation of G_γ was not stimulated by GTPγS (Fig. 6), unlike the carboxyl methylation of CDC42. The carboxyl methylation of another protein with a molecular weight apparently lower than that of the pure γ subunit of transducin was also observed. This might represent an isoform of γ subunit, because at least six such isoforms have been reported to exist in mammalian cells (29). Further studies (e.g., immunologic identification) are needed to prove conclusively that these labeled proteins indeed represent the γ subunits of trimeric GBPs.

Fatty Acylation of GBPs

As stated at the beginning of this chapter (see *Introduction*), in the case of certain GBPs, fatty acids (typically palmitate, and in some cases, arachidonate) are incorporated posttranslationally via a thioester linkage into the cysteine residues upstream of the prenylated and methylated cysteine (30, 31). This modification may further facilitate the interaction of GBPs with their membrane-bound effectors. Recent data indicate that the α subunits of some trimeric GBPs (e.g., $G_{s\alpha}$) may be acylated; this is regulated acutely in response to receptor activation (32), thereby controlling the subcellular distribution of these α subunits (i.e., membrane vs. cytosolic). Receptor activation may also regulate deacylation (30). A selective blocker of protein fatty acylation (cerulenin; see Ref. 33), reduces nutrient-induced insulin secretion from normal islets (7). However, experimental and structural data indicate that neither CDC42 nor G_γ undergo fatty acylation (34, 35). Therefore, it seems possible that acylation of the α subunits of trimeric GBPs and/or other low molecular mass GBPs may also be required for nutrient-induced insulin secretion. Alternatively, other proteins involved in the exocytotic process (such as SNAP-25) (36) may be critically acylated. Additional studies are needed to demonstrate conclusively a putative role(s) for fatty acylation of GBPs in insulin secretion.

Conclusions and Future Directions

These studies, along with recent data from other laboratories, provide evidence to suggest that both low molecular mass GBPs (e.g., CDC42) and γ subunits of trimeric GBPs undergo posttranslational modifications at their C-terminal cysteine residues. They further demonstrate that prevention of these modifications with relatively selective inhibitors, such as lovastatin or AFC, results in altered subcellular distribution of these proteins. Concomitantly, these pharmacologic probes reduce nutrient-induced insulin secretion from normal rat islets. Thus, these studies provide a strong rationale for the formulation that GBPs play a pivotal role(s) in the normal function of the β cell. However, probes such as lovastatin and AFC may interfere with the functioning of more than one relevant GBP. Therefore, it will be necessary to develop systems for the overexpression of GBPs or application of antisense "knock-out" approaches (37, 38) for specific GBPs (e.g., CDC42 and the γ subunit of trimeric GBPs), in order not only to deduce the physiologic functions of these proteins in modulating insulin secretion, but also to suggest potential therapeutic approaches to states of perturbed insulin release.

Acknowledgments

The authors thank Drs. David Scharp and Paul Robertson/Hui-Jian Zhang for human islets and HIT cells, respectively. We also thank Mary Rabaglia, Scott Seavey, and James Stephens for excellent technical assistance. We express our sincere thanks to Dr. Akio Yamazaki for purified G_γ subunits, Dr. Tony Evans for purified CDC42 and also for antisera directed against CDC42, and Dr. Reinhard Jahn for antisera directed against rab3A. These studies were supported by grants from the Veterans Administration and the National Institutes of Health (DK 37312).

References

1. S. A. Metz, M. E. Rabaglia, and T. J. Pintar, *J. Biol. Chem.* **267,** 12517 (1992).
2. S. A. Metz, M. Meredith, M. E. Rabaglia, and A. Kowluru, *J. Clin. Invest.* **92,** 872 (1993).
3. A. G. Gilman, *Annu. Rev. Biochem.* **56,** 615 (1987).
4. R. P. Robertson, E. R. Seaquist, and T. F. Walseth, *Diabetes* **40,** 1 (1991).
5. Y. Takai, K. Kaibuchi, A. Kikuchi, and M. Kawata, *Int. Rev. Cytol.* **133,** 187 (1992).
6. K. H. Muntz, P. C. Sternweis, A. G. Gilman, and S. M. Mumby, *Mol. Biol. Cell.* **3,** 49 (1992).
7. S. A. Metz, M. E. Rabaglia, J. B. Stock, and A. Kowluru, *Biochem. J.* **295,** 31 (1993).
8. G. Li, R. Regazzi, E. Roche, and C. B. Wollheim, *Biochem. J.* **289,** 379 (1993).
9. A. Kowluru and S. A. Metz, *in* "Molecular Biology of Diabetes" (B. Draznin and D. LeRoith, eds.), Vol. 1, p. 249. Humana Press, Totawa, New Jersey, 1994.
10. A. Kowluru and S. A. Metz, *Biochemistry* **33,** 12495 (1994).
11. A. Kowluru, R. S. Rana, and M. J. MacDonald, *Arch. Biochem. Biophys.* **242,** 72 (1985).
12. A. Kowluru and S. A. Metz, *Biochem. J.* **297,** 399 (1994).
13. A. Kowluru, M. E. Rabaglia, K. E. Muse, and S. A. Metz, *Biochim. Biophys. Acta.* **122,** 348 (1994).

14. K. Shinjo, J. C. Koland, M. J. Hart, V. Narasimhan, D. I. Johnson, T. Evans, and R. A. Cerione, *Proc. Natl. Acad. Sci. U.S.A.* **87,** 9853 (1990).

15. J. B. Gibbs, D. L. Pompliano, S. D. Mosser, E. Rands, R. B. Lingham, S. B. Singh, E. M. Scolnick, N. E. Kohl, and A. Oliff, *J. Biol. Chem.* **268,** 7617 (1993).

16. F. Tamanoi, *Trends. Biochem. Sci.* **18,** 349 (1993).

17. P. L. Crowell, R. P. Chang, Z. Ren, C. E. Elson, and M. N. Gould, *J. Biol. Chem.* **266,** 17679 (1991).

18. S. Clarke, *Annu. Rev. Biochem.* **61,** 355 (1992).

19. E. W. T. Tan and R. R. Rando, *Biochemistry* **31,** 5572 (1992).

20. A. Kowluru, S. E. Seavey, M. E. Rabaglia, R. Nesher, and S. A. Metz, *Diabetes* **44,** 262A (1995).

21. B. Favre, S. Zolnierowicz, P. Turowski, and B. A. Hemmings, *J. Biol. Chem.* **269,** 16311 (1994).

22. H. K. Yamane, C. C. Farnsworth, H. Xie, W. Howland, and B. K.-K. Fung, *Proc. Natl. Acad. Sci. U.S.A.* **87,** 5868 (1990).

23. B. K.-K. Fung, H. K. Yamane, I. Mota, and S. Clarke, *FEBS. Lett.* **260,** 313 (1990).

24. S. Clarke, J. P. Vogel, R. J. Deschenes, and J. Stock, *Proc. Natl. Acad. Sci. U.S.A.* **85,** 4643 (1988).

25. M. Floer and J. Stock, *Biochem. Biophys. Res. Commun.* **198,** 372 (1994).

26. A. Kowluru, M. E. Rabaglia, J. Stock, and S. A. Metz, *Diabetes* **42,** 76A (1993).

27. A. Kowluru, S. E. Seavey, M. E. Rabaglia, and S. A. Metz, submitted (1995).

28. R. Regazzi, A. Kikuchi, Y. Takai, and C. B. Wollheim, *J. Biol. Chem.* **267,** 17512 (1992).

29. J. R. Hepler and A. G. Gilman, *Trends. Biochem. Sci.* **17,** 383 (1992).

30. P. B. Wadengaertner, P. J. Wilson, and H. R. Bourne, *J. Biol. Chem.* **270,** 503 (1995).

31. L. Muszbek and M. Laposata, *J. Biol. Chem.* **268,** 18243 (1993).

32. M. Y. Degtyarev, A. M. Spiegel, and T. L. Jones, *J. Biol. Chem.* **268,** 23769 (1993).

33. S. Omura, *Bacteriol. Rev.* **40,** 681 (1976).

34. S. Munemitsu, M. A. Innis, R. Clark, F. McCormick, A. Ullrich, and P. Polakis, *Mol. Cell. Biol.* **10,** 5977 (1990).

35. J. E. Buss, S. M. Mumby, P. J. Casey, A. G. Gilman, and B. M. Sefton, *Proc. Natl. Acad. Sci. U.S.A.* **84,** 7493 (1987).

36. D. T. Hess, T. M. Slater, M. C. Wilson, and J. H. Pate Skene, *J. Neurosci.* **12,** 4634 (1992).

37. P. R. Albert and S. J. Morris, *Trends Pharmacol. Sci.* **15,** 250 (1994).

38. S. J. Persaud and P. M. Jones, *J. Mol. Cell. Endocrinol.* **12,** 127 (1994).

[21] Epidermal Growth Factor-Mediated Regulation of G Proteins and Adenylylcyclase in Cardiac Muscle

Tarun B. Patel,* Hui Sun, Helen Poppleton, Bipin G. Nair,
Hani M. Rashed, and Yi-Ming Yu

Introduction

Epidermal growth factor (EGF), which was first isolated from mouse submaxillary glands (1), is a 53-amino acid polypeptide with three intramolecular disulfide bonds. Although the submaxillary gland has a high concentration of EGF, the growth factor is produced in a variety of tissues (see Ref. 2 for review). In the intact animal the levels of circulating EGF can vary from 0.2 nM under basal conditions to 20 nM in the presence of α-adrenergic agonists (2). Other than EGF, the EGF receptor binds and is activated by a number of ligands. Included among these are tranforming growth factor-α (TGF-α), amphiregulin, vaccinia virus growth factor, and heparin-binding growth factor (3–6). Because EGF and TGF-α are synthesized and secreted from a number of tissues and cell types (2), in addition to modulation by circulating EGF, the EGF receptor is likely to be regulated in an autocrine and/or paracrine manner.

In almost all cell types studied, two subpopulations of EGF binding sites are present, with apparent K_D values that are 10- to 20-fold different (see Ref. 7 for elaboration). The high-affinity binding sites account for approximately 10% of the total receptor population with a K_D of ~50 pM (8). The mature EGF receptor is a single polypeptide (1186 amino acids) that shares considerable sequence homology with the v-*erbB* oncogene protein (9, 10). The tyrosine kinase activity that copurifies with the receptor (11) is intrinsic to the receptor molecule and is stimulated by binding of EGF to the receptor (7). EGF activates the tyrosine kinase activity by increasing the V_{max} of the enzyme with very little effect on its K_m (12, 13). In addition to phosphorylating tyrosine residues of exogenously added substrates, stimulation of the receptor tyrosine kinase activity phosphorylates the receptor on residues 1173, 1148, 1086, and 1068 (14, 15). Evidence from NIH 3T3 and mouse lung

*To whom correspondence should be addressed.

fibroblast (B82L) cells transfected with EGF receptors devoid of tyrosine kinase activity indicates that the kinase activity of the receptor is important for EGF-elicited increases in DNA synthesis, stimulation of inositol phosphate formation, elevation of cytosolic free Ca^{2+}, activation of Na^+/H^+ exchange, and receptor internalization (16, 17). Likewise, it is now fairly well established that phosphorylation by protein kinase C of the EGF receptor decreases the tyrosine kinase activity of the receptor and attenuates the biological responses elicited by EGF (18–22). Covalent modification of the EGF receptor by protein kinase C has also been reported to either decrease the affinity of the receptor for ligand (reviewed in Ref. 19) or decrease the number of cell surface receptors available for binding with EGF (22, 23).

Besides regulating DNA synthesis and growth, EGF elicits a variety of biological responses. For instance, EGF has been demonstrated to stimulate glycolysis in 3T3 cells (24), increase transport of amino acids into hepatocytes (25), enhance fatty acid synthesis in hepatocytes (26, 27), and increase the activity of glycogen synthase in 3T3 cells and A431 cells (28). Additionally, EGF stimulates the activity of glycogen phosphorylase in rat hepatocytes (29); elevates prostaglandin production from glomerular mesangial (30), Madin–Darby canine kidney (MDCK) epithelial (31), and A431 cells (32); and stimulates Na^+/H^+ exchange and Na^+,K^+-ATPase activity in hepatocytes and A431 cells (19, 33). Some of these effects, such as activation of glycogen phosphorylase (29, 34), may be manifested by elevated cytosolic free Ca^{2+} concentrations due to EGF-elicited activation of the phosphatidylinositol second-messenger system (29, 34, 35). In this respect, EGF also stimulates the breakdown of phosphatidylinositol; increases the cellular content of second messengers, inositol 1,4,5-trisphosphate, and diacylglycerol; and elevates cytosolic free Ca^{2+} in a number of other cells, including A431 cells (20–22, 36), NIH 3T3 cells (37), and a human hepatocellular carcinoma cell line (38). In rat hepatocytes, EGF-elicited stimulation of phospholipase C activity is mediated by a pertussis toxin-sensitive mechanism(s) (35). However, in A431 cells, EGF-mediated activation of phospholipase C is not affected by pertussis toxin treatment of cells (21, 36). These findings imply that the coupling between EGF receptor and phospholipase C in different cells involves different mechanisms. Indeed, studies have shown that in A431 cells, phospholipase $C\gamma$ is phosphorylated on tyrosine residues by the activated EGF receptor and this phosphorylation stimulates the enzyme (39–42). These findings and the observations of Moolenaar et al. (20) indicate that the EGF receptor tyrosine kinase activity is essential for activation of phospholipase $C\gamma$.

It is noteworthy that stimulation of phospholipase C activity and activation of the inositol phosphate signaling system by EGF do not account for a variety of EGF-elicited actions. Included among these are the ability of EGF to stimulate Ca^{2+} influx and augment Na^+/H^+ exchange in A431 cells (19), and exert insulin-like effects on glycogen synthase (29) and fatty acid synthesis (26, 27) in hepatocytes. Therefore, it would appear that certain actions of EGF are mediated via mechanism(s) or second-messenger system(s) other than the phosphatidylinositol signaling pathway. In this context, experimental evidence from our laboratory and those of others, described below, indicate that EGF also modulates the cyclic AMP (cAMP) second-messenger system. In rat hepatocytes, EGF decreases glucagon-stimulated cAMP accumulation without altering the basal (unstimulated) cAMP levels, and this decrease in cellular cAMP is accompanied by decrease in the glucagon-stimulated glycogen phosphorylase activity (29). Similarly, EGF decreases prostaglandin E_1-stimulated cAMP content of fibroblastic cells without affecting the basal cAMP levels (43). On the other hand, in cells that overexpress the EGF receptor (A431, HSC-1, MDA), EGF potentiates cAMP accumulation in response to isoproterenol, cholera toxin, forskolin, and 3-isobutyl-1-methylxanthine (IBMX) (44); in these cells, EGF by itself does not alter cellular cAMP content (44). These findings imply that EGF-elicited regulation of cellular cAMP content in fibroblastic cells or hepatocytes (29, 43) and cells overexpressing the EGF receptor (44) involves different mechanism(s). It is noteworthy that, in contrast to the findings in hepatocytes, fibroblastic cells, and cells overexpressing the EGF receptor (29, 43, 44), our studies demonstrate that, in the heart, EGF elevates cellular cAMP content and also stimulates adenylylcyclase via activation of $G_{s\alpha}$ (45, 46). Employing approaches identical to those described in our initial study (45), Nakagawa *et al.* (47) have shown that EGF also stimulates adenylylcyclase in rat parotid gland via a G-protein-based mechanism. Apparently, therefore, EGF differentially modulates the cAMP second-messenger system in different tissues.

In the various sections of this review, we have described the approaches at various levels, ranging from intact hearts to recombinant proteins, that our laboratory has pursued to elucidate the mechanism(s) involved in the activation of adenylylcyclase by EGF and its receptor.

Methodology

EGF-Mediated Alterations in Cardiac Function: Studies Employing Perfused Rat Heart Model

Because the study of Rabkin *et al.* (48) has demonstrated that EGF produces a chronotropic effect in chick embryo cardiac myocytes in culture, employing the isolated perfused rat heart model we address the hypothesis that EGF increases the cellular accumulation of cAMP in rat hearts and produces chrontropic and/or inotropic actions. (The data from these studies are fully described in Refs. 45 and 49.) All of the experiments are performed with the retrograde perfused, Langendorff rat heart preparation (50). Essentially, after anesthetizing the male Sprague–Dawley rats (150–200 g body weight) with an intraperitoneal injection of pentobarbital sodium, the hearts are excised and washed with ice-cold Krebs–Henseleit bicarbonate (KHB) buffer (51) containing 1.3 mM $CaCl_2$ and 10 mM glucose. The hearts are perfused through the aorta with the same KHB buffer at 37°C equilibrated with O_2/CO_2 (95/5, v/v) and the flow rate is maintained constant at 10 ml/min. A pressure transducer in the aortic perfusion line provides a measure of the coronary perfusion pressure. After the hearts begin to beat rhythmically, a balloon made out of PE-10 tubing is inserted into the left ventricle via a small incision in the apex of the heart. The balloon, connected to a pressure transducer and a 1-ml syringe, is then filled with approximately 50 μl of fluid (KHB buffer) so that the contraction of the ventricular muscle is detected by the pressure transducer. The precise volume of the fluid in the balloon is adjusted so that arrhythmic contractions of the ventricles do not occur. The input to the Grass polygraph recorder from the pressure transducer connected to the ventricular balloon catheter provides the heart rate as well as the pressure that develops in the ventricle. The first derivative of the developed pressure is utilized to monitor the dp/dt, or the rate of contractility. Prior to exposure of the hearts to EGF or to any other agent that would modulate cardiac function, the hearts are equilibrated for a period of 30 min and control tracings are recorded. This method is described in detail in Ref. 49.

The data presented in Fig. 1 are typical of the tracings of the various parameters that were obtained from rat hearts exposed to EGF. These data demonstrate that on exposure of hearts to EGF, there is a decline in coronary perfusion pressure and an increase in the heart rate as well as contractility (Fig. 1). To determine whether the inotropic and chronotropic actions of EGF are due to accumulation of cAMP, hearts are perfused as described in Fig. 1, except that the ventricular balloon catheter is omitted. During the course of the perfusion with and without EGF, the hearts are freeze-clamped with a pair of Wollenberger aluminum tongs chilled in liquid nitrogen. The frozen hearts are pulverized into a powder by grinding with chilled pestle and mortar and 0.2 g of the heart tissue is homogenized in 2 ml of 10% trichloroacetic acid. The methodology described by Brooker et al. (52) is employed to extract the cAMP and measure its content by radioimmunoassay. The data in Fig. 2 demonstrate that in comparison with controls, in hearts perfused with EGF the accumulation of cAMP is increased by 10-fold. In order to determine whether the increase in contractility and beating rate of hearts in the presence of EGF is the result of elevation in cellular cAMP levels, experiments are performed with the adenosine A_1 receptor

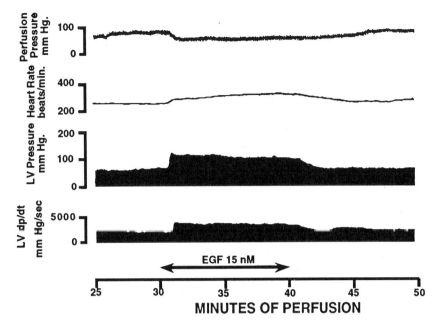

FIG. 1   The effect of EGF on left ventricular cardiac contractility (LV dp/dt), left ventricular pressure (LV pressure), heart rate, and coronary perfusion pressure in isolated perfused rat heart. Following a 20-min equilibration period, the various functional parameters were monitored for a 10-min period prior to continuous infusion of EGF (15 nM) for the 10-min period indicated by the arrow. Recovery of the various parameters was also monitored for a 10-min period following termination of EGF infusion. Details of the perfusion and measurements of various parameters are given in the text. From Ref. 49, with permission.

agonist, $(-)$-N^6-(R-phenylisopropyl)adenosine (PIA). PIA via activation of the adenosine A_1 receptor and activation of the inhibitory G_i protein of adenylylcyclase inhibits the accumulation of cAMP (53). As demonstrated by the data in Fig. 3, in the presence of PIA, EGF-elicited stimulation of contractility and heart rate are markedly attenuated; PIA by itself does not alter the contractility or heart rate. Moreover, in the presence of PIA, the accumulation of cAMP in response to EGF is also markedly diminished (Fig. 2). These and other data described in Ref. 49 demonstrate that EGF increases contractility and beating rate of hearts by elevating the accumulation of cAMP. Although the data from experiments with isolated perfused rat hearts are sufficient to address the hypothesis that EGF elevates the accumulation of cAMP and produces inotropic and chronotropic actions, this model is not

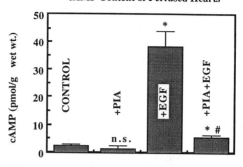

FIG. 2   Effect of EGF on cellular cAMP accumulation in hearts perfused in the presence and absence of $(-)$-N^6-(R-phenylisopropyl)adenosine (PIA). Hearts were perfused under conditions identical to those in Figs. 1 and 3. Just prior to (30-min time point, Figs. 1 and 3) or at the end of the period of EGF infusion (40-min time point, Figs. 1 and 3), hearts were freeze-clamped and the pulverized frozen tissue analyzed for cAMP content as described in the text. Mean ± SEM values of at least three determinations are represented. $*$, $p < 0.005$ as compared with control; $\#$, $p < 0.001$ as compared with corresponding condition without PIA; n.s., not significant. From Ref. 49, with permission.

appropriate to determine whether the effects of EGF on cAMP accumulation occurs in cardiomyocytes or nonmyocytes in the heart. Therefore, further investigations are performed employing primary cultures of cardiac myocytes and nonmyocytes.

Ability of EGF to Stimulate cAMP Accumulation: Experiments with Primary Cultures of Myocytes and Nonmyocytes Isolated from Neonatal Rat Heart Ventricles

As discussed above, because the heart is heterogeneous tissue comprising a variety of cell types, in order to determine whether the effects of EGF on cAMP accumulation are manifested in cardiac myocytes and/or nonmyocytes in the heart, primary cultures of these cells are employed. The methodology employed to isolate cardiac myocytes and nonmyocytes from neonatal rat hearts and culture these cells is derived from the techniques described Harary *et al.* (54) and Libby (55). Essentially, hearts from 20 to 40 Sprague–Dawley rat pups (1–4 days old) are aseptically excised and atria and surrounding tissues are trimmed. The remaining heart tissue, mainly ventricles, is washed with Hanks'–HEPES medium, pH 7.4, containing the following ingredients

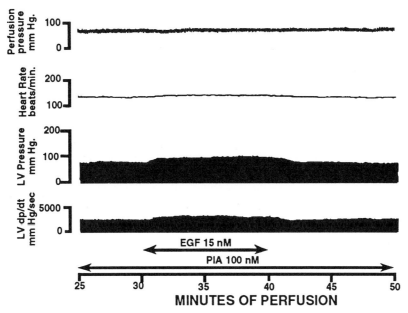

FIG. 3   Attenuation of EGF-mediated alterations of cardiac function by the adenosine
A_1 receptor agonist $(-)$-N^6-(R-phenylisopropyl)adenosine (PIA). Experimental conditions were similar to that described in Fig. 1 except that PIA (100 nM) infusion was
initiated 10 min prior to the infusion of EGF (15 nM). From Ref. 49, with permission.

at the concentrations indicated: NaCl (140 mM), KCl (5.4 mM), MgCl$_2$ (0.8
mM), NaH$_2$PO$_4$ (0.44 mM), Na$_2$HPO$_4$ (0.34 mM), HEPES (20 mM), NaHCO$_3$
(4.2 mM), and glucose (5.5 mM). Digestion is initiated by the addition of a
mixture of Viokase (0.1%, w/v) and DNase (0.002%, w/v) dissolved in the
Hanks'–HEPES medium. The mixture of heart tissue and the enzymes is
stirred at 200 rpm in a humidified incubator (37°C) for 10 min. Following
agitation, the mixture is allowed to stand so that the heart mince settles to
the bottom of the flask. The supernatant from the first digestion step is
decanted and discarded. A fresh batch of the Viokase/DNase mixture is then
added (1 ml/heart) and the digestion procedure repeated. The supernatant
from the second and subsequent six more digests is carefully removed and
placed in 50-ml sterile conical tubes containing an equal volume of CMRL
1066 medium supplemented with 5% horse serum and 5% fetal calf serum
on ice. Following filtration through a Nitex (mesh 80) sieve, cardiac myo-
cyates and nonmyocytes in the supernatant are harvested by centrifugation
(850 g for 10 min). The cell pellets are suspended in CMRL 1066 medium
supplemented with 10% serum components (5% horse serum, 5% fetal calf

serum) and cell density is adjusted to 3×10^6 cells/ml and 2 ml of the cell suspension is placed in 35-mm Falcon 3001 dishes. The cells are left to attach for a period of 1.5 hr at 37°C in a humidified incubator equilibrated with 95% air and 5% CO_2 (v/v). This time is sufficient to allow attachment of nonmyocyte cells but the cardiac myocytes remain unattached, and therefore the supernatant from this preattachment step is enriched with cardiac myocytes (54–58). The supernatant from this preattachment step is combined and any cell aggregates that may be present are dissociated by aspirating the supernatant through a 23-gauge needle into a syringe and filtering through a Nitex (mesh 50) sieve. The density of cells in the filtrate is adjusted to 2×10^6 cells/ml of CMRL 1066 medium containing 10% serum components; 4×10^6 cells (2 ml) are plated in 35-mm dishes (Falcon 3001) and placed in a 37°C humidifed incubator equilibrated with 95% air/5% CO_2. At 18 hr after plating the cells, the medium is changed to remove unattached cells. Under these conditions, cardiomyocytes, which represent >85% of the cell population, beat in a synchronous fashion 48 hr after plating.

For the purposes of culturing nonmyocyte cells, after removing the myocyte-enriched supernatant from the plates in the preattachment step, 2 ml of CMRL 1066 medium containing 5% horse serum and 5% fetal calf serum is added to each of the 35-mm dishes. Under these conditions nonmyocyte cells divide and are confluent 3 days after plating. To ensure minimum contamination of noncardiomyocytes by cardiomyocytes, the procedure of Orlowski and Lingrel (59) is employed. Essentially, cells are harvested with a 0.05% trypsin, 0.02% EDTA solution, and replated at a 1 : 2 split in serum-supplemented medium. By this method, noncardiomyocytes that have been subjected to two passages do not allow any significant growth of cardiac myocytes (<5% of total cell population). Following the second passage, the nonmyocytes are treated identically to the cultured myocytes.

At 24 hr prior to experimentation, medium in the culture dishes is replaced with CMRL 1066 medium containing 0.5% horse serum and 0.5% fetal calf serum. Subsequent to the 24-hr period of incubation in medium containing 1% serum components, the cells are incubated for a period of 2 hr in serum-free CMRL 1066 medium supplemented with 0.5% (w/v) fat-free bovine serum albumin and the inhibitor of cAMP phosphodiesterases, 3-isobutyl-1-methylxanthine (100 μM). Cells are then exposed to EGF and other test agents at the concentrations noted for a period of 10 min unless indicated otherwise. The incubations are terminated by aspirating the medium, adding 1 ml of 1 N NaOH to each dish, and freezing the cells on dry ice. Cyclic AMP content of the 1 : 10 diluted NaOH extracts is determined by the radioimmunoassay method of Brooker et al. (52).

By monitoring the thyroid hormone-elicited reciprocal expression of α- and β-myosin heavy chain mRNA as described previously (59, 60), the ability

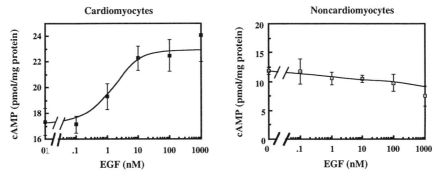

FIG. 4 Concentration dependence of EGF-mediated increase in cellular cAMP accumulation in cultured ventricular cardiomyocytes and noncardiomyocytes. After 4 days in culture, the cells were placed in medium containing 0.5% horse serum and 0.5% fetal calf serum. After 22 hr, the cells were placed in serum-free medium and after 2 hr they were challenged with different concentrations of EGF for 10 min. Control incubations received the vehicle but no EGF. Reactions were terminated and cAMP was measured as described in the text. Data presented are mean ± SEM ($n = 4$) from a representative of three similar experiments. #, $p < 0.05$, *, $p < 0.001$ as compared with controls in the absence of EGF; unpaired Students' t-test analysis. From Ref. 60, Y.-M. Yu, B. G. Nair, and T. B. Patel, *J. Cell. Physiol.,* Copyright © Wiley-Liss, Inc., 1992.

of the cardiomyocytes to respond in a physiologically relevant manner is established. Similarly, by performing Western analyses of the cardiomyocyte and nonmyocytes with anti-β-myosin heavy chain antiserum, it is demonstrated that the nonmyocyte cultures are not contaminated with myocytes (60).

The data presented in Fig. 4 demonstrate that the EGF increases the accumulation of cAMP in cardiac myocytes but not in nonmyocytes derived from rat hearts. However, as determined by the ability of EGF to increase tyrosine phosphorylation of cellular proteins, both myocytes and nonmyocytes express functional EGF receptors (60). Moreover, in both cell types the β-adrenergic receptor agonist isoproterenol increases cAMP accumulation. Therefore, these studies (60) demonstrate that in the heart the elevation of cAMP in response to EGF is due to the effect of EGF on cardiomyocytes and not on nonmyocytes. Moreover, these data (Fig. 4 and Ref. 60) demonstrate that the effects of EGF on cAMP accumulation are cell specific.

Role of $G_{s\alpha}$ and EGF Receptor Protein Tyrosine Kinase in EGF-Elicited Stimulation of Cardiac Adenylylcyclase: Experiments with Isolated Cardiac Membranes

Elevation of cAMP in response to an agonist can be the result of either stimulation of adenylylcyclase or inhibition of cAMP phosphodiesterases.

Because the studies in cardiac myocytes are performed in the presence of the cAMP phosphodiesterase inhibitor, 3-isobutyl-1-methylxanthine, it would appear that EGF elevates cAMP accumulation by stimulating the activity of adenylylcyclase. However, in order to determine directly whether the effects of EGF on cAMP accumulation in perfused hearts and isolated cardiomyocytes are due to activation of adenylylcyclase, further experiments are performed employing isolated cardiac membranes.

Membranes are isolated from hearts of adult male rats (150–180 g body weight). Essentially, after anesthetizing the rats with sodium pentobarbital, the hearts are surgically excised and rinsed in ice-cold buffer containing the following components at the indicated final concentrations: 5 mM Tris-HCl (pH 7.4), 250 mM sucrose, 1 mM phenylmethylsulfonyl fluoride (PMSF), and 1 mM EGTA. The hearts are then minced into fine pieces in the same medium and homogenized by five, 10-sec bursts in a Polytron homogenizer (Virtis) at half-maximal speed. This initial homogenate is further processed in a Potter homogenizer and the volume adjusted so that three rat hearts are represented by 100 ml of the homogenate. The homogenate is centrifuged at 1100 g for 20 min and the pellets are resuspended and homogenized again in the aforementioned medium in a Potter homogenizer. This homogenate is centrifuged at 1100 g for 20 min and the supernatant discarded. The pellets are resuspended in medium without PMSF and the centrifugation process is repeated two more times. It is noteworthy that the center of the pellets, which is granular, should be discarded. The final pellets are resuspended in medium without PMSF at a protein concentration of between 5 and 10 mg/ml and stored in small aliquots at −80°C. Just prior to performance of the adenylylcyclase assay, the membrane suspension is frozen and thawed twice with liquid N_2 and water (room temperature), respectively.

Adenylylcyclase assays are performed essentially as described by Salomon et $al.$ (61). Membrane protein (20–30 μg) is added to a reaction mixture that consists of 50 mM Tris-HCl (pH 7.4), 5.0 mM MgCl$_2$, 12.0 mM phosphocreatine, 1 mg/ml of creatine phosphokinase, 1 mM IBMX, and 0.1 mM [α-^{32}P]ATP (200 dpm/pmol). EGF, Gpp(NH)p, GDPβS, isoproterenol, propranolol, sodium fluoride, aluminum chloride, and other test reagents are added to the assay at the desired final concentration. ^{3}H-Labeled cAMP (15,000–18,000 dpm/assay) is also added to the assay mixture for determination of recovery of ^{32}P-labeled cAMP after column chromatography. The final reaction volume is 250 μl and activities are determined over a 30-min period. The reactions are terminated by transferring 100 μl of incubation mixture to tubes containing 100 μl of 2% SDS, 1 mM ATP, 1.4 mM cAMP, and 50 mM Tris-HCl, pH 7.4, and freezing the mixture in dry ice. ^{32}P-Labeled cAMP is then separated according to the method of Salomon et $al.$ (61). Recovery of cyclic AMP ranges from 70 to 80%. All assays are performed in quadruplicate and activity is linear with respect to membrane protein concentration and time.

Employing the techniques described above, initial experiments demonstrate that in the presence of the GTP analog, Gpp(NH)p, EGF stimulates cardiac adenylylcyclase in a concentration-dependent manner. The effects of EGF on adenylylcyclase are dependent on the presence of GTP analog, and are abolished in the presence of the GDP analog, GDPβS (10 μM), which cannot be phosphorylated (see Ref. 45 for details). Moreover, in the presence of cholera toxin, when adenylylcyclase is activated, EGF does not stimulate the enzyme. This, coupled with the observation that despite complete ADP-ribosylation of G_i by pertussis toxin, EGF still stimulates cardiac adenylylcyclase, suggests that the effects of EGF on adenylylcyclase are mediated via G_s or G_s-like GTP binding protein (see Ref. 45 for details).

In order to elucidate whether the α subunit of G_s is involved in mediating the actions of EGF on cardiac adenylylcyclase, experiments are performed with the anti-$G_{s\alpha}$ antiserum CS1. This antiserum is employed for two main reasons. First, anti-$G_{s\alpha}$ antiserum is generated against the carboxy-terminal decapeptide of the protein. Because the carboxy terminus of $G_{s\alpha}$ is conserved in all alternately spliced variants of the protein (62, 63), the CS1 antiserum recognizes all isoforms of $G_{s\alpha}$ (see, e.g., Ref. 46). Indeed, by Western analysis, in the heart, the CS1 antiserum demonstrates the presence of two isoforms of $G_{s\alpha}$, the 52-kDa species being the most prominent (46). The second reason for employing CS1 antiserum in our experiments is that, because previous studies had indicated that the carboxy terminus of $G_{s\alpha}$ is important in coupling with the receptors (64, 65), we reasoned that the manipulation of the carboxy terminus of $G_{s\alpha}$ with an antibody against this region would abolish receptor/G_s interactions and, therefore, obliterate effects at the level of adenylylcyclase. In our experimental approaches, to study the functional effects of CS1 antiserum, cardiac membranes (50 μg protein) are preincubated with varying amounts (0 to 3 μg protein) of either CS1 antiserum or nonimmune serum in a final volume of 60 μl for a period of 90 min prior to assay for adenylylcyclase activity. Adenylylcyclase activity in these membranes is assayed by adding the assay ingredients such that the final volume of the assay is 250 μl. The concentrations of CS1 antiserum and nonimmune serum in the figures are represented to reflect the final concentrations in the adenylylcyclase assay mixture.

In functional activity assays, neither nonimmune serum nor the CS1 antiserum affects basal adenylylcyclase activities. Similarly, aluminum fluoride- and forskolin-stimulated activities are also not altered by either the nonimmune serum or the CS1 antiserum (Fig. 5). Moreover, in functional assays of G_i activity, CS1 antiserum does not inhibit the interactions between G_i and adenylylcyclase (see Ref. 46 for details). These findings indicate that

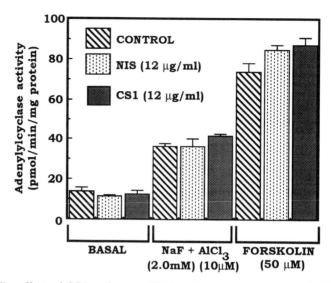

F IG. 5   The effect of CS1 antiserum (CS1) and nonimmune serum (NIS) on basal, aluminum fluoride- and forskolin-stimulated adenylylcyclase activity. As described in the text, cardiac membranes (50 μ protein) were incubated for 90 min either with CS1 or NIS prior to adenylylcyclase activity determination in the absence of Gpp(NH)p. The final concentration of CS1 and NIS in the adenylylcyclase assay was 12.0 μg protein/ml. Adenylylcyclase activity (pmol/min/mg protein) is presented as the mean ± SEM of four determinations. From Ref. 46, with permission.

the CS1 antiserum does not nonspecifically alter either the interactions between G_s or G_i and adenylylcyclase (aluminum fluoride effects) or the activity of the adenylylcyclase catalytic subunit (basal and forskolin stimulated). On the other hand, the CS1 antiserum, but not the nonimmune serum, obliterates the ability of both EGF (Fig. 6) and the β-adrenoreceptor agonist isoproterenol (46) to stimulate cardiac adenylylcyclase. These data, which are described and discussed in detail in Ref. 46, indicate that the effects of EGF on adenylylcyclase in the heart are mediated via activation of G_s.

In another series of experiments, using the isolated cardiac membrane preparation and EGF receptor protein tyrosine kinase-selective inhibitors, tyrphostins, we investigate whether or not the EGF receptor protein tyrosine kinase activity is important for EGF-mediated stimulation of cardiac adenylylcyclase (66). For the synthesis of the tyrphostins the reader is referred to our previous publication (66). In control experiments, the tyrphostins do not alter either the basal or the isoproterenol-stimulated adenylylcyclase activity (66). However, the tyrphostins inhibited both the EGF receptor protein

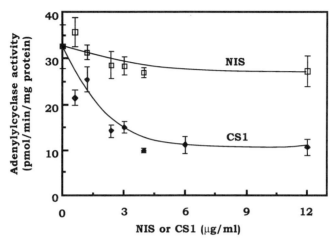

FIG. 6   The effect of varying concentrations of CS1 antiserum (CS1) and nonimmune serum (NIS) on EGF-stimulated adenylylcyclase activity. Cardiac membranes (50 μg protein) were incubated for 90 min with varying amounts of CS1 or NIS. At the end of the preincubation period, the adenylylcyclase assay mixture was added to the membranes and activities in the absence and presence of 10 nM EGF were determined. The Gpp(NH)p concentration in the assay mixture was 10 μM. The final concentrations of CS1 and NIS in adenylylcyclase assay are represented. The mean $\pm$ SEM values from four adenylylcyclase activity determinations are presented. From Ref. 46, with permission.

tyrosine kinase activity and the ability of EGF to stimulate cardiac adenylylcyclase in a concentration-dependent manner (see, e.g., Fig. 7). These data (66) demonstrate that the EGF receptor protein tyrosine kinase activity is important for EGF to stimulate adenylylcyclase.

Investigations into Mechanisms of EGF Receptor and G$_s$ Interactions: Experiments with Purified G$_s$ Components and Peptides

As discussed above, our experimental evidence indicates that the EGF receptor protein tyrosine kinase activity and G$_s$ are important in mediating the actions of EGF on cardiac adenylylcyclase. In order to further determine the mechanism(s) involved in EGF-mediated stimulation of adenylylcyclase and to incorporate the involvement of G$_s$ and EGF receptor protein tyrosine kinase activity, we formulated a hypothesis: on binding of EGF to its recep-

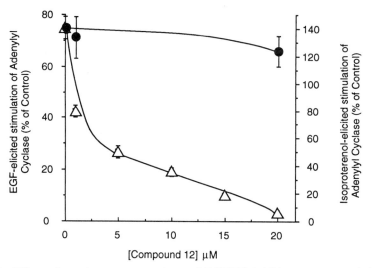

FIG. 7   Effect of varying concentrations of EGFRK inhibitor, compound 12, on the ability of isoproterenol (100 nM, ●) and EGF (10 nM, Δ) to stimulate cardiac adenylylcyclase activity. Cardiac membranes (50 μg of protein) were assayed for adenylylcyclase activity in the presence of isoproterenol and EGF with varying concentrations (1–20 μM) of compound 12. Data presented are the mean ± SEM (n = 4 experiments) of percent agonist (EGF/isoproterenol)-mediated stimulation over corresponding controls performed in the presence of the various indicated concentrations of inhibitor. In the presence of Gpp(NH)p (10 μM), unstimulated control adenylylcyclase activity that remained unaffected by compound 12 was 19.6 ± 0.6 pmol/min/mg protein. Wherever not seen, error bars are within the symbol size. Reprinted from Ref. 66, *Biochem. Pharmacol.* **46**, B. G. Nair and T. B. Patel, Regulation of cardiac adenylylcyclase by EGF: Role of the EGF receptor tyrosine kinase activity. 1239–1245, copyright 1993, with kind permission from Elsevier Science Ltd, The Boulevard, Langford Lane, Kidlington OX5 1GB, UK.

tor, activation of protein tyrosine kinase, and resultant alteration in receptor conformation, a region of the EGF receptor is exposed to interact with and activate G_s. Although activation of the EGF receptor tyrosine kinase, autophosphorylation, and resultant alteration in conformation are established (14, 15 and 67), the possibility that a region of the EGF receptor can interact with and activate G_s remains to be examined. The latter possibility is based on the findings of others with receptors that couple to G proteins. For instance, Okamoto *et al.* (68) have shown that a 15-amino acid peptide corresponding to the β_2-adrenergic receptor can activate G_s. Similarly, a 14-amino acid peptide corresponding to a region in the IGF-II receptor has also been shown to stimulate the inhibitory GTP-binding regulatory protein of adenylylcyclase, G_i (69). From their studies with regions of receptors that

TABLE I   Peptide Sequences and Location of Corresponding Sequences in β_2-Adrenergic Receptor or Rat EGF Receptor

Peptide sequence	Receptor	Location within receptor	Abbreviation
RRSSKFCLKEHKALK	β_2-Adrenergic receptor	Arg^{259}-Lys^{273}	βIII-2
RRREIVRKRTLRR	Rat EGF receptor	Arg^{646}-Arg^{658}	EGFR-13
HLRILKETEFKKIK	Rat EGF receptor	His^{680}-Lys^{693}	EGFR-14
RRLTRKRVIERRR	—	—	revEGFR-13

activate G proteins (68, 69), Okamoto *et al.* described the following three criteria for a region of the receptor to be a candidate for G protein activator: (1) the presence of two basic residues in the N terminus, (2) the presence of the sequence B-B-X-B or B-B-X-X-B (B, basic residues; X, any residue) in the C terminus, and (3) the region (or peptide) should be approximately 14–18 residues long. Within the cytosolic domain of the rat EGF receptor we identified one region resembling the peptide sequence in the β_2-adrenergic receptor that activates G_s (70). This 13-amino acid region Arg^{656}-Arg^{658} (sequence: RRREIVRKRTLRR; EGFR-13) immediately follows the transmembrane region (Ile^{623}-Met^{645}) and meets all of the criteria for a G_s activator described by Okamoto *et al.* (68). Interestingly this 13-amino acid region is also conserved in the human EGF receptor (10). Another region His^{680}-Lys^{693} (sequence: HLRILKETEFKKIK; EGFR-14) fulfills only two (length and C-terminus requirement) of the three criteria mentioned above. Therefore, employing peptides corresponding to these sequences in the EGF receptor, and purified $G_{s\alpha}$ and G protein $\beta\gamma$ subunits, we investigated whether region(s) in the cytosolic domain of the EGF receptor can interact with, and activate, G_s. We also studied the interactions of these regions with G_i.

The α subunit of G_s is purified from bacteria transformed to express this protein. Essentially, the BL21-DE3 strain of *Escherichia coli* transformed with plasmid pQE-60 containing cDNA encoding the 45-kDa form of bovine $G_{s\alpha}$ is obtained (from Dr. Alfred Gilman, University of Texas Southwestern Medical Center, Dallas, TX). Expression of the recombinant $G_{s\alpha}$ from its cDNA subcloned into pQE-60 plasmid is induced in the log phase of growth by the addition of isopropyl-β-D-thiogalactopyranoside and the protein is purified essentially as described by Graziano *et al.* (71). Bovine brain $\beta\gamma$ subunits are purified to homogeneity as described by Mumby *et al.* (72) and Neer *et al.* (73). Heterotrimer G_s is reconstituted by mixing $G_{s\alpha}$ and $\beta\gamma$ subunits ($G_{s\alpha}$: $\beta\gamma$ ratio of 1 : 5) and incubating for 1 hr at 4°C. The peptides EGFR-13, EGFR-14, βIII-2, and rev-EGFR-13 (Table I) are synthesized by the solid-phase system employing an Applied Biosystems synthesizer (Model 430). The peptides are recovered from the resin by HF cleavage and extraction with 5% (v/v) acetic acid. They are desalted by gel filtration with Sephadex G-25 columns using 0.1 M acetic acid and 10% acetonitrile followed by purification by HPLC on a Vydac 218TP54 column. A linear gradient starting with 10% acetonitrile in 0.05% trifluoroacetic acid advancing to 60% acetonitrile in 0.05% trifluoroacetic acid is used in the final purification. The threonine-phosphorylated analog of peptide EGFR-13 (phospho-EGFR-13, Table I) is custom synthesized by SynPep Corp. (Dublin, CA). All synthetic peptides are greater than 95% pure as assessed by HPLC analysis.

In our approaches, changes in G protein activity are monitored by measuring the following three parameters: GTPase activity, GTPγS binding, and the ability of G_s to stimulate adenylylcyclase activity in S49 cyc^- cells, which are mutant mouse lymphoma cells that do not express $G_{s\alpha}$ (74).

[^{35}S]GTPγS binding to G_s is monitored by slight modification of the method of Northup et al. (75). The reconstituted G_s protein ($G_{s\alpha}$: $\beta\gamma$ molar ratio, 1 : 5) is incubated with or without the peptides of interest for a period of 5 min prior to initiation of the [^{35}S]GTPγS binding in 100 μl of medium containing, at the final concentration, the following components: 50 mM HEPES (pH 8.0), 120 μM MgCl$_2$, 100 μM EDTA, and 1 mM DTT. [^{35}S]GTPγS binding is initiated by the addition of 100 nM concentration of the radionucleotide. The final concentration of G_s is 10 nM as assessed by maximal binding in the presence of 1 μM [^{35}S]GTPγS and 25 mM MgCl$_2$. The binding reactions are terminated by addition of 2 ml of ice-cold "stop" buffer containing 25 mM Tris-HCl (pH 7.4),10 mM NaCl, and 25 mM MgCl$_2$. The bound [^{35}S]GTPγS is separated from unbound nucleotide by rapid filtration through BA85 nitrocellulose filters. The filters are washed twice more than 2 ml of the stop buffer. Nonspecific binding is monitored under identical conditions but in the presence of excess (100 μM) unlabeled GTPγS. In controls performed with peptides alone, none of the peptides employed binds any [^{35}S]GTPγS.

In order to monitor GTP hydrolysis, the method described by Brandt et al. (76) is employed with the following modifications. Essentially, the reconstituted G_s protein is incubated with and without the peptides of interest at 25°C in medium containing the following components at the final concentration: 25 mM HEPES–NaOH (pH 8.0), 110 μM EDTA, 200 μM MgSO$_4$, and 1 mM DTT. The final concentration of G_s is 10 nM. At various times after initiation of the incubation with 100 nM [^{35}S]GTPγP, aliquots are withdrawn and transferred into tubes containing ice-cold 5% (w/v) Norit A in 50 mM NaH$_2$PO$_4$ (pH 8.0). Following centrifugation of the mixture in the cold, the supernatant is decanted and counted for [^{32}P]phosphate content. The rates of GTPase activity are linear under all experimental conditions and GTPase activity for each condition is monitored in triplicate. As with [^{35}S]GTPγS binding, in control experiments, the peptides by themselves do not demonstrate any GTPase activity.

The functional activity of G_s is measured by the ability of G_s to activate adenylylcyclase activity in mouse lymphoma S49 *cyc*− membranes. S49 *cyc*− cell membranes are isolated as described by Iyengar *et al.* (77). Reconstituted G_s is incubated in the presence or absence of peptides and GTPγS (100 nM) for 1 hr as described above in the GTPγS binding protocols. Thereafter, 0.1 pmol (10 μl) of G_s is added to 10 μg of S49 *cyc*− cell membrane protein in the presence of 1 nmol of GDPβS. Adenylylcyclase reactions (100 μl final volume) are initiated by the addition of 80 μl of reaction mixture, described previously (45, 46), to the *cyc*− membranes/G_s complex, except that the Mg^{2+} concentration is 1 mM. Assays are conducted for 15 min in triplicates. In control experiments performed with peptides alone (no G_s), the adenylylcyclase activity is not altered.

The data from the experiments employing peptides corresponding to different regions within the cytosolic domain of the EGF receptor, and the methodology described above, are fully detailed in Ref. 70. Briefly, the peptide EGFR-13, corresponding to the juxtamembrane region of the receptor, stimulates GTPγS binding to G_s in a concentration-dependent manner; maximal and half-maximal effects are observed at 10 μM and 700 nM concentrations of EGFR-13, respectively (Fig. 8). EGFR-13 increases both the rate and the extent of GTPγS binding to G_s. The 14-amino acid peptide EGFR-14, which only partially satisfies the criteria for a G_s stimulator, does not alter the binding of GTPγS to G_s, except at very high concentrations (Fig. 8). Moreover, a peptide with sequence irrelevant to the EGF receptor also does not affect GTPγs binding to G_s (Fig. 8B). Similar to the findings with GTPγS binding, EGFR-13 also stimulates the GTPase activity of G_s (see Fig. 9); rEGFR-14 stimulates GTPase activity only at high concentrations (not shown). In order to determine whether the peptide EGFR-13 specifically alters the GTPγS binding to, and GTPase activity of, $G_{s\alpha}$, experiments similar to those described above are performed with $G_{i\alpha2}$. Interestingly, neither EGFR-13 nor EGFR-14 alters GTPγS binding to $G_{i\alpha2}$ (Fig. 9A). However, the GTPase activity of G_i is markedly augmented by EGFR-13 (Fig. 9C). These findings indicate that although EGFR-13 stimulates G_s activity (increases GTPγS binding and GTPase activity), this peptide, by selectively stimulating GTPase activity (no change in GTPγS binding), inactivates the G_i component. Such an activation of G_s and inactivation of G_i would serve to amplify the signal at the level of the effector adenylylcyclase.

An interesting feature of the peptide EGFR-13 is that it includes the threonine residue (Thr-655 in rat receptor; Thr-654 in human receptor), which is the site of protein kinase C phosphorylation. Phosphorylation of the EGF receptor on this Thr residue has been shown to decrease the affinity of the receptor for EGF (18), increase the endocytosis of the receptor (22, 23), and decrease the protein tyrosine kinase activity of the receptor (18). Thus activation of protein kinase C attenuates or abolishes the biological actions

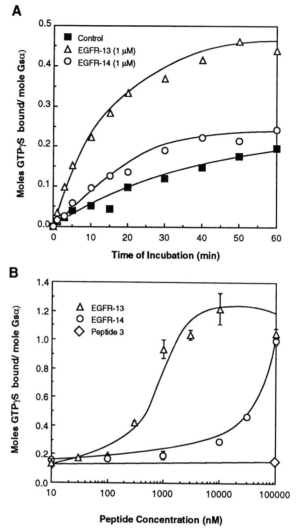

FIG. 8   Effect of peptides, EGFR-13 and EGFR-14, corresponding to regions within the cytosolic domain of the EGF receptor on GTPγS binding to G_s. Purified recombinant $G_{s\alpha}$ (1 pmol) was reconstituted with 5 pmol of purified bovine brain βγ subunits as described in the text; final concentrations of $G_{s\alpha}$ and βγ in the assay were 10 and 50 nM, respectively. [^{35}S]GTPγS binding assays were performed in a final volume of 100 μl of reaction mixture with and without the indicated peptides. (A) Time course of [^{35}S]GTPγS binding to reconstituted G_s in the presence of 1 μM each of

of EGF (19–21). Indeed, previously we have also demonstrated that activation of protein kinase C obliterates the ability of EGF to stimulate adenylylcyclase activity in cardiac membranes (45). Therefore, to determine whether phosphorylation of the threonine residue in EGFR-13 alters the ability of the peptide to activate G_s and inhibit G_i, experiments are performed employing the phosphothreonine form of EGFR-13 (phospho-EGFR-13). The data in Fig. 9 demonstrate that, as compared to its counterpart, phospho-EGFR-13 stimulates GTPγS binding to G_s and GTPase activity of this protein to a lesser extent (Fig. 9, A and B). Moreover, phospho-EGFR-13 activates GTPase activity of G_i to a lesser extent (Fig. 9C) without affecting the binding of GTPγS to G_i. Most importantly, although EGFR-13 augments the ability of EGFR-13 to stimulate adenylylcyclase activity in S49 cyc^- cells (Fig. 9D), neither phospho-EGFR-13 nor the control peptide EGFR-14 alters the functional activity of G_s (Fig. 9D). These data demonstrate that phosphorylation of the threonine residue in EGFR-13 inactivates the peptide with respect to its ability to activate G_s and stimulate adenylylcyclase. These observations are entirely consistent with our previous observation that protein kinase C activation down-regulates the ability of EGF to stimulate cardiac adenylylcyclase (45).

The studies described above and in Ref. 70 have identified the juxtamembrane region in the cytosolic domain of the EGF receptor as an activator of G_s and inhibitor of G_i function. Moreover, the protein kinase C phosphorylation site modulates the interactions of this region with G proteins. The definition of the G protein-interacting sequence of the EGF receptor adds to the number of functionally important regions within the cytosolic domain of this receptor. Thus, in addition to the well-characterized tyrosine kinase domain (78) and the importance of regions around the autophosphorylation sites in the carboxy terminus for recognition of proteins containing Src homology-2 region (79–81), the juxtamembrane region of the receptor is important for interactions with G proteins coupled to adenylylcyclase. Hence, the pleiotropic actions of EGF may be explained by the ability of the

---

EGFR-13 (sequence: RRREIVRKRTLRR) ($\triangle$) and EGFR-14 (sequence: HLRIL-KETEFKKIK) ($\bigcirc$) corresponding to Arg^{646}-Arg^{658} and His^{680}-Lys^{693} of the rat EGF receptor, respectively. Controls ($\blacksquare$) performed in the absence of peptides are also shown. A representative experiment of three is presented. (B) Dose–response curve of [^{35}S]GTPγS binding to reconstituted G_s in the presence of varying concentrations of EGFR-13 ($\triangle$) and EGFR-14 ($\bigcirc$); 100 μM concentration of peptide 3 (sequence: CIASTTTFGG) ($\diamond$), which has no resemblance to any region of the EGF receptor, on [^{35}S]GTPγS binding to the G_s, is also shown. Data are the mean ± SD of three to five determinations. From Ref. 70, with permission.

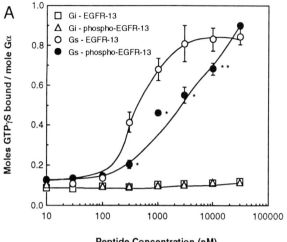

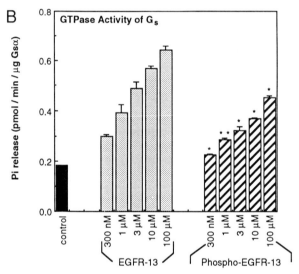

FIG. 9   Modulation by phosphorylation of Thr-655 in EGFR-13 on [^{35}S]GTPγS bind-ing, GTP hydrolyzing activities of G_s and G_i, and ability of G_s to stimulate adenylyl-cyclase activity. (A) [^{35}S]GTPγS binding to G_s (○, ●) and G_i (□, △) was performed in the presence of varying concentrations of EGFR-13 (○, □) and phospho-EGFR-13 (sequence: RRREIVRKRT(PO$_3$H$_2$)LRR (●, △). G_i was reconstituted from purified $G_{i\alpha2}$ expressed in Sf9 cells and purified brain βγ subunits as described in the text. Data are presented as mean ± SD; *, $p < 0.001$; **, $p \leq 0.005$ as compared with identical concentration of EGFR-13. (B) Modulation of GTPase activity of G_s by different concentrations of EGFR-13 (▦) and phospho-EGFR-13 (▨). The experimen-tal conditions were similar to those described in Fig. 3. Data are presented as mean ± SD; *, $p < 0.001$; **, $p \leq 0.005$ as compared with identical concentration of EGFR-13. (C) Modulation of GTPase activity of G_i by different concentrations of EGFR-13 (▦) and phospho-EGFR-13 (▨). The effect of EGFR-14 (10 μM) on GTPase

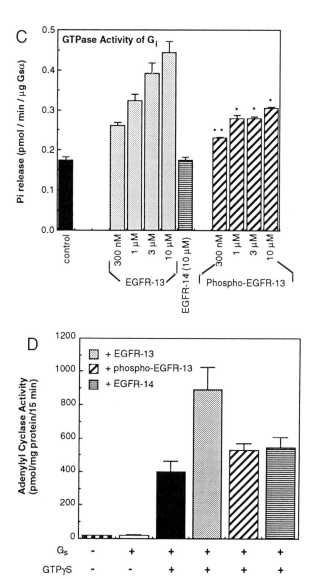

activity of G_i is also shown (▤). Data are presented as mean $\pm$ SD; $*$, $p < 0.001$; $**$, $p \leq 0.005$ as compared with identical concentration of EGFR-13. (D) Stimulation of adenylylcyclase activity by G_s that had been preactivated in the presence or absence of EGFR-13 (▦), phospho-EGFR-13 (◪), and EGFR-14 (▤). Reconstituted G_s was incubated in the presence or absence of peptides and GTPγS (100 nM) for 1 hr as described in the GTPγS binding protocols in the text. Thereafter, 0.1 pmol (10 μl) of G_s was added to 10 μg of S49 cyc^- cell membrane protein in the presence of 1 nmol of GDPβS. Adenylylcyclase reactions (100 μl final volume) were conducted for 15 min. Data are presented as the mean $\pm$ SD of three determinations in two similar experiments. In controls with peptides alone (no G_s), adenylylcyclase activity was not altered (not shown). From Ref. 70, with permission.

multifunctional cytosolic regions of the EGF receptor to activate a number of different types of signaling cascades.

Concluding Remarks

In this review we have presented a number of experimental approaches that have proved to be very useful in elucidating the mechanism(s) involved in EGF-mediated stimulation of cardiac adenylylcyclase. However, there are several additional questions pertaining to the detailed understanding of the mechanism(s) involved in interactions between the EGF receptor and G_s interaction. Thus although the *in vitro* studies demonstrate that the juxtamembrane region of the EGF receptor is important for stimulation of G_s, by performing mutation(s) of key residue(s) in the juxtamembrane region of the EGF receptor, we would like to determine whether the ability of EGF to stimulate adenylylcyclase can be obliterated. Similarly, it is important to determine the region(s) of $G_{s\alpha}$ that interact with the EGF receptor. It is also one of our aims to determine what elements confer specificity to the ability of EGF to stimulate adenylylcyclase. The latter goal poses a particularly important question, because EGF does not elevate cAMP content in all cells that express the EGF receptor, G_s, and adenylylcyclase.

Acknowledgments

We are grateful to the following colleagues for providing various reagents or information as indicated: Dr. Graeme Milligan, anti-$G_{s\alpha}$ antiserum; Dr. Jerome S. Seyer, peptides corresponding to regions on EGF receptor; Dr. Alfred Gilman, bacteria expressing bovine $G_{s\alpha}$; Dr. James Garrison, purified $G_{i\alpha2}$; Dr. H. Shelton Earp, rat EGF receptor sequence; and Dr. Ravi Iyengar, S49 *cyc−* membranes.

NOTE ADDED IN PROOF: We have recently demonstrated that expression of type V isoform of adenylylcyclase is required for EGF to stimulate cAMP accumulation in cells (Chen *et al.*, *J. Biol. Chem.*, **270**, 47 In Press).

References

1. C. R. Savage, Jr., J. H. Hash, and S. Cohen, *J. Biol. Chem.* **248,** 7669–7672 (1973).
2. G. Carpenter and S. Cohen, *Annu. Rev. Biochem.* **48,** 193–216 (1979).
3. H. Marquardt, M. W. Hunkapillar, L. E. Hood, and G. J. Todaro, *Science* **223,** 1079–1082 (1984).
4. P. Stroobant, A. P. Rice, W. J. Gullick, D. J. Cheng, I. M. Kerr, and M. D. Waterfield, *Cell* **42,** 383–393 (1985).

5. M. Shoyab, G. D. Plowman, V. L. McDonald, J. G. Bradley, and G. J. Todaro, *Science* **243,** 1074–1076 (1989).
6. S. Higashiyama, J. A. Abraham, J. Miller, J. C. Fiddes, and M. Klagsbrun, *Science* **251,** 936–939 (1991).
7. G. Carpenter, *Annu. Rev. Biochem.* **56,** 881–914 (1987).
8. M. Faucher, N. Girones, Y. A. Hannun, R. M. Bell, and R. J. Davis, *J. Biol. Chem.* **263,** 5319–5327 (1988).
9. J. Downward, Y. Yarden, E. Mayes, G. Scarce, N. Totty, P. Stockwell, A. Ullrich, J. Schlessinger, and M. D. Waterfield, *Nature (London)* **307,** 521–527 (1984).
10. A. Ullrich, L. Coussens, J. S. Hayflick, T. J. Dull, A. Gray, A. W. Tam, J. Lee, Y. Yarden, T. A. Libermann, J. Schlessinger, J. Downward, E. L. V. Mayes, N. Whittle, M. D. Waterfield and P. H. Seeburg, *Nature (London)* **309,** 418–425 (1984).
11. S. Cohen, H. Ushiro, C. Stoscheck, and M. Chinkers, *J. Biol. Chem.* **257,** 1523–1531 (1982).
12. C. Erneaux, S. Cohen, and G. L. Garber, *J. Biol. Chem.* **258,** 4137–4142 (1983).
13. J. Downward, M. D. Waterfield, and P. Parker, *J. Biol. Chem.* **260,** 14538–14546 (1985).
14. J. Downward, P. Parker, and M. D. Waterfield, *Nature (London)* **311,** 483–485 (1984).
15. B. J. Margolis, I. Lax, R. Kris, M. Dombalagian, A. M. Honnegar, R. Howk, D. Givol, A. Ullrich, and J. Schlessinger, *J. Biol. Chem.* **264,** 10667–10671 (1989).
16. W. H. Moolenaar, A. J. Bierman, B. C. Tilly, I. Verlaan, L. H. K. Defize, A. M. Honegger, A. Ullrich, and J. Schlessinger, *EMBO J.* **7,** 707–710 (1988).
17. J. R. Glenney, W. S. Chen, C. S. Lazar, G. M. Walton, L. M. Zokas, G. M. Rosenfeld, and G. N. Gill, *Cell* **52,** 675–684 (1988).
18. R. J. Davis and M. P. Czech, *Cancer Cells* **3,** 101–108 (1985).
19. I. G. Macara, *J. Biol. Chem.* **261,** 9321–9327 (1986).
20. W. H. Moolenaar, R. J. Aerts, L. G. J. Tertoolen, and S. W. de Laat, *J. Biol. Chem.* **261,** 279–284 (1986).
21. L. J. Pike and A. T. Eakes, *J. Biol. Chem.* **262,** 1644–1651 (1987).
22. M. Wahl and G. Carpenter, *J. Biol. Chem.* **263,** 7581–7590 (1988).
23. L. Beguinot, J. A. Hanover, S. Ito, N. D. Richert, M. D. Willingham, and I. Pastan, *Proc. Natl. Acad. Sci. U.S.A.* **82,** 2774–2778 (1985).
24. J. A. Schneider, I. Diamond, and E. Rozengurt, *J. Biol. Chem.* **253,** 872–877 (1978).
25. S. K. Moule and J. D. McGivan, *Biochem. J.* **247,** 233–235 (1987).
26. R. Holland and D. G. Hardie, *FEBS Lett.* **181,** 308–312 (1985).
27. W. J. Vaartjes, C. G. M. de Haaas, and S. G. van den Bergh, *Biochem. Biophys. Res. Commun.* **162,** 135–142 (1985).
28. C. P. Chan and E. G. Krebs, *Proc. Natl. Acad. Sci. U.S.A.* **82,** 4563–4567 (1985).
29. F. Bosch, B. Bouscarel, J. Slaton, P. Blackmore, and J. H. Exton, *Biochem. J.* **239,** 523–530 (1986).
30. B. L. Margolis, J. V. Bonventre, S. G. Kremer, J. E. Kudlow, and K. L. Skorecki, *Biochem. J.* **249,** 587–592 (1988).

31. L. Levine and A. Hassid, *Biochem. Biophys. Res. Commun.* **76,** 1181–1187 (1977).
32. A. Berchuck, P. C. MacDonald, L. Milewich, and M. L. Casey, *Mol. Cell. Endocrinol.* **57,** 87–92 (1988).
33. M. Fehlman, B. Canivet, and P. Freychet, *Biochem. Biophys. Res. Commun.* **100,** 254–260 (1981).
34. R. M. Johnson, P. A. Connelly, R. B. Sisk, B. F. Pobiner, E. L. Hewlett, and J. C. Garrison, *Proc. Natl. Acad. Sci. U.S.A.* **83,** 2032–2036 (1986).
35. R. M. Johnson and J. C. Garrison, *J. Biol. Chem.* **262,** 17285–17293 (1987).
36. B. C. Tilly, P. A. Van Paridon, I. Verlaan, S. W. de Laat, and W. H. Moolenaar, *Biochem. J.* **252,** 857–863 (1988).
37. A. Pandiella, L. Beguinot, T. J. Velu, and J. Meldolesi, *Biochem. J.* **254,** 223–228 (1988).
38. A. Gilligan, M. Prentki, M. J. Glennon, and B. Knowles, *FEBS Lett.* **233,** 41–46 (1988).
39. M. I. Wahl, T. O. Daniel, and G. Carpenter, *Science* **241,** 968–970 (1988).
40. M. I. Wahl, S. Nishibe, P.-G. Suh, S. G. Rhee, and G. Carpenter, *Proc. Natl. Acad. Sci. U.S.A.* **86,** 1568–1572 (1989).
41. S. Nishibe, M. I. Wahl, S. G. Rhee, and G. Carpenter, *J. Biol. Chem.* **264,** 10335–10338 (1989).
42. S. Nishibe, M. I. Wahl, S. M. T. Hernandez-Sotomayer, N. K. Tonks, S. G. Rhee, and G. Carpenter, *Science* **250,** 1253–1256 (1990).
43. W. B. Anderson, M. Gallo, J. Wilson, E. Lovelace, and I. Pastan, *FEBS Lett.* **102,** 329–332 (1979).
44. R. L. Ball, K. D. Tanner, and G. Carpenter, *J. Biol. Chem.* **265,** 12836–12845 (1990).
45. B. G. Nair, H. M. Rashed, and T. B. Patel, *Biochem. J.* **264,** 563–571 (1989).
46. B. G. Nair, B. Parikh, G. Milligan, and T. B. Patel, *J. Biol. Chem.* **265,** 21317–21322 (1990).
47. Y. Nakagawa, J. Gammichia, K. R. Purushotham, C. A. Schneyer, and M. G. Humphreys-Beher, *Biochem. Pharmacol.* **42,** 2333–2340 (1991).
48. S. W. Rabkin, P. Sunga, and S. Myrdal, *Biochem. Biophys. Res. Commun.* **146,** 889–897 (1987).
49. B. G. Nair, H. M. Rashed, and T. B. Patel, *Growth Factors* **8,** 41–48 (1993).
50. O. Langendorff, *Pleugers Arch.* **61,** 291–332 (1895).
51. H. A. Krebs and K. Henseleit, *Z. Physiol. Chem.* **210,** 33–36 (1932).
52. G. Brooker, J. F. Harper, W. L. Terasaki, and R. D. Moylan, *in* "Advances in Cyclic Nucleotide Research" (G. Brooker, P. Greengard, and G. A. Robison, eds.), Vol. 10, pp. 1–33. Raven Pres, New York, 1979.
53. G. A. Stiles, *J. Biol. Chem.* **260,** 6728–6732 (1985).
54. I. Harary, F. Hoover, and B. Farley, *in* "Methods in Enzymology" (S. Fleischer and L. Packer, eds.), Vol. 32, pp. 740–745. Academic Press, New York, 1974.
55. P. Libby, *J. Mol. Cell. Cardiol.* **16,** 803–811 (1984).
56. B. Blondel, I. Roijen, and J. P. Cheneval, *Experimentia* **27,** 356–358 (1971).
57. I. S. Polinger, *Exp. Cell Res.* **63,** 78–82 (1970).

58. D. G. Wenzel, J. W. Wheatley, and G. D. Byrd, *Toxicol. Appl. Pharmacol.* **17,** 774–785 (1970).

59. J. Orlowski and J. B. Lingrel, *J. Biol. Chem.* **265,** 3462–3470 (1990).

60. Y.-M. Yu, B. G. Nair, and T. B. Patel, *J. Cell. Physiol.* **150,** 559–567 (1992).

61. Y. Salomon, C. Londos, and M. Rodbell, *Anal. Biochem.* **54,** 541–548 (1974).

62. J. D. Robishaw, M. D. Smigel, and A. G. Gilman, *J. Biol. Chem.* **261,** 9587–9590 (1986).

63. P. Bray, A. Carter, C. Simons, V. Guo, C. Puckett, J. Kamholta, A. Spiegel, and M. Nirenberg, *Proc. Natl. Acad. Sci. U.S.A.* **83,** 8893–8897 (1986).

64. K. A. Sullivan, R. T. Miller, S. B. Masters, B. Beiderman, W. Heideman, and H. Bourne, *Nature (London)* **330,** 758–760 (1987).

65. S. B. Masters, K. A. Sullivan, R. T. Miller, B. Biederman, N. G. Lopez, J. Ramachandran, and H. Bourne, *Science* **241,** 448–451 (1988).

66. B. G. Nair and T. B. Patel, *Biochem. Pharmacol.* **46,** 1239–1245 (1993).

67. D. L. Cadena, C. L. Chan, and G. N. Gill, *J. Biol. Chem.* **269,** 260–265 (1994).

68. T. Okamoto, Y. Murayama, Y. Hayashi, M. Inagaki, E. Ogata, and I. Nishimoto, *Cell* **68,** 723–730 (1991).

69. T. Okamoto, T. Katada, Y. Murayama, M. Ui, E. Ogata, and I. Nishimoto, *Cell* **62,** 709–717 (1990).

70. H. Sun, J. M. Seyer, and T. B. Patel, *Proc. Natl. Acad. Sci. U.S.A.* **92,** 2229–2233 (1995).

71. M. P. Graziano, M. Freissmuth, and A. G. Gilman, *in* "Methods in Enzymology" (R. A. Johnson and J. D. Corbin, eds.), Vol. 195, pp. 192–215. Academic Press, San Diego, 1991.

72. S. Mumby, I.-H. Pang, A. G. Gilman, and P. C. Sternweis, *J. Biol. Chem.* **263,** 2020–2026 (1988).

73. E. J. Neer, J. M. Lok, and L. G. Wolf, *J. Biol. Chem.* **259,** 14222–14229 (1984).

74. P. C. Sternweis and A. G. Gilman, *J. Biol. Chem.* **254,** 3333–3340 (1979).

75. J. K. Northup, M. D. Smigel, and A. G. Gilman, *J. Biol. Chem.* **257,** 11416–11423 (1982).

76. D. R. Brandt, T. Asano, S. E. Pedersen, and E. M. Ross, *Biochemistry* **22,** 4357–4362 (1983).

77. R. Iyengar, M. K. Bhat, M. E. Riser, and L. Birnbaumer, *J. Biol. Chem.* **256,** 4810–4815 (1981).

78. S. M. Hernandez-Sotomayer and G. Carpenter, *J. Membrane Biol.* **128,** 81–89 (1992).

79. C. A. Koch, D. Anderson, M. F. Moran, C. Ellis, and T. Pawson, *Science* **252,** 668–674 (1991).

80. M. J. Fry, G. Panayotou, G. W. Booker, and M. D. Waterfield, *Protein Sci.* **2,** 1785–1797 (1993).

81. P. van der Geer and T. Hunter, *Annu. Rev. Cell Biol.* **10,** 251–337 (1994).

[22] G Protein Dependence of α_1-Adrenergic Receptor Subtype Action in Cardiac Myocytes

Vitalyi Rybin, Hyung-Mee Han, and Susan F. Steinberg

Introduction

α_1-Adrenergic receptor stimulation is increasingly appreciated as a mechanism for the rapid modulation of cardiac contractile function, as well as the long-term induction of several genetic and morphologic features of the cardiac hypertrophic response (reviewed in Ref. 1). Studies in a number of laboratories have probed the mechanisms underlying these α_1-receptor responses. These studies provide ample evidence that the rapid effects of α_1-receptor agonists to influence cardiac pacemaker function, repolarization, and myocyte contractility, occurring within minutes following stimulation by α_1-receptor agonists, result from the simultaneous activation of one or more GTP-binding protein (G protein)-coupled intracellular response pathways. The nature of these responses to α_1-receptor agonists changes importantly during normal growth and development of the heart. In contrast, more chronic effects of α_1-receptor agonists to influence myocyte structure and function primarily reflect G-protein-dependent increases in intracellular lipid-derived signaling molecules and the phosphorylation of key intracellular target substrates in both neonatal and adult ventricular myocytes (2, 3).

This chapter reviews our understanding of G protein expression and function in α_1-receptor response pathways in cardiac tissue. In particular, we will emphasize the evidence that establishes the importance of sympathetic innervation in the functional acquisition of a pertussis toxin-sensitive G protein linking the α_1-adrenergic receptor to inhibition of automaticity in the mature, innervated heart. The effects of α_1-agonists to modulate contractile function will be discussed within the context of the current understanding of α_1-adrenergic receptor subtypes and their linkage to G proteins in cardiac myocytes. From the rather correlative nature of the data currently available, it will become apparent that despite the early identification of a functional linkage between an α_1-receptor subtype, a pertussis toxin-sensitive G protein, and an inhibitory chronotropic response, the molecular identity and/or the precise biochemical events that control α_1-receptor signaling through this pathway have yet to be uncovered. Thus, the final section of this chapter will

Methods in Neurosciences, Volume 29

pose questions that should constitute the challenge for future investigation in this area.

Methods

Membrane Preparation

Tissues are trimmed of fat and connective tissue, weighed, minced, and homogenized twice for 10 sec in 4 volumes (w/v) of ice-cold homogenization buffer (1 mM EDTA, 0.1 mM phenylmethylsulfonyl fluoride, 5 μM pepstatin A, 0.25 M sucrose, and 0.03 M histidine, pH 7.4) with a Polytron (Brinkman Instruments, Inc., Westbury, NY). The crude homogenate is centrifuged at 1500 g for 15 min to remove large tissue fragments, nuclear debris, and cellular organelles. The supernatant is recentrifuged at 43,000 g for 45 min at 4°C and the pellet is resuspended in homogenization buffer at a protein concentration of 3–5 mg/ml. Membranes are stored in aliquots at −70°C. The average membrane protein yield is approximately 0.15–0.30% of the initial tissue wet weight.

$[^{32}P]$ADP-Ribosylation Assays

Assays of pertussis toxin (PTX)-sensitive G proteins measure the incorporation of $[^{32}P]$ADP-ribose from $[^{32}P]$NAD into the appropriate molecular weight membrane proteins as described previously (4). Labeling with PTX is accomplished by incubating 1 or 2 μg of membrane in 20 μl of a 50 mM Tris-Cl (pH 8.0) buffer containing 2 mM $MgCl_2$, 1 mM EDTA, 10 mM DTT, 0.1% Lubrol PX, 10 mM thymidine, 10 μM $[^{32}P]$NAD (1.5 μCi/assay), and 20 μg/ml PTX for 1 hr at 37°C. For each sample, reactions are linear with protein concentration under these assay conditions. Reactions are terminated by addition of SDS-PAGE sample buffer and boiling for 5 min. Electrophoresis is performed on vertical slab gels [resolving gel, 12% acrylamide; stacking gel, 4% acrylamide (w/v)]. Following proportional counting of the gels using a Betascan, PTX-sensitive G proteins are quantified by relating the number of counts in the band specifically labeled to the specific activity of the $[^{32}P]$NAD and the protein concentration.

Immunoblotting

Samples (25 μg/lane) are electrophoresed on a 12% SDS-polyacrylamide gel and transferred to nitrocellulose. Prestained molecular weight markers are electrophoresed in parallel. Four G protein subunit-specific antisera are used:

anti-$G_{\alpha \text{ common}}$, a polyclonal antiserum, previously characterized as antiserum 1398 (5), which is strongly reactive against all PTX-sensitive α subunits; anti-G/oα a polyclonal antiserum directed against the amino terminus of $G_{q\alpha}$; anti-$G_{q\alpha 11}$, a polyclonal antiserum directed against the common C-terminal sequence of $G_{q\alpha}$ and $G_{\alpha 11}$; and anti-β, a polyclonal antiserum raised against an internal decapeptide sequence of the human β-subunit that recognizes both the 35- and 36-kDa forms of the β-subunit. Although these antisera are raised against rat or human G protein subunits, there is a high degree of species cross-reactivity, presumably due to substantial sequence homology of G proteins in mammalian cells. The nitrocellulose is incubated in 5% dry milk, 50 mM Tris, pH 7.5, 200 mM NaCl, and 0.05% Nonidet P-40 (blocking buffer) for 1 hr at room temperature to block nonspecific binding and then probed with a 1 : 200 dilution of G protein subunit-specific antiserum in 5% bovine serum albumin, 50 mM Tris, pH 7.5, 200 mM NaCl, 0.05% Nonidet P-40, and 0.02% NaN_3 overnight at 4°C. The nitrocellulose is then washed five times, 5 min each, with 50 mM Tris, pH 7.5, 200 mM NaCl, 0.05% Nonidet P-40, and then incubated in blocking buffer for 30 min at room temperature. To detect bound primary antibody, blots are incubated for 1 hr at room temperature with ^{125}I-labeled goat antirabbit IgG F(ab')$_2$ fragment at a final dilution of 0.67 μCi/ml in blocking buffer. The nitrocellulose is washed seven times as described above, dried, and autoradiographed with Kodak XAR film with intensifying screens at −70°C. In each case, the density of specific immunoreactive bands on the autoradiogram increases linearly with the amount of protein loaded. Accordingly, the relative abundance of individual proteins identified is quantified by scanning densitometry.

Materials

[^{32}P]NAD is purchased from Dupont-New England Nuclear (Boston, MA) and pertussis toxin (PTX) is purchased from List Biological Company (Campbell, CA). Polyclonal anti-$G_{o\alpha}$ is purchased from Dupont-New England Nuclear (Boston, MA). Polyclonal anti-$G_{\alpha \text{ common}}$, anti-$G_{q\alpha 11}$, and anti-$\beta$-subunit antisera are the generous gift of Dr. David Manning (University of Pennsylvania, Philadelphia, PA). All other chemicals are reagent grade.

Results and Discussion

G Protein Structure and Function

The G proteins represent a fundamental and widespread mechanism for the transmission of signals from agonist-occupied cell surface receptors to a number of membrane-bound effector mechanisms, including enzymes (adenylylcyclases and phospholipases) and ion channels (reviewed in Ref. 6). The G proteins are $\alpha\beta\gamma$ heterotrimers that are classified based on the identity of their α subunit (although multiple forms of the β and γ subunits also have been identified; see below). All α subunits contain GTP-binding and GTPase-activating domains (7) that critically regulate the signal-generating process. In the basal state, the GTP-binding site of the α subunit is occupied by GDP. Agonist-activated receptors promote the exchange of bound GDP for GTP. GTP binding to the α subunit induces conformational changes that promote rapid α dissociation from $\beta\gamma$; both the GTP-bound α subunit as well as the freed $\beta\gamma$ complex can interact with effector molecules. Ultimately, each cycle of activation is terminated by a GTPase activity intrinsic to the α subunit, which hydrolyzes GTP to GDP, leads to reformation of the $\alpha\beta\gamma$ heterotrimer, and returns the system to the basal state. Thus, subunit dissociation (α–$\beta\gamma$) and guanine nucleotide exchange cycles critically regulate G protein action.

Many G protein α subunits contain sites for their covalent modification by bacterial toxins, such as cholera toxin (CTX) or pertussis toxin (PTX). This reaction, termed ADP-ribosylation, involves the transfer of adenosine diphosphate (ADP)-ribose from nicotinamide-adenine dinucleotide (NAD) onto an arginine (CTX) or cysteine (PTX) residue of the G_α protein. Toxin-catalyzed ADP-ribosylation reactions have provided a powerful approach to study G protein expression and function. For example, G protein expression has been estimated by performing ADP-ribosylation reactions with [^{32}P]NAD, labeled on the ADP-ribose moiety of the molecule. Because this reaction results in the covalent incorporation of 1 mol of [^{32}P]ADP-ribose per mole G_α protein, this method provides the means to tag G_α proteins, identify these proteins on electrophoretic gels, and quantify their expression in various tissues (although see Ref. 8 for caveats related to the use of this method for rigorous quantification of G_α proteins). Toxin-dependent ADP-ribosylation also result in characteristic functional alterations that can be extremely useful to delineate G protein actions in cells. ADP-ribosylation by CTX results in persistent α-subunit activation as a result of inhibition of the protein's intrinsic GTPase activity. In contrast, ADP-ribosylation by PTX blocks G_α protein–receptor interactions. This results in functional inactivation of signal transduction pathways mediated by PTX-sensitive G_α proteins and thereby provides a powerful means to identify signaling pathways that are controlled by this class of G protein.

On the basis of their distinct α subunits, G proteins have been broadly grouped into four closely related classes (Table I). The G_s class includes $G_{s\alpha}$ and $G_{\alpha\,olf}$; these G proteins, which are substrates for CTX-dependent ADP-ribosylation, activate adenylylcyclase and increase cAMP accumulation. Whereas $G_{s\alpha}$ expression is ubiquitous, $G_{\alpha\,olf}$ expression is restricted to neurons in olfactory epithelium. The G_i class includes three forms of $G_{i\alpha}$ ($G_{i\alpha1}$,

TABLE I   G Protein Subunits

G_α	β	γ
$G_{s\alpha}$	β_1	γ_1
$\quad G_{s\alpha}$	β_2	γ_2
$\quad G_{\alpha\,olf}$ (olfactory)	β_3	γ_3
$G_{i\alpha}$	β_4	γ_4
$\quad G_{i\alpha1}$, $G_{i\alpha2}$, $G_{i\alpha3}$		γ_5
$\quad G_{o\alpha}$		γ_6
$\quad G_{\alpha\,tr}$, $G_{\alpha\,tc}$ (transducins)		γ_7
$\quad G_{z\alpha}$		
$G_{q\alpha}$		
$\quad G_{q\alpha}$		
$\quad G_{\alpha11}$		
$\quad G_{\alpha14}$		
$\quad G_{\alpha15}$		
$\quad G_{\alpha16}$		
$G_{\alpha12}$		
$\quad G_{\alpha12}$		
$\quad G_{\alpha13}$		

$G_{i\alpha2}$, and $G_{i\alpha3}$), $G_{o\alpha}$, two forms of transducin ($G_{\alpha\,tr}$ and $G_{\alpha\,tc}$), and G_z. $G_{i\alpha}$ is expressed by a variety of tissues and cell lines (although the specific isoform expressed frequently varies) (9) whereas the two forms of transducin ($G_{\alpha\,tr}$ and $G_{\alpha\,tc}$) are highly localized to retinal rod and cone cells, respectively. With the exception of G_z, the G_i class of α subunits is distinguished by its sensitivity to ADP-ribosylation by PTX. The G_q class includes $G_{q\alpha}$, $G_{\alpha11}$, $G_{\alpha14}$, $G_{\alpha15}$, and $G_{\alpha16}$. These G proteins are not substrates for CTX- or PTX-catalyzed ADP-ribosylation. Although G_q and G_{11} are detected in many tissues, expression of other members of this class is restricted to stromal and epithelial ($G_{\alpha14}$) or hematopoietic ($G_{\alpha15}$ and $G_{\alpha16}$) cells. Studies indicate that this class of G protein couples receptors to PTX-insensitive activation of phospholipase C and the generation of IP_3 and diacylglycerol (DAG) (10–12). Finally, the G_{12} class, which includes $G_{\alpha12}$ and $G_{\alpha13}$, is yet another PTX-insensitive class of G protein whose function is the least well understood currently (12a, b).

The $\beta\gamma$ complex was traditionally thought to contribute to receptor signaling indirectly by interacting with and deactivating the G protein α subunit, thereby preventing activation of effector molecules. However, this notion appears to be an oversimplification in view of recent evidence that $\beta\gamma$ dimers can directly regulate the activity of several types of effector mechanisms, including the adenylylcyclase enzyme, phospholipases, and ion channels (6, 13–16). Currently, there are four known species of β subunit and seven different forms of the γ subunit; β subunits are highly homologous at the amino acid level, whereas γ subunits are structurally quite divergent (6, 17). Theoretically 28 different $\beta\gamma$ subunit combinations are possible, although there appear to be preferred associations between β and γ subunits such that G proteins purified from different tissues differ with respect to their γ subunit (18). This knowledge has led to the hypothesis that different γ subunits distinguish between different α and β subunits and that the γ species can impart functional specificity to G protein action. In support of this hypothesis, individual γ-subunit subtypes have been reported to lend specificity to the receptor signaling pathway by influencing the ability of an individual G protein to discriminate between individual receptors and effector mechanisms (19).

Sympathetic Innervation as Key Determinant of α_1-Adrenergic Receptor Effects on Automaticity

α_1-Receptor agonists tend to increase impulse initiation in the neonatal heart and to inhibit it in the adult heart. Noting that the developmental conversion of the α_1-adrenergic chronotropic response from excitatory to inhibitory coincides temporally with sympathetic innervation of the heart, our laboratory, in collaboration with Dr. Richard Robinson, previously examined the role of sympathetic innervation in the developmental conversion of the α_1-adrenergic chronotropic response.

The most rigorous evidence to support the conclusion that sympathetic innervation critically modulates α_1-receptor responsiveness comes from studies in cultured myocytes. Myocytes cultured from the newborn rat ventricle, which is not yet innervated by sympathetic neurons, respond to α_1-receptor agonists with an increase in automaticity (20). These myocytes, when cocultured with neurons dissociated from sympathetic ganglia, become innervated. The structural and functional evidence for *in vitro* innervation includes the demonstration of close apposition between muscle and nerve terminal membranes as well as the demonstration that tyramine increases the automaticity of myocytes in nerve–muscle cocultures (20). Sympathetic innervation of rat ventricular myocytes *in vitro* is associated with the appearance of an inhibitory α_1-adrenergic chronotropic response [i.e., it mimics the developmental change in α_1-receptor responsiveness in the intact heart (20, 21)]. The decisive observation regarding the role of a PTX-sensitive G protein in the inhibitory α_1-adrenergic chronotropic response was made in this culture paradigm (21). Specifically, innervated myocyte cultures were found to contain significantly more substrate for PTX-dependent ADP ribosylation than noninnervated myocyte cultures. Control experiments indicated that this difference could not be attributed to the small amount of PTX-sensitive G protein α subunits derived from neuronal material in the nerve–muscle cocultures, but rather must reflect enhanced PTX-sensitive G protein α subunits in the ventricular myocytes. Because ADP-ribosylation by PTX functionally uncouples G protein α subunits from their cognate receptor molecules, and thereby inhibits responses that are dependent on these G proteins, we used PTX to establish the functional link between the PTX-sensitive G protein, the α_1-adrenergic receptor, and the decrease in automaticity. PTX did not influence the excitatory α_1-adrenergic chronotropic response in pure muscle cultures. In contrast, PTX converted the inhibitory α_1-adrenergic chronotropic response of nerve–muscle cocultures to a stimulatory response characteristic of the pure muscle cultures. These studies established the critical role of sympathetic innervation in the functional acquisition of a PTX-sensitive G protein linking the α_1-adrenergic receptor to an inhibitory chronotropic response. However, given the knowledge that PTX ADP-ribosylates a family of related G protein α subunits, these experiments did not permit a precise molecular identification of the G protein involved in α_1-receptor-dependent inhibition of automaticity.

Studies in the *in situ* heart provided additional evidence that sympathetic innervation plays a pivotal role in the functional acquisition of a PTX-sensitive G protein linking the α_1-adrenergic receptor to an inhibitory chronotropic response (22). In newborn rats, acceleration of innervation *in vivo* with nerve growth factor (NGF) increased the amount of PTX-sensitive G protein detected in ADP-ribosylation assays and hastened the appearance of the α_1-adrenergic inhibitory chronotropic response. Conversely, delay of sympathetic innervation using antibody to NGF reduced the amount of PTX-sensitive G protein. These rats retained an excitatory α_1-adrenergic chronotropic response.

Although studies in cultured rat ventricular myocytes and in intact rat and dog cardiac tissues have provided consistent evidence that a PTX-sensitive G protein(s) is critical for α_1-receptor inhibition of automaticity, these studies also suggest that factors in addition to the absolute level of PTX-sensitive G protein expression influences α_1-receptor responses. This is best illustrated by the data presented in Fig. 1, which provides an analysis of the relationship between PTX-dependent inhibition of the α_1-adrenergic decrease in automaticity and PTX-dependent ADP-ribosylation of G_α subunit(s) in nerve–muscle coculture. As previously reported (20, 21), the α_1-adrenergic receptor agonist phenylephrine decreases automaticity in the majority of nerve–muscle cocultures (65%). This α_1-receptor response is mediated by a PTX-sensitive pathway in that it is inhibited progressively as the concentration of PTX in the pretreatment culture medium is increased from 0.1 to 5 ng/ml. As previously reported (4, 21), nerve–muscle cocultures have significantly more PTX-sensi-

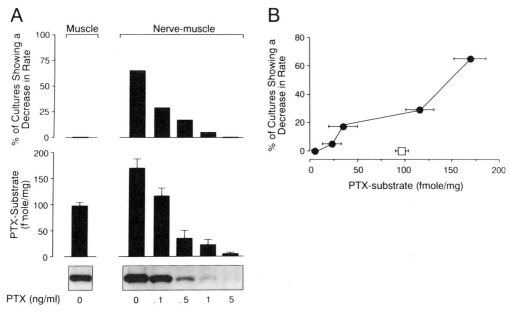

FIG. 1   Relationship between the fractional response to phenylephrine and the detectable PTX substrate in muscle cultures and nerve–muscle cocultures. (A) Studies were performed on pure muscle cell cultures pretreated with saline or nerve–muscle cocultures that were pretreated with saline or increasing concentrations of PTX (0.1, 0.5, 1, and 5 ng/ml) for 24–36 hr. (A, *Upper*) The effect of 10^{-8} M phenylephrine to influence spontaneous contractile rate was determined according to methods published previously (20). (A, *Lower*) The level of substrate for PTX-catalyzed ADP-ribosylation was assayed (see text). The autoradiogram is from one of the four experiments on separate cultures illustrated by the data. (B) The relationship between the level of PTX-sensitive substrate and the percentage of cultures retaining an inhibitory chronotropic response to phenylephrine in saline and PTX-pretreated nerve–muscle cocultures (●) is compared to similar parameters in saline-pretreated pure muscle cell cultures (□).

tive substrate than pure muscle cultures (170 ± 17 vs. 97 ± 7 fmol/mg protein, respectively). Exposure of nerve–muscle cocultures to PTX results in ADP-ribosylation of the G_α protein(s) by endogenous unlabeled cellular NAD to an extent that is dependent on the concentration of PTX in the pretreatment interval. This is evident from the bottom panel of Fig. 1A, where membranes are subsequently exposed to PTX in the presence of exogenous labeled [^{32}P]NAD and only PTX-sensitive G_α proteins not previously ADP-ribosylated are available for *in vitro* ADP-ribosylation (i.e., there is an inverse relationship between the extent of ADP-ribosylation during the initial pre-treatment interval and the amount of radioactivity that can be incorporated in the subsequent *in vitro* ADP-ribosylation reaction). However, this comparison of the dose–response relationships for PTX actions reveals an important and as yet unexplained observation. Whereas treatment with 0.5–1 ng/ml PTX reduces the PTX-sensitive substrate in nerve–muscle cocultures to a level well below that measured in pure myocyte cultures, an inhibitory α_1-adrenergic chronotropic response persists in a fraction of these cultures. This is best highlighted by the comparison of PTX-sensitive substrate levels and the α_1-adrenergic receptor chronotropic response in pure muscle and nerve–muscle cultures, illustrated in Fig. 1B. These data argue that the absolute level of PTX-sensitive G_α is not sufficient to dictate α_1-receptor responsiveness and that additional molecular mechanisms also must be operative.

Distinct α_1-Receptor Chronotropic Responses in Newborn and Adult Myocytes Reflect Actions at Distinct α_1-Receptor Subtypes Linked to Different G Proteins

α_1-Receptor subtypes can be distinguished by their sensitivity to relatively low concentrations of WB4101 or to the irreversible antagonist chloroethyl-clonidine (CEC). Although these α_1-receptor subtypes have been referred to as α_{1a} and α_{1b}, respectively, recent studies reveal a family of closely related α_1-adrenergic receptor subtypes (23–27). Indeed, the mRNAs for three distinct α_1-receptor subtypes, α_{1B}, α_{1D}, and α_{1C}, have been identified in neonatal and adult rat ventricular myocardium (28). However, the precise relationship between the pharmacologically defined cardiac α_1-receptor subtypes and the identified clones is as yet uncertain. In the absence of an assignment of physiologic functions to specific cloned α_1-receptor subtypes, this chapter will continue to define operationally the α_1-receptor subtypes on the basis of their sensitivity to inhibition by WB4101 and CEC.

In the newborn heart, the α_1-receptor subtype that increases automaticity is inhibited by low concentrations of WB4101. This receptor subtype acts through the PTX-insensitive G protein, $G_{q\alpha}$, to stimulate phosphoinositide hydrolysis and enhance the intracellular accumulation of IP_3 and DAG in neonatal rat ventricular myocytes (Fig. 2). Although the precise cellular actions of IP_3 in cardiac tissue remain uncertain, the weight of recent evidence suggests that IP_3 can induce calcium release from cardiac sarcoplasmic reticulum (SR). Although the kinetics of the response are probably too slow to play a role in excitation–contraction coupling, IP_3 enhances spontaneous

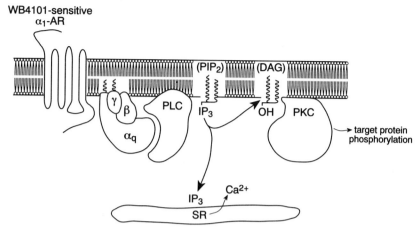

FIG. 2   Schematic diagram depicting WB4101-sensitive α_1-adrenergic receptor sub-type action in cardiac myocytes. α_1-AR, α_1-Adrenergic receptor; PLC, phospholipase C; PIP$_2$, phosphatidylinositol 4,5-bisphosphate; DAG, diacylglycerol; IP$_3$, inositol trisphosphate; SR, sarcoplasmic reticulum; PKC, protein kinase C. Responses activated by the α_1-adrenergic receptor: positive chronotropic response and hypertrophy ($G_{q\alpha}$). Expression is in neonatal and adult heart. See text for details.

calcium release from calcium overloaded SR, suggesting that the biologic actions of IP$_3$ may assume importance in various pathologic states (29). Similarly, DAG is the endogenous activator of protein kinase C, a serine/threonine kinase that has been implicated in the modulation of numerous cellular functions in cardiac myocytes, including the regulation of membrane ionic conductances (30, 31), intracellular calcium and pH (32–34), contractile function (30, 32, 33,35), and the induction of myocardial cell hypertrophy (36, 37). It should be noted that this α_1-receptor subtype, via the $G_{q\alpha}$ protein, also has been linked to the entire repertoire of the hypertrophic phenotype (2, 3).

The signal transduction pathway linking α_1-agonists to a decrease in automaticity in the adult heart involves a different α_1-receptor subtype (which is irreversibly alkylated by CEC) and a PTX-sensitive G protein (which becomes functionally linked to the α_1-receptor following sympathetic innervation of the heart; Fig. 3). Activation of this pathway leads to activation of the Na$^+$,K$^+$-ATPase (38), a decrease in intracellular sodium (39) hyperpolarization, and slowing of pacemaker activity in the adult heart. Although the CEC-sensitive α_1-receptor also is present in neonatal ventricular myocardium, it has not been linked to function in that tissue (28, 40). It should be noted that a WB4101-sensitive stimulatory α_1-adrenergic positive chronotropic response persists in the adult heart and is revealed following treatment

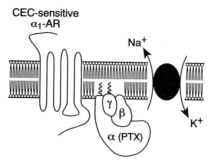

FIG. 3   Schematic diagram depicting CEC-sensitive α_1-adrenergic receptor subtype action in cardiac myocytes. α_1-AR, α_1-Adrenergic receptor. This receptor activates negative chronotropic responses only in adult heart and sympathetically innervated neonatal myocytes.

with PTX. Thus, the integrated response to α_1-receptor agonists reflects the balance of the cellular actions of divergent response mechanisms and is dictated, in large part, by the degree of maturation and sympathetic innervation of the cardiac tissue.

Finally, it should be noted that the conclusions of most studies examining α_1-receptor subtype action in cardiac myocytes are internally consistent. However, there is one notable exception in that an α_1-receptor subtype that is inhibited by CEC couples to phosphoinositide hydrolysis and a positive inotropic response in rabbit ventricular myocardium. (41). This may represent an important species-dependent difference in α_1-receptor subtype action and requires further study.

G-Protein Expression in Cardiac Myocytes

A detailed analysis of G protein subunit expression in cardiac myocytes is not currently available. Indeed, the concept that cardiac myocytes express only certain G proteins may be somewhat naive, as there is evidence for regional, developmental, and/or species-dependent differences in G protein subunit expression. For example, there is evidence that $G_{o\alpha}$ is highly enriched in atrial tissue relative to ventricular tissue (42, 43) (see Fig. 4). Although $G_{o\alpha}$ is detected at only very low levels in intact ventricular myocardium, $G_{o\alpha}$ is abundant in cultured neonatal ventricular myocytes (43, 44). Moreover, $G_{o\alpha}$ expression in cultured myocytes is influenced by factors such as sustained membrane depolarization with high K^+ as well as cAMP, suggesting that G protein subunit expression may be a highly regulated process in cardiac

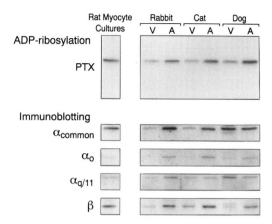

FIG. 1   G protein subunit expression in membranes from cultured neonatal rat ventricular myocytes as well as atrial and ventricular myocardial tissues from rabbit, cat, and dog hearts. ADP-ribosylation and immunoblot analysis with the indicated antisera were performed as described in the text.

myocytes (43, 44). Thus, nonuniformity of G protein subunit expression could provide the basis for regional variations in receptor-dependent responses in the heart. Developmental changes in G protein subunit expression also have been reported. For example, the principal form of G_i in ventricular myocardium, $G_{i\alpha2}$, decreases substantially during the developmental maturation of the rat ventricle (42, 45). Yet the functional consequences of this change are not yet understood. Species-dependent differences in G protein expression also may be important but have not been adequately considered.

Accordingly, Fig. 4 examines G protein subunit expression in membranes prepared from cultured neonatal rat ventricular myocytes as well as atrial and ventricular myocardial tissue from adult rabbit, cat, and dog hearts. These preparations were chosen for illustrative purposes because most of the available information regarding α_1-receptor subtype action, and the role of G proteins in this process, has been obtained in these species (21, 38–41, 46–49). The analysis is limited to those species of G protein α subunits that are plausible candidate regulators of α_1-receptor responsiveness, namely, PTX-sensitive G protein α subunits and $G_{q\alpha11}$.

PTX-catalyzed ADP-ribosylation is illustrated in the top panel of Fig. 4. The highest levels of PTX-sensitive substrate were observed in cultured neonatal rat ventricular myocytes and atrial preparations from each species, although distinct G_α subunit species could not be resolved. PTX-dependent ADP-ribosylation is influenced by the amount of G protein α-subunit protein substrate as well as the degree of endogenous ADP-ribosylation and the availability of $\beta\gamma$ subunits, necessary cofactors for the ADP-ribosylation of G_i (50). Therefore, we used immunoblot analysis to further examine G protein α-subunit expression in these preparations. Membranes were probed with a panel of polyclonal antisera generated against synthetic peptides derived from the distinct sequences of individual G protein subunits. Using a polyclonal antiserum that recognizes all PTX-sensitive α subunits ($G_{\alpha \text{ common}}$), we detected immunoreactive protein in all preparations, with the greatest amounts of immunoreactivity in cultured neonatal rat ventricular myocytes and atrial myocardial tissues. A similar pattern was observed using the antiserum that recognizes $G_{o\alpha}$ (i.e., $G_{o\alpha}$ was found to be enriched in neonatal rat ventricular myocytes and atrial myocardial tissue). However, it is noteworthy that the amount of $G_{\alpha \text{ common}}$ and $G_{o\alpha}$ immunoreactivity in dog ventricular myocardium is comparable to that detected in dog atrial myocardium and is substantially greater than that detected in ventricular myocardium from the rabbit or cat. These immunoblot analyses begin to distinguish the individual members of the $G_{i\alpha}$ family and constitute an important extension of studies using PTX-catalyzed ADP-ribosylation. Comparisons of G protein α-subunit expression among the various preparations establishes important differences in their relative abundance that could have substantial impact on receptor-mediated functions. These comparisons are valid, because the preparations were analyzed under identical experimental conditions (dilutions of antisera; exposure time for autoradiographs). However, it should be emphasized that our immunoblotting method does not incorporate corrections for potential differences in titer and/or hybridization efficiency between individual antisera. Therefore, precise quantification of G protein α-subunit expression in a given tissue is not possible.

The PTX-insensitive $G_{q\alpha11}$ subunit also was detected in similar abundance in all of the preparations examined. The presence of $G_{q\alpha11}$ in cultured neonatal rat ventricular myocytes is consistent with the recent observation that $G_{q\alpha11}$ is expressed by cultured neonatal rat ventricular myocytes, where it couples the α_1-adrenergic receptor to the PTX-insensitive stimulation of phosphinositide hydrolysis and the formation of IP_3 (3).

Finally, G protein β-subunit immunoreactivity was detectable in each preparation. It is noteworthy that the abundance of the β subunit was greatest in preparations from cultured rat ventricular myocytes and atrial myocardium, which also are enriched in PTX-sensitive α subunits (primarily $G_{o\alpha}$). The mechanism for this association is uncertain because, to the best of our knowledge, coordinate regulation of G protein subunit expression has not been described. However, these results suggest that the increase in substrate for PTX-catalyzed ADP-ribosylation in cultured neonatal rat ventricular myocytes and atrial myocardial tissues likely reflects increased G_α subunit expression as well as increased availability of $\beta\gamma$ subunits, which serve as necessary cofactors for PTX-catalyzed ADP-ribosylation (50). Althought β-subunit expression was measured in this study with an antiserum that does not discriminate between the different isoforms of this protein, our results are consistent with recent evidence that β_1 and β_2 subunits are expressed at higher levels in rat atrial than in rat ventricular myocardium (β_3 subunits are not detected in either preparation; 51a). Moreover, using a panel of antisera capable of discriminating individual members of the γ subunit family, Hanson *et al.* recently presented evidence that γ-subunit expression in the heart is restricted to certain isotypes and that there is regional and developmental variations in γ-subunit expression. Thus, changes in the β and/or γ subunits of the G protein also could contribute to regional, species-dependent, or developmental differences in α_1-receptor responsiveness in cardiac myocytes.

Although the cultures used in these studies are enriched in neonatal rat ventricular myocytes, it should be emphasized that intact heart preparations are composed of cardiomyocytes as well as other contaminating noncardio-myocyte cell populations (fibroblasts, endothelial cells, and neurons) that may influence the results. Studies on membrane preparations do not permit unambiguous identification of G protein subunit expression in cardiac myocytes. Ultimately, immunodetection techniques with intact tissues will be required for that purpose.

Future Directions

Do changes in the identity of G protein β and/or γ subunits contribute to changes in α_1-receptor signaling in cardiac myocytes? Although most studies have concentrated on G protein α subunits, differences in G protein β and/or γ subunits also may exist and may contribute to differences in receptor function. Yet, a careful analysis of the molecular identity of the β and γ subunits, to determine whether particular combinations of α, β, and γ subunits exist in cardiac myocytes, is not available. Future studies must consider the possibility that each G protein subunit imparts specificity to α_1-receptor–effector coupling. First, it is possible that the $\beta\gamma$ complex plays a direct role in α_1-receptor-mediated signaling and that all $\beta\gamma$ complexes are not functionally equivalent. According to this hypothesis, developmental changes in β- and/or γ-subunit expression could influence their ability to transduce signals from agonist-occupied α_1-receptors to effector molecules. Recent evidence for selectivity in $\beta\gamma$ interactions (51, 52) as well as reports that structurally distinct $\beta\gamma$ complexes differ in their ability to activate effector molecules (53–55) provide the precedent for this notion. A second intriguing possibility is that the identity of the γ subunit influences the α_1-receptor–G protein interaction. According to this hypothesis, developmental changes in γ-subunit expression may critically influence G protein–receptor targeting and thereby constitute an additional mechanism for molecular diversity in α_1-receptor signaling pathways.

Do posttranslational modifications of G protein subunits contribute to changes in α_1-receptor signaling in cardiac myocytes? Experimental data argue that the precise level of α-subunit expression is not sufficient to define α_1-receptor responsiveness. Additional factors must contribute to the control of α_1-receptor responsiveness. The notion that covalent lipid modifications that increase protein hydrophobicity may act to target G protein subunits to the membrane, thereby increasing their effective concentrations for interactions with receptor and effector molecules, has not been tested. Two lipid modifications of G_α proteins have been described (reviewed in Ref. 56). Members of the $G_{i\alpha}$ class are myristoylated (57, 58). This is thought to be a cotranslational and irreversible modification that appears to play an important role in localizing these proteins to the plasma membrane. Many G_α subunits, including $G_{s\alpha}$, $G_{i\alpha}$, $G_{o\alpha}$, $G_{q\alpha}$, $G_{z\alpha}$, and $G_{\alpha12}$, also are palmitoylated. This dynamic posttranslational lipid modification of G_α proteins has been shown to be critical for membrane attachment and normal signaling and it has been speculated that regulated palmitoylation might influence G_α–receptor interactions. This speculation is particularly intriguing given the current remaining uncertainties regarding the mechanism(s) underlying the developmental change in α_1-receptor–G protein coupling. Finally, γ subunits also undergo posttranslational modification. The γ subunits are prenylated at a cysteine that is part of the COOH-terminal CAAX motif (where A is an aliphatic amino acid and X is any amino acid). This involves the enzymatic transfer of a 15-carbon farnesyl or a 20-carbon geranylgeranyl group to the cysteine residue of the γ subunit, depending on the identity of the amino acid residue at the X position. This modification of γ does not appear to be required for the interaction between β and γ. However, carboxyl-terminal processing facilitates $\beta\gamma$ complex incorporation into the membrane and critically influences interactions of $\beta\gamma$ with α and/or effector molecules such as adenylylcyclase. Accordingly, future studies of this phenomenon, in the context of cardiac tissues and/or development, potentially are of great importance.

Are other G proteins involved in the α_1-receptor-dependent contractile and/or hypertrophic responses in cardiac myocytes? A G protein, termed G_h, has been shown to copurify with rat liver α_1-adrenergic receptors (59, 60). This G protein contains high molecular mass α (74- to 80-kDa) and β ($\sim$50-kDa) subunits and has been identified in a number of tissues, including the heart (61). Studies in a reconstitution system indicate that G_h activates phospholipase C (62). Moreover, $G_{h\alpha}$ expressed in COS-1 cells mimics $G_{q\alpha}$ in its ability to couple the agonist-occupied α_{1B}-adrenergic receptor to stimulation of phosphoinositide hydrolysis. However, $G_{h\alpha}$ is a multifunctional G protein in that it also contains a transglutaminase activity that is regulated by GTP and receptor activation (63). The relationship between these two actions of $G_{h\alpha}$ and the potential role of $G_{h\alpha}$ in α_1-adrenergic receptor action in the heart requires further study.

Acknowledgments

This work was supported by USPHS–NHLBI Grants HL-28958 and HL-49537. The authors wish to thank Drs. Richard B. Robinson and John P. Bilezikian for their helpful discussions during the work on this project, Drs. Richard B. Robinson and Michael R. Rosen for critically reading this manuscript, and Mrs. Ema Stasko for her skillful technical assistance.

References

1. A. Terzic, M. Puceat, G. Vassort, and S. M. Vogel, *Pharm. Rev.* **45,** 147–175 (1993).
2. K. U. Knowlton, M. C. Michel, M. Itani, H. E. Shubeita, K. Ishihara, J. H. Brown, and K. R. Chien, *J. Biol. Chem.* **268,** 15374–15380 (1993).
3. V. J. LaMorte, J. Thorburn, D. Absher, A. M. Spiegel, J. H. Brown, K. R. Chien, J. R. Feramisco, and K. U. Knowlton, *J. Biol. Chem.* **269,** 13490–13496 (1994).
4. H. M. Han, R. B. Robinson, J. P. Bilezikian, and S. F. Steinberg, *Circ. Res.* **65,** 1763–1773 (1989).
5. K. E. Carlson, L. F. Brass, and D. R. Manning, *J. Biol. Chem.* **264,** 13298–13305 (1989).
6. M. I. Simon, M. P. Strathmann, and N. Gautam, *Science* **252,** 802–808 (1991).
7. D. W. Markby, R. Onrust, and H. R. Bourne, *Science* **262,** 1895–1901 (1993).
8. L. A. Ransnas and P. A. Insel, *J. Biol. Chem.* **263,** 9482–9485 (1988).
9. D. T. Jones and R. R. Reed, *J. Biol. Chem.* **262,** 14241–14249 (1987).
10. C. H. Lee, D. Park, D. Wu, S. G. Rhee, and M. I. Simon, *J. Biol. Chem.* **267,** 16044–16047 (1992).
11. S. Gutowski, A. Smrcka, L. Nowak, D. Wu, M. Simon, and P. C. Sternweis, *J. Biol. Chem.* **266,** 20519–20524 (1991).
12. D. Wu, C. H. Lee, S. G. Rhee, and M. I. Simon, *J. Biol. Chem.* **267,** 1811–1817 (1992).
12a. T. Voyno-Yasenetskaya, B. R. Conklin, R. L. Gilbert, R. Hooley, H. R. Bourne, and D. L. Barber, *J. Biol. Chem.* **269,** 4721–4724 (1994).
12b. A. M. Butte, N. L. Johnson, N. Dhanasekaran, and G. C. Johnson, *J. Biol. Chem.* **270,** 24631–24634 (1995).
13. W. J. Tang and A. G. Gilman, *Science* **254,** 1500–1504 (1991).

14. D. Park, D. Jhon, C. Lee, K. Lee, and S. G. Rhee, *J. Biol. Chem.* **268,** 4573–4576 (1993).
15. C. L. Jelsema and J. Axelrod, *Biochemistry* **84,** 3623–3627 (1987).
16. D. E. Clapham and E. J. Neer, *Nature* (*London*) **365,** 403–406 (1993).
17. J. J. Cali, E. A. Balcueva, I. Rybalkin, and J. D. Robishaw, *J. Biol. Chem.* **267,** 24023–24027 (1992).
18. N. Gautam, J. Northup, H. Tamir, and M. I. Simon, *Proc. Natl. Acad. Sci. U.S.A.* **87,** 7973–7977 (1990).
19. C. Kleuss, H. Scherubl, J. Heschler, G. Schultz, and B. Wittig, *Science* **259,** 832–834 (1993).
20. E. D. Drugge, M. R. Rosen, and R. B. Robinson, *Circ. Res.* **57,** 415–423 (1985).
21. S. F. Steinberg, E. D. Drugge, J. P. Bilezikian, and R. B. Robinson, *Science* **230,** 186–188 (1985).
22. G. Malfatto, T. S. Rosen, S. F. Steinberg, P. C. Ursell, L. S. Sun, S. Daniel, P. Danilo, and M. R. Rosen, *Circ. Res.* **66,** 427–437 (1990).
23. J. W. Lomasney, S. Cotecchia, W. Lorenz, W. Y. Leung, D. A. Schwinn, T. L. Yang-Feng, M. Brownstein, R. J. Lefkowitz, and M. G. Caron, *J. Biol. Chem.* **266,** 6365–6369 (1991).
24. S. Cottechia, D. A. Schwinn, R. R. Randall, R. J. Lefkowitz, M. G. Caron, and B. K. Kobilka, *Proc. Natl. Acad. Sci. U.S.A.* **85,** 7159–7163 (1988).
25. D. A. Schwinn, J. W. Lomasney, W. Lorenz, P. J. Szklut, R. T. Fremeau, T. L. Yang-Feng, M. G. Caron, R. J. Lefkowitz, and S. Cotecchia, *J. Biol. Chem.* **265,** 8183–8189 (1990).
26. D. M. Perez, M. T. Piascik, and R. M. Graham, *Mol. Pharmacol.* **40,** 876–883 (1991).
27. C. S. Ramarao, J. M. K. Denker, D. M. Perez, R. J. Gaivin, R. P. Riek, and R. M. Graham, *J. Biol. Chem.* **267,** 21936–21945 (1992).
28. A. F. R. Stewart, D. G. Rokosh, B. A. Bailey, L. R. Karns, K. C. Chang, C. S. Long, K. Kariya, and P. C. Simpson, *Circ. Res.* **75,** 796–802 (1994).
29. A. M. Vites and A. J. Pappano, *Am. J. Physiol.* **258,** H1745–H1752 (1990).
30. A. Dosemeci, R. S. Dhallan, N. M. Cohen, W. J. Lederer, and T. B. Rogers, *Circ. Res.* **62,** 347–357 (1988).
31. G. N. Tseng and P. A. Boyden. *Am. J. Physiol.* **261,** H364–H379 (1991).
32. K. T. MacLeod and S. E. Harding, *J. Physiol.* **444,** 481–498 (1991).
33. M. C. Capogrossi, T. Kaku, C. R. Filburn, D. J. Pelto, R. G. Hansford, H. A. Spurgeon, and E. G. Lakatta, *Circ. Res.* **66,** 1143–1155 (1990).
34. M. Puceat, O. Clement-Chomienne, A. Terzic, and G. Vassort, *Am. J. Physiol.* **264,** H310–H319 (1993).
35. S. Yuan, F. A. Sunahara, and A. K. Sen, *Circ. Res.* **61,** 372–378 (1987).
36. P. M. Dunnmon, K. Iwaki, S. A. Henderson, A. Sen, and K. R. Chien, *J. Mol. Cell. Cardiol.* **22,** 901–910 (1990).
37. S. N. Allo, L. L. Carl, and H. E. Morgan, *Am. J. Physiol.* **263,** C319–C325 (1992).
38. A. Shah, I. S. Cohen, and M. R. Rosen, *Biophys. J.* **54,** 219–225 (1988).
39. A. Zaza, R. P. Kline, and M. R. Rosen, *Circ. Res.* **66,** 416–426 (1990).
40. U. delBalzo, M. R. Rosen, G. Malfatto, L. M. Kaplan, and S. F. Steinberg, *Circ. Res.* **67,** 1535–1551 (1990).

41. M. Takanashi, I. Norota, and M. Endoh, *Naunyn-Schmiedeberg's Arch. Pharmacol.* **343,** 669–673 (1991).

42. C. W. Luetje, K. M. Tietje, J. L. Christian, and N. M. Nathanson, *J. Biol. Chem.* **263,** 13357–13365 (1988).

43. K. A. Foster, P. J. McDermott, and J. D. Robishaw, *Am. J. Physiol.* **259,** H432–H441 (1990).

44. K. A. Foster and J. D. Robishaw, *Am. J. Physiol. Suppl.* **261,** 15–20 (1991).

45. S. R. Holmer, S. Stevens, and C. J. Homcy, *Circ. Res.* **65,** 1136–1140 (1989).

46. D. J. Sheridan, P. A. Penkoske, B. E. Sobel, and P. B. Corr, *J. Clin. Invest.* **65,** 161–171 (1980).

47. G. P. Heathers, K. A. Yamada, E. M. Kanter, and P. B. Corr, *Circ. Res.* **61,** 735–746 (1987).

48. P. B. Corr, J. A. Shayman, J. B. Kramer, and R. J. Kipnis, *J. Clin. Invest.* **67,** 1232–1236 (1981).

49. J. H. Lee, S. F. Steinberg, and M. R. Rosen, *J. Pharm. Exp. Ther.* **258,** 681–687 (1991).

50. E. J. Neer, J. M. Lok, and L. G. Wolf, *J. Biol. Chem.* **259,** 14222–14229 (1984).

51. C. J. Schmidt, T. C. Thomas, M. A. Levine, and E. J. Neer, *J. Biol. Chem.* **267,** 13807–13810 (1992).

51a. C. A. Hansen, A. G. Schroering, and J. D. Robishaw, *J. Mol. Cell. Cardiol.* **24,** 471–484 (1995).

52. N. Ueda, J. A. Iniguez-Lluhi, E. Lee, A. V. Smrcka, J. D. Robishaw, and A. G. Gilman, *J. Biol. Chem.* **269,** 4388–4395 (1994).

53. D. E. Logothetis, D. Kim, J. K. Northup, E. J. Neer, and D. E. Clapham, *Proc. Natl. Acad. Sci. U.S.A.* **85,** 5814–5818 (1988).

54. J. L. Boyer, S. G. Graber, G. L. Waldo, T. K. Harden, and J. C. Garrison, *J. Biol. Chem.* **269,** 2814–2819 (1994).

55. A. Fawzi, D. S. Fay, E. A. Murphy, H. Tamir, J. J. Erdos, and J. K. Northup, *J. Biol. Chem.* **266,** 12194–12200 (1991).

56. P. B. Wedegaertner, P. T. Wilson, and H. R. Bourne, *J. Biol. Chem.* **270,** 503–506 (1995).

57. T. L. Z. Jones, W. F. Simonds, J. J. Merendino, M. R. Brann, and A. M. Spiegel, *Proc. Natl. Acad. Sci. U.S.A.* **87,** 568–572 (1990).

58. S. M. Mumby, R. O. Heukeroth, J. I. Gordon, and A. G. Gilman, *Proc. Natl. Acad. Sci. U.S.A.* **87,** 728–732 (1990).

59. M. J. Im and R. M. Graham, *J. Biol. Chem.* **265,** 18944–18951 (1990).

60. M. J. Im, P. Riek, and R. M. Graham, *J. Biol. Chem.* **265,** 18952–18960 (1990).

61. K. J. Baek, T. Das, C. Gray, S. Antar, G. Murugesan, and M. J. Im, *J. Biol. Chem.* **268,** 27390–27397 (1993).

62. M. J. Im, C. Gray, A. J. Rim, *J. Biol. Chem.* **267,** 8887–8894 (1992).

63. H. Nakaoka, D. M. Perez, K. J. Baek, T. Das, A. Husain, K. Misono, M. J. Im, and R. M. Graham, *Science* **264,** 1593–1596 (1994).

[23] G$_{s\alpha}$ Mutations and Pituitary Adenomas

Joseph M. Alexander

Introduction

This chapter outlines the various methodologies used to characterize G$_{s\alpha}$ mutations from human tissues. Activating mutations of the stimulatory α subunit of heterotrimeric G proteins (G$_{s\alpha}$) have been extensively documented in tumors of the pituitary gland (1), as well as in patients with McCune–Albright syndrome (2, 3). Several studies have shown that up to 40% of growth hormone-secreting human pituitary adenomas harbor such mutations. To date, activating mutations have been localized to two sites within G$_{s\alpha}$, i.e., codon 201 and codon 207. Both mutations result in a constitutively activated stimulatory G protein trimeric complex, which leads to increased adenylylcyclase activity, and elevated intracellular cAMP levels. Depending on the tissue studied, this constitutive activation of G protein signaling pathways can lead to hypersecretion of hormones (such as that documented in somatotroph adenomas) and/or increased cellular proliferation.

Due to advances in molecular biological techniques such as the polymerase chain reaction (PCR) and single-strand conformational polymorphism (SSCP) analysis, G protein mutations can be readily detected in human tissues from a variety of sources. Because of the wide applicability of PCR analysis of genomic mutations in human tumors, many molecular biology companies offer prepared reagents and methods kits for many of the manipulations discussed in this chapter. I have encouraged their use when possible because I believe they are more time- and cost-effective than preparing reagents in the laboratory. Specific companies are references as sources of prepared reagents and kits at many points in the text. I offer these only as information, and not as an endorsement of any product over another.

Gene Structure of G$_{s\alpha}$

Figure 1 indicates the gene structure of human G$_{s\alpha}$, as well as the placement of oligonucleotide primers used for PCR amplification, SSCP analysis, and screening for mutations. It also demarcates the sites of known activating mutations of human G$_{s\alpha}$. The human G$_{s\alpha}$ gene is composed of 13 exons (4, 5). Mutations of codons 201 and 227 that activate G$_{s\alpha}$ are located on exons

Methods in Neurosciences, Volume 29

8 and 9, respectively. Arginine-201 is the site of covalent ADP-ribosylation by the pathogenic *Vibrio cholerae* exotoxin, thus activating $G_{s\alpha}$ by blocking its GTPase activity (6); Gln-227 is within the critical GTP-binding region. Disruption of this protein domain inhibits GTPase activity of $G_{s\alpha}$ and constitutively activates G protein signal transduction (6). Thus, both mutations constitutively activate $G_{s\alpha}$ by inhibiting its intrinsic ability to hydrolyze GTP to guanosine 5'-diphosphate, transition, and return to its inactive conformation.

DNA Extraction from Fresh and Fixed Tissue

The success of any PCR amplification is dependent on the quality of genomic DNA from the tissue of interest. PCR will reliably amplify the human $G_{s\alpha}$ gene sequence from high molecular weight genomic DNA extracted from fresh tissue and surgical specimens. For some samples, such as formalin-fixed and paraffin-embedded DNA with moderate covalent cross-linking and degradation, amplification can be more problematic and may require additional PCR cycles and/or "nested" PCR. However, amplification of $G_{s\alpha}$ gene sequences from even the poorest quality genomic DNA preparations is usually achieved because of the exponential amplification of intact DNA.

DNA Extraction from Fresh Tissue

This protocol is based on a method described by Gross-Bellard and colleagues (7). Fresh or frozen tissue first is minced or ground, respectively, and is then placed in TE9 (10 mM Tris, 1 mM ethylenediaminetetraacetic acid, pH 9.0). As a general consideration, tissue volume should be approximated, and 3 to 5 volumes of TE9 added. Sodium dodecyl sulfate (SDS) is then added to a final volume of 1% (v/v, from a 10% stock solution) followed by 500 μg/ml proteinase K (Boehringer Mannheim, Indianapolis, IN). Tissue is then incubated at 48°C until tissue is entirely digested (usually 3 hr to overnight, depending on the tissue type). NaCl then is added to a final concentration of 200 mM after tissue digestion. Next, an equal volume of phenol/isoamyl alcohol/chloroform (PIC; 25 : 24 : 1, v/v) that has been preequilibrated with TE7.5 (10 mM Tris, 1 mM EDTA, pH 7.5) is added to the aqueous tissue preparation. (Although these reagents can be obtained separately and prepared as an organic extraction solution, several manufacturers offer premixed and equilibrated organic DNA extraction solutions that are reasonably inexpensive and quality controlled.) The organic extraction is vortexed for 1 min and centrifuged at 10,000 g for 10 min at room temperature. Following centrifugation, the upper aqueous phase should appear clear or slightly cloudy, and a white protein precipitate will separate the aqueous and organic phases. The upper aqueous phase should be removed without disturbing the protein precipitate at the interface, transferred to a new tube, and the extraction procedure repeated twice with chloroform/isoamyl alcohol (49 : 1), or until the remainder of the white protein precipitate at the interface has been removed. Following organic extraction with PIC, the volume of the aqueous phase is estimated and 0.25 volume of 10 M ammonium acetate (NH$_4$Ac) is added. If the amount of tissue sample is limited (e.g., paraffin-embedded tissue removed from a single slide), glycogen (1 μl of a 10 mg/ml stock solution) may be added to nucleate the DNA precipitation reaction and visualize the DNA pellet after centrifugation. Ethanol (2.5 volumes, 100%) is then added to precipitate genomic DNA. The preparation should be vortexed briefly or the tube repeatedly inverted until a visible precipitate forms (typically, high molecular weight DNA will appear as a fibrous material precipitate; however, degraded DNA from archival samples may precipitate as a cloudy material, or, in low amounts, form no visible precipitate). The precipitated DNA then is centrifuged at 10,000 g for 20 min at 4°C. The supernatant should be carefully decanted and the pellet rinsed with 70% ethanol to remove excess NH$_4$Ac. The DNA precipitate then is air or vacuum dried to remove the remaining 70% ethanol. The pellet then is resuspended in TE (pH 7.5). If DNA does not solubilize in a short period (1–2 hr), it should be stored at 4°C overnight to allow for complete DNA solubility.

In order to determine DNA concentrations for each sample, UV absorption at OD$_{260}$ and OD$_{280}$ is measured. The extinction coefficient for double-stranded DNA is 50 μg(ml^{-1})/OD unit. Calculate the ratio of OD$_{260}$: OD$_{280}$. Organically extracted DNA will have an OD$_{260}$: OD$_{280}$ ratio of 2.00 ± 0.20. If OD measurements lie outside that range, a second organic extraction should be considered.

DNA Extraction from Formalin-Fixed, Paraffin-Embedded Tissue

This protocol is based on a method described by Rogers *et al.* (8). Archival samples require additional preparation prior to the above outlined protocol for extraction of DNA. The paramount concern is the complete removal of paraffin from tissue samples. After a tissue specimen has been scraped from an archival slide or carefully excised from a paraffin-embedded block, it should be placed in xylene to dissolve the remaining paraffin. While the tissue specimen should remain intact during xylene extraction, it may be necessary to centrifuge the sample between xylene extraction and ethanol rehydration steps. The xylene treatment should be repeated until the white or translucent paraffin coat has disappeared from the periphery of the tissue sample. After xylene has been completely removed, the tissue sample should be rehydrated gradually through a series of ethanol steps. This is an extremely critical step. If samples are rehydrated too rapidly, solubilized parafin will reprecipitate, and genomic DNA recovery will be minimal. The rehydration steps are as follows: two treatments with 100% ethanol, 5 min; two treatments with 95% ethanol, 5 min; two treatments with 90% ethanol, 5 min; one treatment with 70% ethanol, 2 min; one treatment with 50% ethanol, 1 min. Finally, samples should be briefly air or vacuum dried and resuspended in TE9. At this point, genomic DNA extraction proceeds as described above.

Avoiding Tissue DNA Cross-Contamination

Because of the powerful amplification properties of thermostable DNA polymerase, extreme care must be taken to avoid contamination of DNA samples with DNA from other specimens that are being prepared simultaneously. Only aerosol-resistant pipette tips should be utilized for transfer of DNA samples. These prevent contamination of the pipetter barrel with DNA, and minimize contamination between samples during DNA isolation. Whenever possible, only single-use plastic Eppendorf and clinical centrifuge tubes should be used to manipulate and store genomic DNA samples. Because these are designed for single-use, cross-contamination of sample DNAs is less likely. In addition, DNA tends to be less adherent to plastic (or silicon-treated glass), and for this reason plasticware usually gives better genomic DNA yields. Bleach and/or UV light treatment of equipment and glassware is also effective at removing sources of DNA contamination. It is also advisable to keep PCR equipment and genomic DNA samples in a separate area away from subcloning and sequencing of G$_{s\alpha}$ gene DNA (for example, a pipetter set that is designated solely for PCR of genomic DNA should be used throughout the procedure, and not one that is routinely used for subcloning and sequencing of G$_{s\alpha}$ clones).

Oligonucleotide Primers

Table I lists oligonucleotide primers for PCR that have been demonstrated to amplify G$_{s\alpha}$ gene sequences from genomic DNA (1, 9), along with their sequence location in the G$_{s\alpha}$ gene. The nucleotide location of each primer is based on the GenBank accession file M21142. This file contains the genomic sequence from intron 6 to the 3′ end of the primary (unprocessed) RNA

TABLE I   Sequence of Oligonucleotide Primers Used for PCR Amplification of Human $G_{s\alpha}$ Gene

Oligonucleotide number[a]	Oligonucleotide sequence	Sequence location[b]
1	GTG ATC AAG CAG GCT GAC TAT GTG	258
2	GCT GCT GGC CAG CAC GAA GAT GAT	792
3	CCC CTC CCC ACC AGA GGA CTC TGA	304
4	AGA GCG TGA GCA GCG ACC CTG ATC	737
5	TCC CTC TGG AAT AAC CAG CTG T	561
6	ACA GCT GGT TAT TCC AGA GGG A	582
7	CTA CTC CAG ACC TTT GCT TTA G	342

[a] See Fig. 1.
[b] GenBank Acc# M21142.

transcript. Primers 1 and 2 are used for primary PCR of genomic DNA, and generate a 534-bp $G_{s\alpha}$ gene fragment that includes a portion of intron 7, exons 8 and 9 (as well as intron 8), and a portion of intron 9. Primers 3 and 4 are immediately internal to the first set, and are used for "nested" secondary PCR (see below). This reaction yields a 433-bp $G_{s\alpha}$ gene fragment. Nested PCR is sometimes necessary to amplify DNA from specimens that have low yields of genomic DNA and/or have a high degree of covalent cross-linking due to formalin fixation. Primers 5 and 7 are used for single-stranded conformational polymorphism (SSCP) analysis of codon Arg-201 mutations, whereas primers 6 and 4 are used for SSCP analysis of codon Gln-227 mutations. Oligonucleotides 8 and 9 are degenerate oligonucleotides used to detect wild-type alleles as well as all potential activating mutations of Arg-201 and Gln-227, respectively. Construction of oligonucleotides 8 and 9 are described in more detail in Table II.

Numerous companies provide oligonucleotide synthesis services, and many research institutions have central oligonucleotide synthesis fee-for-service facilities. When requesting oligonucleotide synthesis, the sequence is written 5' to 3', and a routine desalting of the oligonucleotide primer should be requested. This is fairly standard, and desalting after synthesis usually is available at no extra charge. More elaborate purification (such as polyacrylamide gel electrophoretic purification) usually is not necessary. Before attempting to amplify $G_{s\alpha}$ gene sequences from experimental samples (particularly samples that are highly degraded and/or covalently cross-linked), PCR primers should be tested on control high molecular weight DNA. Human blood leukocyte DNA is an excellent and readily available control.

TABLE II   Sequences of Degenerate Oligonucleotide Probes Directed against
Codons 201 and 227 of the Human G$_{s\alpha}$ Gene

Codon	Sequence	Amino acid
Arg-201 TTCGCTGCCG**CGT**CCTGACT		
Probe 1	TGT	Cysteine
	AGT	Serine
	GGT	Glycine
Probe 2	CAT	Histidine
	CCT	Proline
	CTT	Leucine
Gln-227 GTGGGTGGC**CAG**CGCGATGA		
Probe 1	AAG	Lysine
	GAG	Glutamine
Probe 2	CTG	Leucine
	CCG	Proline
	CGG	Arginine
Probe 3	CAC	Histidine
	CAT	Histidine

PCR Conditions for Amplification of G$_{s\alpha}$ Genes

PCR reaction conditions for amplification of G$_{s\alpha}$ gene sequences are fairly straightforward. *Thermus aquaticus* thermostable polymerase (*Taq*) can be purchased alone, or PCR kits with *Taq* polymerase, 10× PCR buffer, MgCl$_2$, deoxynucleotides, control DNA, and primers can also be purchased from Perkin-Elmer (Norwalk, CT). Typically, PCR reactions are carried out in a total volume of 100 μl, containing 50 mM KCl, 10 mM Tris-HCl (pH 9.0), 3.5 mM MgCl$_2$, 0.1% Triton X-100, 200 μM dNTPs, and 0.5 U of *Taq*1 polymerase. For primary PCR, 50 pmol of oligonucleotide primers 1 and 2 is added, as well as ≤200 ng of genomic DNA from experimental samples. PCR thermocycling conditions are as follows: 1 min at 94°C (to denature double-stranded genomic DNA), 2 min at 57°C (to allow the primers to anneal to G$_{s\alpha}$ gene sequences), and 3 min at 72°C (extension period for *Taq* polymerase), for a total of 30 cycles.

For DNA samples that are highly degraded and/or covalently cross-linked, secondary, or "nested," PCR may be required. The strategy involves using primers designed to anneal within the sequences amplified by the previous primer set. PCR amplification is an exponential reaction, and the nested primer strategy affords greater specificity, offers greater amplification of genomic sequences, and can usually successfully generate G$_{s\alpha}$ gene sequences from even the most degraded or cross-linked genomic DNA sample. After primary PCR, amplified $G_{s\alpha}$ gene sequences could be separated from oligonucleotide primers and PCR reaction reagents. This can be done quickly by DNA affinity chromatography. For example, Magic PCR Preps (Promega Corp., Madison WI) and GeneClean (Bio101, LaJolla, CA) are two of many commercially available kits for purifying PCR products. The protocols for this step are straightforward and are included in the product literature with the reagents. Once the primary PCR products are isolated, 50 pmol of nested primers 3 and 4 should be utilized following the same PCR reactions detailed for the primary PCR reaction.

Visualization of PCR Products

Typically, agarose gel electrophoresis is used to analyze PCR products (10). Mix 10 μl of each PCR reaction sample with 2 μl of 6× gel loading buffer [0.25% (w/v) xylene cyanol, 0.25% (w/v) bromphenol blue, 1% (w/v) Ficoll in H_2O]. A gel solution containing 2% agarose/1× TBE buffer (20× TBE: 1 M Tris base, 1 M boric acid, 20 mM EDTA) should be thoroughly heated until all agarose is dissolved, and allowed to cool to 70°C. Add 1 μl of a 10 mg/ml ethidium bromide stock just before pouring the gel. (*Caution:* Ethidium bromide is a carcinogen! Do not breathe vapor containing ethidium bromide, and avoid contact with skin.) Appropriate DNA size markers should be run with PCR products to determine their size. Several size markers of the appropriate length are available from companies that specialize in molecular biology products. After electrophoresis, PCR bands may be visualized with UV light to confirm that the amplification was successful and the PCR products are of the correct size. The size of the PCR products should be 534 and 433 bp after primary and nested PCR, respectively.

Generating Positive and Negative Controls for Oligonucleotide-Specific Hybridization

In order to optimize hybridization conditions for each wild-type and degenerate probe, appropriate positive and negative control sequences should be generated and included on each nitrocellulose blot. DNA from patient tissues with normal $G_{s\alpha}$ and known mutations may be used. However, genomic DNA harboring activating $G_{s\alpha}$ mutations is often not available. It is possible to synthesize controls using PCR utilizing degenerate oligonucleotide 8 (for

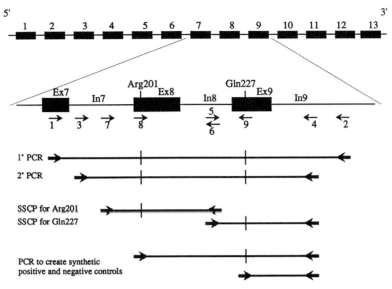

FIG. 1   The human $G_{s\alpha}$ gene. A schematic of $G_{s\alpha}$ gene exon structure is shown, as well as a more detailed schematic of the area of activating mutations. The position of each oligonucleotide is shown below the map, as well as the expected size of each PCR product.

codon 201) or 9 (for codon 227) and PCR primer 4 from Fig. 1. In this strategy, mutations are incorporated into DNA amplification reactions using the oligonucleotide probes that will later be used to probe for mutations in experimental tissues (see Fig. 1). These controls are essential to the interpretation of oligonucleotide-specific hybridization of degenerate and wild-type oligonucleotide probes. The PCR conditions outlined above should be used to create synthetic controls for hybridization.

Immobilization of PCR Products on Nitrocellulose Membranes

After PCR of $G_{s\alpha}$ sequences and confirmation of the PCR product size, DNA samples should be immobilized on nitrocellulose membranes and probed with specific oligonucleotides designed to detect point mutations at codons 201 and 227. Because the PCR reaction product encompasses both mutation sites, the entire mutational screening can usually be carried out with a single nitrocellulose blot. Typically, DNA samples are applied to nitrocellulose membranes using a vacuum manifold blotting apparatus. For example, the

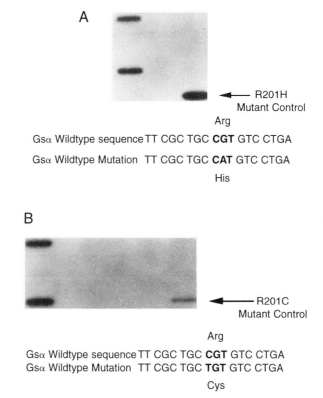

A

← R201H
Mutant Control

Arg

Gsα Wildtype sequence TT CGC TGC **CGT** GTC CTGA

Gsα Wildtype Mutation TT CGC TGC **CAT** GTC CTGA

His

B

← R201C
Mutant Control

Arg

Gsα Wildtype sequence TT CGC TGC **CGT** GTC CTGA
Gsα Wildtype Mutation TT CGC TGC **TGT** GTC CTGA

Cys

Fɪɢ. 2   G$_{sα}$ mutations in somatotroph adenomas. PCR-amplified DNA from tumors was applied to slot-blots and hybridized with the indicated probes. (A) Two of four tumors hybridized to degenerate oligonucleotide 2 (Table II), which sequencing analysis revealed to be a G to A missense mutation altering codon 201 from normal arginine to constitutively active histidine. Note the R201H positive control. Other negative synthetic controls are above it on the blot. (B) Two of six tumors hybridized to degenerate oligonucleotide 1 (Table II), which sequencing analysis revealed to be a C to T missense mutation altering codon 201 from normal arginine to constitutively active cysteine. Note the R201C positive control. Other negative synthetic controls are above it on the blot.

data shown in Fig. 2 were generated using a slot-blot apparatus (Schleicher and Schuell, Keene, NY). However, PCR reactions may be applied to membranes individually with a pipetter without such an appartus and allowed to air dry.

Many different brands of membranes are available, and most are equivalent in their affinity for DNA and low background under specific hybridization conditions. It is recommended that nylon-backed nitrocellulose be used because of ease of handling for multiple reprobing with degenerate probes; pure nitrocellulose is quite fragile and difficult to handle after one or two hybridizations. The data shown in Fig. 2 were generated using Genescreen nylon-backed membranes (New England Nuclear-DuPont, Boston, MA). Latex gloves should always be used when handling any nitrocellulose membrane. If DNA is applied manually without a blotting apparatus, the coordinants of the samples should be marked with a pencil, and the membrane placed on Whatman chromatography paper. Heat 10 μl of each PCR reaction (as well as positive and negative synthetic controls) to 95°C, place on ice, and apply to the membrane and allow to air dry. Membranes then are placed between two pieces of Whatman paper, wrapped in aluminum foil, and baked at 80°C for 2 hr in a vacuum oven. This procedure irreversibly binds DNA samples to the membrane for subsequent hybridization.

Radiolabeling of Oligonucleotide Probes with T4 Polynucleotide Kinase

Degenerate oligonucleotide probes listed in Table II can be radioactively labeled to a very high specific activity at their 5' terminal phosphate with T4 polynucleotide kinase (PNK) and [γ-^{32}P]ATP. The specific activity of [γ-^{32}P]ATP is very high, and extreme caution should be used to minimize exposure to any potential source of contamination. A $\frac{1}{4}$-inch Plexiglas shield should always be utilized when handling radioactive solutions. The PNK reaction is as follows: 10 pmol of oligonucleotide in 1 μl of H$_2$O is used as a substrate in the reaction, 1 μl of [γ-^{32}P]ATP (3000 Ci/mmol at 10 mCi/ml; New England Nuclear-DuPont, Boston, MA), 1 μl of 10× kinase buffer (0.5 M Tris, pH 7.6, 0.1 M MgCl$_2$, 0.05 M dithiothreitol, 1 mM spermidine, 1 mM EDTA), 8–10 U (1 μl) of PNK (Promega Corp., Madison, WI), and 6 μl of H$_2$O are added to a total volume of 10 μl. The reaction is incubated at 37°C for 20 min, followed by the addition of 1 μl of 0.5 M EDTA to terminate the reaction. The labeled oligonucleotide should be purified from unincorporated [γ-^{32}P]ATP using a prepared Sephadex G-25 size-exclusion column according to manufacturer's recommendations (NAP-5 column, Pharmacia Biotech, Inc., Piscataway, NJ). The labeling efficiency of the kinase reaction can be measured by scintillation counting of 1 μl of the total volume of radiolabeled oligonucleotide in 5 ml of scintillation fluid (Econoflor, New England Nuclear-DuPont, Boston, MA). Efficient labeling will result in a specific activity of 5×10^6 cpm/pmol oligonucleotide.

Detection of Point Mutations in Human $G_{s\alpha}$ Using Oligonucleotide-Specific Hybridization

Prehybridization, hybridization, and washing steps are typically carried out in plastic heat-seal bags in volumes of 10–20 ml, depending on the number of membranes being hybridized. Alternatively, these steps can be carried out in airtight sealed plastic trays. Washes are typically carried out in plastic trays due to the greater volumes required. After membranes have been baked to immobilize DNA, they should be prehybridized in 10 ml of the following buffer for 3 hr at 42°C; 1.8 M NaCl/0.1 M NaH$_2$PO$_4$/0.01 M EDTA (10× SSPE), 0.5% sodium dodecyl sulfate (SDS), 5× Denhardt's solution (1% w/v Ficoll, polyvinylpyrrolidone, fraction V BSA in H$_2$O), 100 μg/ml salmon sperm DNA (salmon sperm can be obtained from Sigma Chemical Corp., St. Louis, MO; a 10 mg/ml stock is prepared by placing 1 g of DNA in 100 ml of H$_2$O and autoclaving for 30 min). Membranes then are placed in 10 ml fresh hybridization buffer preheated to 55°C consisting of 0.05 M Tris, pH 8.0, 3 M TMAC (trimethylammonium chloride; Sigma Chemicals, St. Louis, MO), 0.02 M EDTA, 0.1% SDS, 100 μg/ml salmon sperm DNA, and 5× Denhardt's solution). Add 1 × 10^5 cpm/ml of the oligonucleotide probe and hybridize for a minimum of 4 hr to overnight at 55°C. After hybridization, membranes are washed briefly twice at room temperature in 50 ml of 2× SSPE/0.1% SDS. This removes the vast majority of the oligonucleotide probe from the membrane. Membranes are then washed three times for 15 min each in TMAC wash buffer consisting of 0.05 M Tris, pH 8.0, 3 M TMAC, 0.02 M EDTA, 0.1% SDS at 50–55°C.

To visualize hybridized oligonucleotide probe, membranes are drained of excess liquid but not allowed to dry completely. Membranes then are sealed in heat-seal bags, and exposed to X-ray film overnight. As long as membranes are not allowed to dry completely, they may be rewashed under more stringent conditions after autoradiography. Continue to wash the membranes in TMAC wash buffer at increasingly higher temperatures (in steps of 5°C) until synthetic negative controls are free of bound oligonucleotide probe, and while synthetic positive controls continue to show a strong autoradiographic signal (see Fig. 2).

Single-Strand Conformational Polymorphism Analysis of $G_{s\alpha}$ Mutations

Single-strand conformational polymorphism analysis of genetic mutations relies on the differential migration in nondenaturing gel electrophoresis of DNA duplexes with single base pair mismatches (11, 12). The advantages of this technique include the ability to rapidly screen large numbers of samples for all potential mutations at G$_{s\alpha}$ codons 201 or 227 (rather than hybridizing degenerated oligonucleotide probes individually) and a more limited exposure to high levels of [^{32}P]dATP. Its drawbacks are (1) the potential to not resolve mutations in a subset of samples, and (2) only small areas of the genome can be analyzed (<250 bp). In the case of G$_{s\alpha}$, two PCR reactions are required for a complete analysis (Fig. 1) because the 600-bp fragment generated by PCR for oligonucleotide-specific hybridization is too large to resolve on SSCP gels.

PCR reactions are identical to the previously described protocol with the following exceptions: (1) due to the limited resolution of SSCP gels, codons 201 and 227 must be amplified separately with the oligonucleotide primers listed in Fig. 1, (2) deoxynucleotide concentrations should be lowered by 10-fold (from 200 to 20 μM) to give added specificity to the amplification reaction, and (3) PCR products are visualized by the addition of [^{32}P]dCTP (100 nCi/reaction) to PCR reactions, which then incorporates into elongating DNA strands. For both codon 201 (primers 6 and 7, Fig. 1) and codon 227 (primers 4 and 5, Fig. 1), PCR recations are carried out in a total volume of 100 μl, containing 50 mM KCl, 10 mM Tris-HCl (pH 9.0), 3.5 mM MgCl$_2$, 0.1% Triton X-100, 40 mM dNTPs, 0.1 μl of [^{32}P]dCTP, and 0.5 U of Taq1 polymerase using 200 ng of genomic DNA from experimental samples as substrate. PCR thermocycling conditions are as follows: 1' 94°C (to denature double-stranded genomic DNA), 2' 57°C (to allow the primers to anneal to G$_{s\alpha}$ gene sequences), and 3' 72°C (extension period for Taq polymerase) for a total of 30 cycles; 30 cycles is typically sufficient to amplify most genomic DNA sequences and has proved adequate for G$_{s\alpha}$ PCR reactions for SSCP analysis. For very degraded or highly cross-linked DNA samples, a nested PCR strategy may be necessary, first using primary PCR with unlabeled (nonradioactive) oligonucleotide primers 1 and 2 to amplify genomic DNA, followed by SSCP PCR as discussed above.

For SSCP gel electrophoresis, a standard sequence gel apparatus (30 × 40 × 0.04 cm) is sealed with tape and/or metal clips on three sides to limit leakage during gel casting (10). A single glass plate may be siliconized with SigmaCote (Sigma Chemicals, St. Louis, MO) to facilitate removal of the gel from the surface of the plate after electrophoresis. Gel solution should be prepared using 6% acrylamide (Protogel; 30% acrylamide/bisacrylamide gel solution,) with 10% glycerol (w/v) and 1× TBE. For a 100-ml gel, 20 ml of Protogel (National Diagnostics, Atlanta, GA), 10 g of glycerol, 5 ml of 20× TBE, 800 μl of ammonium persulfate (10% stock solution, w/v), and 80 μl TEMED (N,N,N',N'-tetramethylethylenediamine; Sigma Chemicals, St. Louis, MO) are sequentially added. The addition of glycerol to gels offers greater resolution and conformational polymorphic patterns are more evident. The acrylamide gel solution must be poured immediately, followed by insertion of the gel comb to form loading wells. Acrylamide gels should polymerize for at least 30 min before beginning electrophoresis.

Mix 10 μl of PCR reaction products with 10 μl of loading buffer [0.25% (w/v) xylene cyanol, 0.25% (w/v) bromphenol blue, 15% (w/v) Ficoll in H$_2$O]; samples are then heated to 95°C for 2 min, and put on ice. The sequencing gel apparatus is filled with 1× TBE buffer, and samples are loaded with a standard P20 Eppendorf pipetter. Gels typically are run at 3 W (constant power setting) for approximately 15 hr at room temperature. After SSCP gel electrophoresis, the sequencing appartus is disassembled and the gel glass plates are separated (remember that the TBE tank buffer is radioactive and must be handled appropriately due to unincorporated ^{32}P-labeled oligonucleotide primers that have migrated off the end of the gel). The removal of 0.04-cm acrylamide gels from glass sequencing plates can be somewhat problematic, but can be facilitated with a few simple steps. First, place the gel flat on a bench surface, with the siliconized glass plate facing up. Second, a temperature differential between the glass plates and the gel often aids in disassembly. For this reason, spreading flaked ice over the plate to cool the siliconized glass, allowing it to chill for 5 min, is sometimes helpful in cleanly separating glass plates. Third, the top plate then is gently lifted away from the gel, beginning at one corner. The combined effects of the siliconized surface and temperature differential should allow the top plate to be removed easily. Fourth, cut a piece of Whatman paper slightly larger than the gel, place it on top of the gel, and gently smooth it over the area of the gel, making certain it is in complete contact with the acrylamide. Fifth, flip the bottom gel plate over, and gently lift the plate away from the Whatman paper. The gel should lift away easily from the glass. The Whatman paper/acrylamide gel is placed on a gel dryer and the entire area of the acrylamide is covered with plastic wrap and is dried under vacuum at 80°C for 1 hr. Typically, dried gels should be exposed to film for 6–24 hr, depending on the efficiency of the oligonucleotide labeling and PCR amplification. Tumor DNA samples that harbor G$_{s\alpha}$-activating mutations typically exhibit an additionally labeled band on the autoradiographic exposure (see Fig. 3), and should be processed further for sequence analysis to verify G$_{s\alpha}$-activating mutation.

Subcloning and Sequencing of DNA Samples That Harbor Activating Mutations of the G$_{s\alpha}$ Gene

It is essential that any putative mutation discovered by SSCP or oligonucleotide-specific hybridization be verified by DNA subcloning and sequencing. DNA subcloning and Sanger sequencing are completely standardized, and

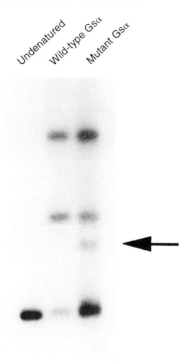

Fig. 3 SSCP analysis of $G_{s\alpha}$ mutations. DNA from a tumor sample known to have an activating mutation at R201 was amplified and fractionated by SSCP gel electrophoresis. The arrow indicates the anomalous migration pattern of the mutant $G_{s\alpha}$ allele. (Gel autoradiograph courtesy of Dr. Wen Yi Cai, Neuroendocrine Unit, Massachusetts General Hospital.)

there are numerous molecular biological kits and resource materials available that detail these protocols. However, there are specific methodological concerns that pertain to subcloning and sequencing of PCR products. For mutated DNA samples that are detected using oligonucleotide-specific hybridization, DNA may be subcloned from the original PCR reactions. For SSCP-analyzed mutations, tissue DNA samples must be reamplified using unlabeled primers in order to continue with subcloning and sequence analysis. Many of the steps are not discussed in detail here, but Fig. 4 provides a flow chart that overviews the process. [For detailed protocols concerning the basic principles and methodologies of subcloning (i.e., ligation of DNA into plasmid vectors, preparation of competent *Escherichia coli* cells, bacterial transformation, plasmid ''miniprep'' protocols, and sequencing) see Ref. 10.]

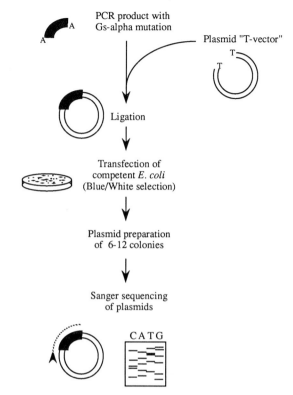

FIG. 4   Flow chart of DNA subcloning and sequencing to verify $G_{s\alpha}$ mutations in human tissues.

Subcloning of PCR Products

The polymerization of DNA in PCR reactions using thermostable *Taq* polymerase presents a unique problem for efficient subcloning of DNA. Double-stranded DNA generated by PCR has an additional adenosine nucleotide at the 3' end of each strand. This single-nucleotide overhang markedly reduces the efficiency of blunt-end ligation into plasmid vectors. Accordingly, plasmid vectors have been commercially developed that have a 5'-thymidine overhang to accommodate ligation of PCR products. Such "T-vectors" are available from several sources, including Invitrogen Corporation (T/A Cloning Kit) and Promega Corporation (pGEM-T vectors). These plasmids, in addition to efficiently subcloning PCR products, contain standard ampicillin-resistance markers and β-galactosidase (blue/white) screening sequences.

Once PCR products have been ligated into the appropriate plasmid vector, plasmid is transfected into competent *E. coli* cells [prepared competent cells are available from BRL (Bethesda, MD) or from the Promega Corporation (Madison, WI)]. *Escherichia coli* cells are plated on Luria broth/agarose plates that have been supplemented with X-galactosidase (Promega Corporation, Madison, WI) as a substrate for β-galactosidase (blue/white) screening. Plates then are incubated at 37°C overnight for bacterial growth.

Bacterial colonies are then selected for insertion (white colonies), and small-scale bacterial preparations are grown overnight. Because virtually all DNA samples with G$_{s\alpha}$ mutations contain wild-type G$_{s\alpha}$ gene sequences as the other allele, 6 to 12 plasmid preparations should be analyzed from each tumor DNA sample. Plasmid minipreparation kits are available from numerous molecular biology supply companies. After obtaining subcloned DNA samples, DNAs should be sequenced using a Sequenase sequencing kit (U.S. Biochemicals, Cleveland, OH), and analyzed for activating point mutations at codons 201 and 227. The details of DNA Sanger sequencing are available from many resource materials, and from Sequenase product literature.

Summary

This chapter details the essential protocols for the detection and analysis of activating mutations of the G$_{s\alpha}$ gene from human tissues using oligonucleotide-specific hybridization and SSCP. Both strategies have inherent advantages and drawbacks. Oligonucleotide-specific hybridization, although more thorough in its ability to detect mutations, is more time-consuming and requires more manipulation and exposure to ^{32}P radioactivity. SSCP is a more rapid screening method; however, mutations in some tissue samples may be overlooked on SSCP gel electrophoresis. Once an activating mutation is suspected, verification by subcloning and DNA Sanger sequencing is necessary. These techniques are standard and have been optimized in recent years both by the continued refinement of molecular biology techniques and by the availability of commercial reagents. The analysis of activating mutations of G$_{s\alpha}$ and other candidate oncogenes should continue to become more and more routine in molecular pathology laboratories, and will facilitate the molecular pathogenetic diagnosis of a growing number of human tumor types.

References

1. C. A. Landis, S. B. Masters, A. Spada, A. M. Pace, H. R. Bourne, and L. Vallar, *Nature (London)* **340,** 692 (1989).
2. L. S. Weinstein, P. V. Gejman, E. Friedman, *et al., Proc. Natl. Acad. Sci. U.S.A.* **87,** 8287 (1990).

3. L. S. Weinstein, A. Shenker, P. V. Gejman, M. J. Merino, E. Friedman, and A. M. Spiegel, *N. Engl. J. Med.* **325,** 1688 (1991).
4. R. Mattera, J. Codina, A. Crozat, V. Kidd, S. L. Woo, and L. Birnbaumer, *FEBS Lett.* **206,** 36 (1986).
5. T. Kozasa, H. Itoh, T. Tsukamoto, and Y. Kaziro, *Proc. Natl. Acad. Sci. U.S.A.* **85,** 2081 (1988).
6. H. R. Bourne, D. A. Sanders, and F. McCormick, *Nature (London)* **349,** 117 (1991).
7. M. Gross-Bellard, P. Oudet, and P. Chambon, *Eur. J. Biochem.* **36,** 32 (1973).
8. B. B. Rogers, L. C. Alpert, E. A. Hine, and G. J. Buffone, *Am. J. Pathol.* **136,** 541 (1990).
9. P. E. Harris, J. M. Alexander, H. A. Bikkal *et al., J. Clin. Endocrinol. Metab.* **75,** 918 (1992).
10. T. Maniatis, E. F. Fritsch, and J. Sambrook, "Molecular Cloning: A Laboratory Manual." Cold Spring Harbor Laboratory, Cold Spring Harbor, New York, 1982.
11. M. Orita, H. Iwahana, H. Kanazawa, K. Hayashi, and T. Sekiya, *Proc. Natl. Acad. Sci. U.S.A.* **86,** 2766 (1989).
12. Y. Suzuki, M. Orita, M. Shiraishi, K. Hayashi, and T. Sekiya, *Oncogene* **5,** 1037 (1990).

[24] Molecular Methods for Analysis of Genetic Polymorphisms: Application to the Molecular Genetic Study of Genes Encoding β_2-Adrenoceptor and Stimulatory G Protein α Subunit

Charles W. Emala and Michael A. Levine*

Introduction

Identification of specific genes that are responsible for disease phenotypes has gained rapid momentum due to improved methodologies for screening and detecting mutations and polymorphisms in DNA sequence. Neutral changes in DNA sequence are commonly used in genetic linkage studies in which candidate genes are suspected, or in whole genome strategies (e.g., positional cloning) to implicate specific genes as responsible for a disease. Subsequently, sequence analysis of specific genes may be undertaken to determine whether defects are present that can explain the disease phenotype. These methodologies can be categorized as those evaluating large changes in gene sequence (insertions, deletions, rearrangements) and those based on small (i.e., single base) changes, which will be the focus of this review.

Identification of changes in gene sequence can serve two purposes. A specific sequence change can result in altered expression or function of an encoded protein and thereby represent the basis for disease. Alternatively, changes in DNA sequence may not cause disease, but can serve as genetic markers that facilitate genome-wide searches (i.e., positional cloning) or linkage analysis with specific candidate genes. The inheritance of specific genes or chromosomal regions can be analyzed to determine linkage with specific traits to localize the area of the genome that is responsible for a particular disease. In pedigree analysis, transmission of particular gene alleles can be rapidly compared to the inheritance of disease phenotypes or traits to assess genetic linkage. Neutral variations in DNA sequence can serve as useful markers of a region within the genome, and include single-base changes

*To whom correspondence should be addressed.

and variable number of tandem repeats (VNTR), which are short (11–60 bp) and generally benign repeats of nucleotide sequence occurring frequently in noncoding regions of DNA (1). Although VNTRs, which are highly heterozygous, are very useful in genetic linkage studies and facilitate positional cloning strategies, polymorphic base changes within known candidate genes can be used for both linkage studies, and in themselves may potentially represent disease-causing mutations (2). This review is limited to more commonly used techniques involved in the detection and analysis of small base changes in DNA.

In this chapter the term mutation refers to changes in gene sequence that result in the disease phenotype, either through a deleterious change in amino acid sequence, mRNA splicing or stability, or gene transcription. "Neutral" polymorphisms refer to changes in nucleotide sequence that do not affect gene expression. For example, a nucleotide substitution in an exon may not alter the encoded amino acid or could result in a conservative amino acid replacement that does not change the function of the protein.

Molecular studies of the genes encoding the β_2-adrenergic receptor (β_2AR) and the α subunit of the G_s protein ($G_{s\alpha}$) have employed genetic techniques to identify potential defects that may have disease causality. Additionally, in the case of the β_2AR gene, polymorphisms have been used to perform linkage analysis to determine if the β_2AR gene is a candidate gene for asthma, and in several instances specific amino acid substitutions have been evaluated for their effects on receptor protein function.

Methods Employed for Detection of Single-Base DNA Sequence Changes

Restriction Fragment Length Polymorphism

Restriction fragment length polymorphisms (RFLPs) result from small mutations that create or abolish a recognition sequence for a restriction endonuclease. These base changes result in generation of restriction fragments of variable size that can be detected by Southern blot analysis of genomic DNA or by polymerase chain reaction (PCR)-based strategies (Fig. 1). RFLPs may be identified using a panel of restriction enzymes to digest genomic DNA (or cDNA) from multiple unrelated subjects; which is then size fractionated by gel electrophoresis and transferred to a membrane substrate. The digested DNA is then hybridized to a labeled cDNA or genomic DNA corresponding to the candidate gene (*i.e.*, Southern blotting). Each restriction enzyme will yield a particular pattern of fragments for each gene; variations in the size or number of fragments often indicate the presence of a polymorphic base

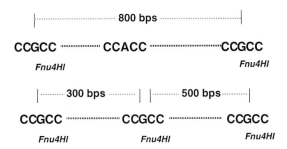

FIG. 1   An example of a restriction fragment length polymorphism. The endonuclease *Fnu4H*I recognizes the sequence CCGCC. Digestion of the upper allele would yield a DNA fragment of 800 bp. A polymorphic base change in the lower allele creates an additional endonuclease recognition site such that digestion of this allele with *Fnu4H*I yields DNA fragments of 300 and 500 bp.

sequence that has created or abolished a recognition site for the restriction enzyme. Analysis of genomic DNA from many subjects is required to detect an RFLP unless it is very common. RFLP analysis can be further refined when the nucleotide sequence or restriction map of the gene is known, because the expected sizes of DNA fragments for any restriction enzyme can be predicted. This may permit more informed selection of the restriction enzymes to be analyzed, enhancing the chances that a polymorphism will be detected. Although this technique allows rapid testing of a large number of restriction enzymes, only a small percentage of single-base changes result in the creation or abolition of a restriction site, and thus this method has a low sensitivity for single-base changes.

A more rapid approach for screening a gene for RFLP can be accomplished using a pool of genomic DNA from 10–15 unrelated individuals. Samples of DNA from the pool are digested with various enzymes and the resulting gene fragments are then analyzed by Southern blotting. The presence of multiple bands of varying intensity often indicates the presence of an RFLP, because DNA from some but not all subjects in the pool will generate alternative gene fragments. This method will not identify polymorphisms of low frequency, because bands with low intensity may not be detected. Confirmation of the RFLP, and determination of the relative allelic frequency, can then be accomplished by restriction of genomic DNA from unrelated subjects.

Skolnick and White (3) have concluded that for a polymorphism to be generally useful as a genetic marker the probability that an offspring is informative (i.e., the polymorphism information content, PIC value) must be at least 0.15. Thus, many polymorphisms will not be sufficiently heterozygous to be of significant clinical utility, and screening methods that facilitate more rapid identification of base changes have been developed. Several of these techniques are based on the observation that single-nucleotide substitutions change the melting point of a DNA segment. These melting polymorphisms offer several important advantages over RFLPs, particularly when applied to linkage analysis studies. Although RFLPs are very useful for distinguishing two alleles at chromosomal loci, they can be detected only when DNA polymorphisms are present in the recognition sequences for the corresponding restriction endonucleases or when deletion or insertion of a short DNA sequence is present within the region detected by a particular probe. However, most polymorphisms (and mutations) are not directly detectable by restriction endonuclease digestion (4). In many instances the base change does not alter the vulnerability of DNA to restriction endonuclease cleavage. Moreover, certain mutations can affect a restriction endonuclease site, but the resulting fragments may be either too small or too similar in size to an already existing fragment to be detected. Finally, the occurrence of similar genes in clusters (repeated copies) may mask or confuse the results of the DNA restriction fragment studies because one radioactive probe may identify fragments from all of the similar genes (4). By contrast, mobility shift analysis of DNA amplified by PCR is an efficient and rapid method for detecting DNA polymorphisms that result from single-base alterations, and can distinguish multiple alleles at chromosomal loci without the use of radioactivity.

Denaturing Gradient Gel Electrophoresis

The denaturing gradient gel electrophoresis (DGGE) technique satisfies many of the requirements for a screening method that can rapidly identify point mutations in defined regions of DNA. DGGE is capable of detecting single-base substitutions in DNA fragments 100–1000 bp in length on the basis of sequence-dependent melting properties of double-stranded nucleic acid molecules (4–6). By attaching a GC-rich clamp to the test DNA sample, the test segment becomes relatively GC poor and will be a more homogeneous melting domain (7) (Fig. 2). In practice, amplified DNA fragments that differ by only one nucleotide will usually exhibit differences in their electrophoretic migration through polyacrylamide gels containing concentration gradients of two denaturants, formamide and urea. Accordingly, the GC clamp increases the number of potential mutations detectable by DGGE from 50% of all possible single-base changes to close to 100% (7).

Double-stranded DNA undergoes an abrupt transition from the totally helical state to a partially melted state when it migrates to a specific denaturant concentration in the gel; the denaturant concentration required for this

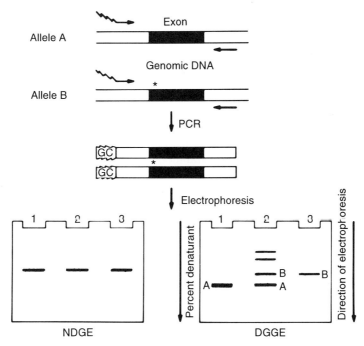

FIG. 2   Analysis of genomic DNA for point mutations. Genomic DNA is amplified by polymerase chain reaction (PCR) using primer pairs in which one oligonucleotide had a 40-base GC extension at the 5' end (GC clamp). PCR products are then subjected to electrophoresis at 60°C using polyacrylamide gels that contain a linearly increasing gradient of formamide and urea. Homologous alleles result in a single band (denoted A or B), whereas heterologous (A and B) alleles yield two homoduplexes and two more slowly migrating heteroduplex bands. DGGE, Denaturing gradient gel electrophoresis; NDGE, nondenaturing gradient gel electrophoresis. Reprinted from Ref. 81, with permission.

transition is dependent on the nucleotide sequence. DNA fragments differing by as little as a single base substitution will melt at slightly different denaturant concentrations because of differences in the stacking interactions between adjacent bases in each DNA strand. Accordingly, wild-type and mutant alleles of amplified DNA fragments will migrate differently within the electrophoretic gel. To optimize detection of nucleotide changes by DGGE, one oligonucleotide of each primer pair can be synthesized with the GC-rich clamp at the 5' end, and the position of the primers can be designed using a mathematical algorithm such as MELTMAP 87 (8). These two modifications serve to ensure amplification of a DNA fragment that contains a single melting domain.

The mixture of normal DNA with mutant samples to ensure heteroduplex formation has also increased the sensitivity of this method (9, 10). Other modifications to this technique include (1) melting of the duplexes in solution using step-wise increase in denaturant concentration prior to standard polyacrylamide electrophoresis, (2) constant denaturant gel electrophoresis (CDGE) whereby separation is done in a gel containing a concentration of denaturant predetermined to be the melting domain of the fragment analyzed (11), and temperature gradient gel electrophoresis (TGGE) rather than using a denaturant gradient (12). In temperature gradient gel electrophoresis gels are placed on a horizontal metal support that generates a temperature gradient. When a double-stranded DNA fragment migrates into the gel region at its melting temperature, denaturation of the strand begins and migration of the DNA strand is arrested.

Typically, DGGE analysis of amplified PCR products is performed using a 7.5% polyacrylamide gel that contains a linearly increasing gradient of the denaturants formamide and urea. Denaturing gels are prepared from stock solutions of 0% denaturant mix [6.5% acrylamide in Tris/acetic acid/EDTA (TAE) buffer] and 100% denaturant mix (6.5% acrylamide, 40% formamide, 7 M urea in TAE buffer). The melting point of a DNA fragment may be determined using computer-based formulas, but empirical testing is often necessary. A 20–80% denaturant gel provides a good starting point for analysis of DNA fragments of 200–700 bp, and perpendicular gels may be used to determine the melting point of the DNA fragment experimentally (8). Amplified DNA fragments are typically electrophoresed at 75–100 V for 16–20 hours at 60°C. DNA is visualized by ultraviolet light after gels are stained with ethidium bromide; other stains, such as silver stain, can provide enhanced sensitivity. Once the melting point of a given DNA fragment is determined, fragments may be analyzed using gels containing a denaturant gradient extending from 15% above the melting point to 15% below. For example, if a DNA fragment is found to denature at 50% denaturant in a preliminary gradient extending from 10–90%, subsequent gels to analyze this DNA fragment would utilize a denaturant gradient of 35–65%. Narrowing the range of denaturant concentration allows optimal focusing of the DNA fragment and permits resolution of DNA fragments that contain base substitutions that change the melting point and thereby alter the electrophoretic mobility of the fragment.

Modifications of the GC clamp and heteroduplex formation allow detection of a large percentage of single-base changes in DNA fragments up to 1000 bp. However, some fragments may contain multiple melting domains in addition to the high-melting domain created by the GC clamp (7). Once denaturation occurs in the fragment's lowest melting point, gel migration stops and potential base changes in areas of higher melting domains would not be detected. Despite these limitations, which do not ensure detection of all base changes, a high percentage of base changes are detected. In addition, the ability to recover the resolved DNA fragments from the gel for subsequent sequencing or subcloning has led to the application of DGGE for direct mutation detection in many disease-candidate genes, including factor VIII deficiency (10), rhodopsin mutations (13), β-thalassemia (14), α-1-antitrypsin deficiency (15), porphyria carriers (16), and asthma (17), as well as linkage analysis studies of the preproPTH gene (18).

Single-Strand Conformational Polymorphisms

The method for detection of single-base changes, single-strand conformational polymorphism (SSCP), takes advantage of defined secondary structure that is assumed by single-stranded DNA molecules under nondenaturing conditions. This inherent secondary structure is altered when one of the bases is changed, resulting in a change in the migration pattern of the fragment in a nondenaturing gel. This method of DNA fragment analysis is analogous to other techniques based on differences in melting temperature, but unlike DGGE, it does not require the use of a gel containing a denaturant gradient, and unlike TGGE, it does not require a special apparatus to form a temperature gradient. SSCP analysis typically is performed using a sequencing gel apparatus, in 8 M urea denaturant polyacrylamide gels containing 0 or 10% glycerol. Sensitivity is further enhanced when gels are run at either 4°C or room temperature. As in DGGE, less than 100% of single-base changes can be detected, which has motivated the development of further modifications to enhance sensitivity. To enhance sensitivity, smaller fragments of DNA may be analyzed by digesting DNA with frequent-cutting restriction enzymes. This not only results in smaller fragments but also improves detection of base changes due to different fragment contexts for the base change. Single-base changes in PCR products or digested genomic DNA can be detected by this method (19). The recent application of SSCP to analysis of the β_3-adrenergic receptor gene has facilitated detection of a novel Trp[64] → Arg polymorphism that is highly associated with the time to onset of noninsulin dependent diabetes mellitus (2).

RNA SSCP may have greater sensitivity than DNA SSCP. RNA assumes more elaborate and greater numbers of conformational complexes, thus a single-base change can have greater impact on its secondary structure. However, RNA SSCP requires the additional step of *in vitro* transcription of a DNA template, and yet still does not ensure detection of all base changes. Nonetheless, SSCP has rapidly gained favor as a screening method for single-base changes due to its simplicity and lack of specialized equipment.

Chemical or Enzymatic Cleavage of Mismatched Bases

Whereas SSCP and DGGE identify base sequence changes on the basis of altered mobility of intact DNA fragments in a gel, other methods rely on chemical or enzymatic cleavage of hybrid molecules at the site of nucleotide mismatch to generate molecules of novel sizes, which may then be resolved by electrophoresis in a denaturing polyacrylamide gel. Chemical and enzymatic cleavage techniques require synthesis of a labeled single-stranded probe of defined length. Following hybridization of the probe with the target sequence, chemical or enzymatic digestion is performed. If no mismatched bases are present, the target fragment "protects" the full-length probe sequence from digestion. By contrast, base substitutions, deletions, or additions in the target sequences will result in nucleotide mismatch that allows digestion of the hybrid molecule to two or more smaller labeled fragments. Thus the presence and approximate location of a base change can be determined by these methods.

Mismatched C bases and mismatched T bases are more reactive with hydroxylamine and osmium tetroxide, respectively, than are matched base pairs. Further reaction with piperidine cleaves the strand containing the mismatched base (20). Mismatched G and A bases are detected with a probe of the opposite sense that would seek mismatched C and T bases, respectively. Moreover, use of mutant as well as wild-type probes accords two opportunities to detect a base change and thereby further enhance the sensitivity of this technique.

Single-base changes may also be identified using enzymes such as RNase A, which cleaves DNA:RNA or RNA:RNA duplexes at sites of base mismatches. Although S1 nuclease was initially found to cleave DNA:DNA duplexes at mismatched sites, and would theoretically be useful for identifying mismatches, S1 nuclease has a low level of cleavage at mismatched sites and many mismatched sites are not cleaved at all (21). RNase A has been exploited for its higher sensitivity of cleaving mismatched pyrimidine residues (C or T or U) in DNA:RNA or RNA:RNA hybrids. A ^{32}P-labeled RNA probe generated by *in vitro* transcription is hybridized with denatured DNA or RNA in solution followed by digestion with RNase A. The size of the protected labeled RNA probe is then determined by gel electrophoresis. If no pyrimidine base mismatches occur within the length of the probe sequence, a full-length probe will be visualized. If a pyrimidine mismatch occurs, as in detection of products of chemical cleavage, two or more smaller band fragments will be visualized. The limitations of this enzymatic cleavage with RNase A are that only about 70% of mismatches are detected, because mismatched purines are usually not cleaved (22), and that *in vitro* synthesis of a ^{32}P-labeled probe is required.

TABLE I   Noncoding Region Polymorphisms of the Human β_2-Adrenergic
           Receptor Gene

Restriction enzyme	Allele size (kb)	Frequency	Detection method[a]	Ref.
*Ban*I	3.7, 3.4	0.25, 0.75	RFLP	27
*Bso*FI (*Fnu4H*I)	0.65, 0.6	0.35, 0.65	RFLP	23

[a] RFLP, Restriction fragment length polymorphism.

Direct Sequencing

The most direct approach to identify polymorphisms within a gene of interest
is DNA sequencing. Some knowledge of the gene or mRNA sequence is
required to design an appropriate sequencing primer. This methodology is
perhaps the most laborious approach to identify polymorphisms, because
large regions of DNA from many individuals may need to be sequenced to
identify even a single polymorphic site. In the absence of an automated
sequencer, direct DNA sequencing is performed manually and is therefore
typically reserved for identification of specific nucleotide changes that have
been detected using screening methods such as RFLP, DGGE, or SSCP.

Detection of Single-Base Changes in the β_2-Adrenergic Receptor Gene

Interest in polymorphic variants and potential mutations within the β_2-adren-
ergic receptor gene is based on the potential role of β_2-adrenergic receptor
dysfunction in asthma. The gene for the human β_2-adrenergic receptor is
contained within a single exon and has been mapped to the q31–q32 region
of chromosome 5. Multiple polymorphic single-base substitutions, many of
which alter the predicted amino acid sequence, have been identified within
this gene and have been evaluated for their potential effects on receptor
function. Three methods have been used to analyze the β_2-adrenergic receptor gene (Tables I and II). RFLP analysis has identified polymorphic restriction enzyme sites for *Ban*I and *Fnu4H*I (*Bso*FI). The identification of these
polymorphic sites, which occur in noncoding regions of the gene, was subsequently followed by the identification of nine polymorphic nucleotide sites
within the coding region of the gene using temperature gradient gel electrophoresis and direct DNA sequencing.

The first allelic polymorphism of the human β_2AR gene to be described
was detected by RFLP analysis. Forty-eight different restriction enzymes
were used to digest genomic DNA isolated from 20 unrelated North American

TABLE II   Coding Region Polymorphisms of the Human β_2-Adrenergic
Receptor Gene

Codon substitution	Nucleic acid position	Frequency	Detection method[a]	Ref.
Neutral polymorphisms				
Leu-84	252	0.22	TGGE	17
Arg-175	523	0.20	TGGE	17
Gly-351	1053	0.27	TGGE	17
Tyr-366	1098	0.005	TGGE	17
Leu-413	1239	0.34	TGGE	17
Amino acid substitutions				
Arg[16] $\rightarrow$ Gly	46	0.72	Sequencing	17
Gln[27] $\rightarrow$ Glu	79	0.49	Sequencing	17
Val[34] $\rightarrow$ Met	100	0.005	Sequencing	17
Thr[164] $\rightarrow$ Ile	491	0.01	TGGE	17

[a] TGGE, Temperature gradient gel electrophoresis.

caucasians. The enzyme *Ban*I identified a biallelic polymorphism that resulted in generation of bands of either 3.4 kb (A) or 3.7 kb (B). The frequency of these alleles in this population was 0.75 for A and 0.25 for B. This indicates that 25% of alleles contain a sequence alteration that results in the loss of a *Ban*I recognition site, which enables hybridization of the probe to a 3.7-kB *Ban*I fragment generated by enzymatic digestion at a *Ban*I site that is 300 bases downstream (23).

The association of the *Ban*I polymorphism with atopy was first evaluated in a study of 72 South African atopic and nonatopic individuals. In this study no association was found between a β_2-adrenergic receptor allele and atopy (24). Interestingly, the gene frequencies for the B and A alleles were 0.45 and 0.55 in the South African population as compared to 0.25 and 0.75 in the North American population, emphasizing the variability of gene frequencies in different populations. The relationship between the *Ban*I polymorphism and reactive airway disease (asthma or methacholine airway hyperresponsiveness) has been evaluated in 56 members of four Japanese families. Although a higher prevalence of symptomatic asthma occurred in individuals who lacked the B allele, there was no significant association between the lack of the B allele and methacholine airway responsiveness (a commonly used objective measure of a tendency toward hyperreactive airway responses) or between the lack of the B allele and atopy (25). In a followup to this study, the relative frequency of β_2-adrenergic receptor alleles was no different in 77 Japanese asthmatic patients as compared to control patients (26).

RFLP analysis identified a second polymorphism in the noncoding region of the human β_2-adrenergic receptor gene. *BsoF*I (*Fnu4H*I) identified a biallelic polymorphism resulting in DNA fragments of 600 (P) and 650 (Q) bp with frequencies of 0.65 (P) and 0.35 (Q) in 46 chromosomes from 23 unrelated Caucasians (27).

The largest number of polymorphisms within the β_2-adrenergic receptor gene and the frequencies of these polymorphisms have been reported by Reishaus *et al.* (17) (Table II). Genomic DNA from 51 unrelated asthmatics and 56 unrelated nonasthmatics was amplified by PCR using five pairs of primers. The PCR products were evaluated by TGGE and direct dideoxy DNA sequencing. Nine polymorphisms were identified within the coding region of the gene. Five base substitutions were natural polymorphisms, which did not change the encoded amino acid. By contrast, four polymorphisms resulted in amino acid changes ($Arg^{16} \rightarrow Gly$; $Gln^{27} \rightarrow Glu$; $Val^{34} \rightarrow Met$; $Thr^{164} \rightarrow Ile$) (Fig. 3). The frequencies of these polymorphisms were similar in the asthmatic and nonasthmatic groups, suggesting that these polymorphic changes alone could not account for asthma in the majority of randomly selected asthmatics. Subsequently, three of the polymorphisms that result in amino acid replacements were evaluated by stable expression in mammalian cells to evaluate the functional role, if any, of the change in β_2-adrenergic receptor sequence. The $Arg^{16} \rightarrow Gly$ (R16G) substitution leads to enhanced agonist-mediated down-regulation of the receptor (28), with normal agonist binding and normal stimulation of adenylyl cyclase. The R16G variant has been found to be significantly increased in a subset of asthmatics that exhibit nocturnal asthma (29). In contrast, the Q27E variant receptor lacks down-regulation (28), and individuals who are homozygous for this allele have *less* airway reactivity to methacholine compared to individuals who have the Q27 receptor (30). The T164I substitution results in a receptor with impaired agonist binding, impaired ability to stimulate adenylyl cyclase, and an impaired rate of receptor sequestration following agonist binding (31). Although the frequency of these variants is not significantly increased in randomly selected asthmatics (17), the functional defects of these expressed receptor variants suggest that single DNA base changes in the β_2-adrenergic receptor gene may influence receptor function and the ultimate expression of the asthmatic phenotype.

Polymorphisms within the β_2-adrenergic receptor gene have been used to perform linkage analysis of this gene with asthma. As many as six polymorphic markers have been used in multiplex families with a familial incidence of asthma. Figure 4 is a representative example of the analysis of polymorphic markers for its linkage to asthma in one such family. In this family the maternal allele, designated BQCREU, was inherited by three of her children, which is discordant with the inheritance of asthma in only two of the three children. This independence of the inheritance of the β_2-adrenergic receptor

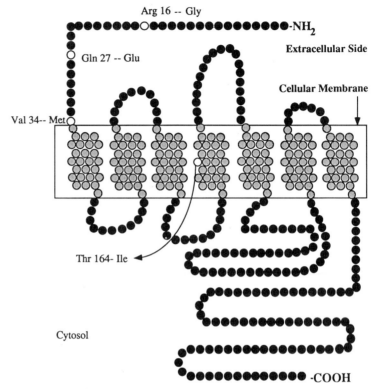

FIG. 3   The predicted protein structure and membrane topology of the human β_2-adrenergic receptor, depicting amino acid substitutions secondary to DNA polymorphic sites ($\bigcirc$).

allele and asthma suggests no linkage of asthma to this candidate gene in this family.

Methods Employed for Detection of DNA Sequence Changes in the $G_{s\alpha}$ Subunit Gene

The heterotrimeric ($\alpha\beta\gamma$) guanine nucleotide-binding proteins (G proteins) are members of a superfamily of GTPase "molecular switches." Members of this superfamily utilize the hydrolysis of GTP to control diverse cellular processes, including protein translation (e.g., EF-Tu), vesicular transport (e.g., rab) (32), the actin cytoskeleton (e.g., rho) (33), growth and differentia-

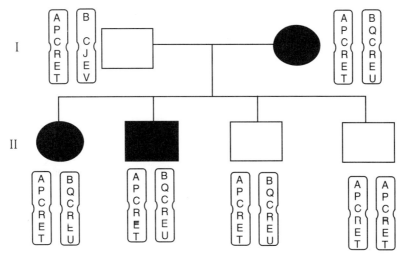

FIG. 4   A representative asthmatic family pedigree. Linkage analysis was performed using six polymorpic markers within the β_2-adrenergic receptor gene. ●, Female with asthma; ■, males with asthma; □, males without asthma. Letter designations for polymorphic markers: A/B, *Ban*I RFLP; P/Q, *Bso*FI RFLP; C/M, polymorphism characterized by DGGE but not sequenced; R/J, Arg[16] → Gly; T/U, Gln[27] → Glu; E/K, Leu-84. The mother in generation I has asthma and three children in generation II inherit the same allele from her. However, only two of the three children who inherit this allele develop asthma. These results suggest that either (1) there is no linkage between the β_2-adrenergic receptor and asthma in this family, or (2) that inheritance of this allele is necessary but not sufficient for the development of asthma.

tion (e.g., ras) (34), and transmembrane signal transduction (e.g., G proteins) (35). G proteins have been identified in a variety of a species, ranging from unicellular organisms to mammals. G proteins subserve a broad range of signal transduction functions in their ability to couple receptors for light, odorants, hormones, growth factors, and neurotransmitters to signal generating enzymes as well as ion channels. To fulfill these diverse roles many different G proteins have evolved. At least 16 genes encoding mammalian G protein α subunits have been identified by biochemical purification or molecular cloning. In many instances, molecular cloning has revealed the existence of more than one protein that might subserve the role of a specific G protein that was first identified by function (e.g., four species of α_s are known to be generated by alternative splicing, and three separate genes encode species of α_i). The α subunits associate with at least five distinct β subunits (36, 37) and perhaps as many as 10 distinct γ subunits (38, 39). The β and γ subunits exhibit combinatorial specificity and cell-specific distribution (36, 39, 40). Moreover, specific $\beta\gamma$ dimers exhibit preferential association with specific α subunits, an effect that is determined primarily by sequences in the γ subunit (40). This combinatorial specificity results in a large number of unique G proteins that can couple specific receptors to multiple signal effector enzymes and ion channels (41, 42).

Mutations and polymorphisms have been identified in the gene that encodes the α chain of the G protein that stimulates adenylyl cyclase ($G_{s\alpha}$). The human α_s protein is encoded by the *GNAS1* gene located on chromosome 20q13.1 → 13.2 (43). It spans approximately 20 kb and contains at least 13 coding exons (44). The α_s mRNA is alternatively spliced to give four biologically active products, by including or excluding exon 3, and by utilizing alternative splice acceptor sites in intron 3 to include or exclude a single Ser residue (45). In addition, there appear to be at least two alternative first exons in α_s transcripts. In one case, an alternative promoter produces a transcript with a novel exon 1 that does not contain an initiator ATG; thus a truncated, nonfunctional α_s protein is translated from an in-frame ATG in exon 2 (46). In the other case, a transcript has been identified that encodes a larger α_s isoform ($XL\alpha_s$), in which a novel 51-kDa protein is spliced to exons 2 through 13 of α_s. The function of the large (92-kDa) α_s isoform is unknown, but the protein appears to be associated with the trans Golgi network (47).

Concordant with the important role of adenylyl cyclase and the second messenger cAMP in mediating the actions of many peptide hormones and neurotransmitters, altered expression or activity of α_s has been implicated in the pathophysiology of several endocrine disorders. Mutations in *GNAS1* that impair its ability to hydrolyze GTP lead to constitutive activation of adenylyl cyclase and result in hyperfunction and autonomous growth of endocrine tissues (e.g., sporadic endocrine tumors and McCune–Albright syndrome). By contrast, mutations in *GNAS1* that lead to decreased expression or function of α_s are associated with generalized resistance to hormones that act by stimulating adenylyl cyclase (e.g., pseudohypoparathyroidism type 1a and Albright hereditary osteodystrophy).

A role for α_s in tumorigenesis was first suggested by Vallar *et al.* (48), who found that a subset of growth hormone-secreting pituitary tumors from human subjects with acromegaly had increased adenylyl cyclase activity. Landis *et al.* (49) established the molecular basis for this constitutive activation of adenylyl cyclase as a mutation in *GNAS1* that resulted in replacement of either Arg-201 or Gln-227. Both of these residues are located in the GTP binding site and are important regulators of the intrinsic GTPase of the α chain, and previous studies had suggested that modification of these amino acids could provide a basis for human disease. For example, ADP-ribosylation of Arg-201 of α_s by the exotoxin of *Vibrio cholera* markedly impairs

GTPase activity, and thereby transforms α_s to a constitutively active state that is capable of ligand-independent stimulation of adenylyl cyclase. ADP-ribosylation of $G_{s\alpha}$ in intestinal epithelial cells increases the level of cAMP, which inhibits salt and water reabsorption and leads in part to the watery diarrhea that is characteristic of cholera. In a second example, Gln-227 in α_s is analogous to the cognate amino acid, Gln-61, that is frequently replaced in the protooncogene p21^{ras}, a member of the superfamily of GTPase molecular switches. Point mutations in p21^{ras} that inhibit GTPase activity are present in a variety of human tumors (34) and account for neoplastic transformation in these cells via activation of signal pathways (49). The net result is promotion of both autonomous function and clonal expansion of cells carrying the mutation. Thus, the Arg-201 and Gln-227 mutations in *GNAS1* share some transforming characteristics with p21^{ras} oncogenes, and have been termed *gsp1*. Several studies have now confirmed the presence of *gsp1* mutations in approximately 50% of human pituitary tumors (50–52), and have also identified the *gsp1* mutation in a smaller subset (10–15%) of the thyroid tumors (50–53). There are no obvious differences in the clinical behavior of tumors with or without the *gsp1* mutation (54, 55).

Similar *gsp1* mutations have been identified in affected tissues from patients with the McCune–Albright syndrome, an unusual metabolic disorder characterized by the clinical triad of polyostotic fibrous dysplasia, *café-au-lait* pigmented skin lesions, and endocrine dysfunction, most notably precocious puberty in girls. Subsequent studies have disclosed multiple and diverse endocrinopathies in patients with McCune–Albright syndrome (MAS), including gonadotropin-independent precocious puberty (56), autonomous thyroid nodules (57), growth hormone excess and hyperprolactinemia (58), hypercortisolism (59), and hyperphosphaturic hypophosphatemic rickets or osteomalacia (60). These metabolic disorders occur alone or in combination in individual patients, and range in severity from occult biochemical abnormalities to life-threatening clinical disabilities. Careful clinical observation led to the proposal that MAS is caused by a postzygotic, somatic cell mutation (61). This hypothesis was based on the variable involvement of endocrine glands and bones, and the pattern of the skin lesions (which follow lines of embryologic development), suggesting a mosaic distribution of the mutation. The lack of documented transmission of the syndrome was felt to imply that germ-line inheritance of the gene defect would be lethal (61). Several lines of evidence had suggested that the molecular basis for MAS involved a postreceptor defect that led to accumulation of elevated levels of cAMP: (1) cAMP is the major intracellular second messenger for hormones that regulate cell proliferation and function in the affected tissues, (2) the serum levels of tropic hormones that normally regulate the hyperfunctional endocrine glands are not elevated, (3) each involved cell type expresses a different receptor, suggesting that a single mutation would have to be in a gene encoding a protein expressed in all tissues, such as a G protein, and (4) sporadic mutations that inhibit the intrinsic GTPase activity of α_s (49) and produce hormone-independent activation of adenylyl cyclase can cause sporadic pituitary tumors and thyroid tumors (see above). Weinstein *et al.* (62) and Schwindinger *et al.* (63) subsequently identified mutations in Arg-201 in DNA isolated from pathologic and clinical specimens obtained from patients with MAS. Two specific mutations have been identified: $Arg^{201}(CGT) \rightarrow His(CAT)$ and $Arg^{201}(CGT) \rightarrow Cys (TGT)$. These mutations have been found in affected endocrine tissues (62), skin (63), and bone (64), as well as in tissues not classically involved in MAS, including blood (62, 63), liver (65), heart, and others (66). These mutations are not present in all cells of an individual patient with MAS, and even in an affected organ they are present in more cells in histologically abnormal tissue than in histologically normal tissue. Thus, the mutation exhibits a mosaic distribution, and thereby fulfills the original hypothesis proposed by Happle (61).

DGGE has facilitated identification of a large number of mutations in the $G_{s\alpha}$ gene that led to loss of function in patients with Albright hereditary osteodystrophy (AHO), a syndrome characterized by decreased expression or function of $G_{s\alpha}$ and a constellation of unusual somatic features, including short stature, brachydactyly, and heterotopic ossifications. In addition, most patients with AHO are resitant to parathyroid hormone and other hormones that activate receptors coupled via G_s to the stimulation of adenylyl cyclase, a condition termed pseudohypoparathyroidism (PHP) type 1a (67). Other patients with AHO and $G_{s\alpha}$ deficiency lack hormone resistance, a condition termed pseudopseudohypoparathyroidism (67). Distinct mutations in the *GNAS1* gene, including missense mutations (68–70), point mutations in sequences required for efficient splicing (71), and small deletions (68, 71, 72), have been found in each kindred studied, implying that new and independent mutations sustain this disorder in the population (Fig. 5). Most patients with AHO have genetic defects that impair the synthesis of $G_{s\alpha}$ protein and therefore have $G_{s\alpha}$ deficiency. In other patients mutations in the gene encoding $G_{s\alpha}$ lead to synthesis of dysfunctional proteins (Fig. 5). The first mutation in *GNAS1* that was identified in a patient with AHO altered the initiator ATG codon and led to synthesis of a $G_{s\alpha}$ protein that was truncated at the amino-terminal end (70). This protein presumably lacks the ability to interact with receptors or adenylyl cyclase. In other cases missense mutations may selectively impair a specific function of $G_{s\alpha}$. A point mutation in exon 13 of the $G_{s\alpha}$ gene that results in the replacement of Arg-385 with His near the carboxy terminus of $G_{s\alpha}$ has recently been described in one patient with PHP type 1a (69) (Fig. 5). This mutation is located five amino acids upstream of a mutation (*unc*) in $G_{s\alpha}$ that was previously identified in the S49 murine

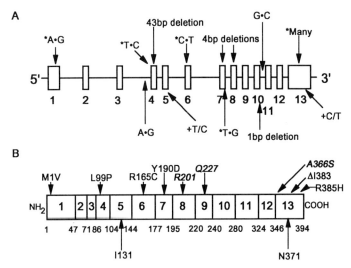

FIG. 5   Mutations in the G$_{s\alpha}$ gene. (A) The human G$_{s\alpha}$ gene, which spans over 20 kbp and contains 13 exons and 12 introns. Thirteen unique mutations that result in *loss of G$_{s\alpha}$ function* have been identified in affected members of 13 unrelated AHO families; missense mutations are denoted by an asterisk. (B) The position of these missense mutations is above the protein structure. There are two polymorphisms (+; A) and the position of the unchanged amino acid is denoted beneath the predicted G$_{s\alpha}$ protein (B). The sites of two missense mutations that result in *gain of function* (replacement of either Arg-201 or Gln-227) in patients with MAS (62–64, 66, 77) or in sporadic tumors (49, 51) are denoted by italics. The mutation in exon 1 eliminates the initiator methionine codon and prevents synthesis of a normal G$_{s\alpha}$ protein (70). The 4-bp deletions in exon 7 (82) and exon 8 (68) and the 1-bp deletion in exon 10 all shift the normal reading frame and prevent normal mRNA and/or protein synthesis. Mutations in intron 3 and at the donor splice junction between exon 10 and intron 10 cause splicing abnormalities that prevent normal mRNA synthesis (83). The mutations indicated with an asterisk represent missense mutations (68, 69, 77, 84); the resultant amino acid substitutions are indicated in the schematic diagram of the G$_{s\alpha}$ protein at the bottom of the figure. Some of these mutations may prevent normal protein synthesis by altering protein secondary structure, but the Arg → His substitution in exon 13 appears to encode an altered protein that cannot couple normally to receptors (69), and the Ala → Ser mutation encodes an activated G$_{s\alpha}$ protein that is unstable at 37°C (77).

lymphoma cell line (73, 74). The *unc* mutation (arg³⁸⁹ → Pro) "uncouples" G$_{s\alpha}$ from cell surface receptors (75). Similarly, expression studies of the Arg³⁸⁵ → His mutation in G$_{s\alpha}$ indicates that this molecule is also unable to couple cell surface receptors to activation of adenylyl cyclase (69).

TABLE III   Neutral Polymorphisms in the Human $G_{s\alpha}$ Gene

Exon	Codon	Frequency	Detection method[a]	Ref.
Exon 5	Ile-131	0.62	DGGE	68
Amplicon exon 4–5	Ile-131	0.53	DGGE/RFLP	78
Exon 13	Asn-371	0.21	DGGE/RFLP	79

[a] DGGE, Denaturing gradient gel electrophoresis; RFLP, restriction fragment length polymorphism.

The wide variety of clinical presentations caused by mutations in α_s is perhaps best illustrated by the cases of two unrelated boys who both had α_s deficiency with gonadotropin-independent precocious puberty and renal resistance to PTH infusion (76). These two boys showed the simultaneous expression of a feature of MAS (excessive α_s function) and a feature of PHP type Ia (α_s deficiency). The molecular basis for this unusual presentation was determined to be a missense mutation in *GNAS1* that replaced Ala-366 with Ser. This mutation had two effects (77): (1) it increased the rate of GDP release from mutant α_s 80-fold, mimicking the action of hormone coupled receptor, and (2) it decreased the thermal stability of the mutant protein, resulting in rapid degradation at 37°C. Thus in the testis, where the temperature is 33°C, the mutant α_s accumulates and causes hormone-independent activation of adenylyl cyclase, while in the rest of the body the mutant α_s is degraded, resulting in decreased stimulation of adenylyl cyclase in response to hormones.

DGGE has been used to identify neutral polymorphisms in exons 5 (68, 78) and 13 (79) (Table III). The neutral polymorphism in exon 5 consists of a T → C substitution that conserves the isoleucine residue at codon 131 (ATT → ATC) (68, 78). This single-base substitution changes the melting point of $G_{s\alpha}$ gene amplicons consisting of either exons 4 and 5 plus intron 4 (68, 78), or exon 5 alone (68), and facilitates resolution of the two different alleles by DGGE. In addition, this nucleotide polymorphism can be detected by digestion of the amplicon with *Fok*I (78). In this case, alleles that contain the *Fok*I site (CATCC) are cleaved to generate two fragments of 126 and 279 bp (using the PCR primers specified by the authors (78)), whereas alleles that lack the *Fok*I site (CATTC) result in an intact fragment of 405 bp. An additional neutral polymorphism has been detected by DGGE in exon 13 of the $G_{s\alpha}$ gene (79). This C → T transition at codon 371 does not replace the normal amino acid Asn, but does abolish a recognition site for *Fok*I (79). The previously described allele (44) is designated A_1; the variant allele, A_2, contains the *Fok*I site. Restriction endonuclease digestion of the 284-bp PCR product from individuals who are homozygous for the A_1 alleles generates two fragments, 120 and 164 bp in length (80).

Summary

The ability to screen DNA rapidly for small changes in base sequence has revolutionized the search for genetic basis of disease. These techniques have led to the identification of specific mutations that account for the disease phenotype in syndromes such as Albright hereditary osteodystrophy and McCune–Albright syndrome, and have provided polymorphic markers for candidate gene and positional cloning studies of inherited airway hyperresponsiveness. These and additional techniques will continue to accelerate the discovery of the genetic cause of many human diseases.

Acknowledgments

This work was supported in part by NIH Grants DK34281 (MAL) and HL45794 (CWE) from the U.S. Public Health Service.

References

1. Y. Nakamura, M. Leppert, P. O'Connell, R. Wolfe, T. Holm, M. Culver, C. Martin, E. Fujimoto, M. Hoff, E. Kumlin, and R. White, *Science* **235**, 1616 (1987).
2. J. Walston, K. Silver, C. Bogardus, W. C. Knowler, F. S. Celi, S. Austin, B. Manning, A. D. Strosberg, M. P. Stern, N. Raben, J. D. Sorkin, J. Roth, and A. R. Shuldiner, *N. Engl. J. Med.* **333**, 343 (1995).
3. M. H. Skolnick and R. White, *Cytogenet. Cell. Genet.* **32**, 58 (1982).
4. R. M. Myers and T. Maniatis, *in* "Molecular Biology of Homo Sapiens," pp. 275–284. Cold Spring Harbor Laboratory, Cold Spring Harbor, New York, 1986.
5. S. G. Fischer and L. S. Lerman, *Proc. Natl. Acad. Sci. U.S.A.* **80**, 1579 (1983).
6. R. M. Myers, S. G. Fischer, T. Maniatis, and L. S. Lerman, *Nucleic Acids Res.* **13**, 3111 (1985).
7. V. C. Sheffield, D. R. Cox, L. S. Lerman, and R. M. Myers, *Proc. Natl. Acad. Sci. U.S.A.* **86**, 232 (1989).
8. L. S. Lerman and K. Silverstein, *in* "Methods in Enzymology" (R. Wu, ed.), p. 482. Academic Press, Orlando, Florida, 1987.
9. S.-P. Cai and Y. W. Kan, *J. Clin. Invest.* **85**, 550 (1990).
10. M. Higuchi, H. H. Kazazian, Jr., C. K. Kasper, J. A. Phillips, and S. E. Antonarakis, *Pediatr. Res.* **27**, 777 (1990).
11. E. Hovig, B. Smith-Sorensen, A. Brogger, and A.-L. Borresen, *Mutation Res.* **262**, 36 (1991).
12. V. Rosenbaum and D. Reissner, *Biophys. Chem.* **26**, 236 (1987).
13. C.-H. Sung, C. M. Davenport, J. C. Hennessey, I. H. Maumenee, S. G. Jacobson, J. R. Heckenlively, R. Nowakowski, G. Fishman, P. Gouras, and J. Nathans, *Proc. Natl. Acad. Sci. U.S.A.* **88**, 6481 (1991).

14. M. Losekoot, R. Fodde, C. I. Harteveld, H. Van Heeren, P. C. Giordano, and L. F. Bernini, *Br. J. Haematol.* **76,** 269 (1990).

15. P. H. Johnson, H. Cadiou, and D. A. Hopkinson, *Ann. Hum. Genet.* **55,** 183 (1991).

16. F. Bourgeois, X. F. Gu, J. C. Deyback, M. P. Te Velde, F. de Rooij, Y. Nordmann, and B. Grandchamp, *Clin. Chem.* **38,** 93 (1992).

17. E. Reishaus, M. Innis, N. MacIntyre, and S. B. Liggett, *Am. J. Respir. Cell Biol.* **8,** 334 (1993).

18. A. Miric and M. A. Levine, *J. Clin. Endocrinol. Metab.* **74,** 509 (1991).

19. M. Orita, H. Iwahana, H. Kanazawa, K. Hayashi, and T. Sekiya, *Proc. Natl. Acad. Sci. U.S.A.* **86,** 2766 (1989).

20. R. G. H. Cotton, N. R. Rodrigues, and R. D. Campbell, *Proc. Natl. Acad. Sci. U.S.A.* **85,** 4397 (1988).

21. T. E. Shenk, C. Rhodes, P. W. J. Rigby, and P. Berg, *Proc. Natl. Acad. Sci. U.S.A.* **72,** 989 (1975).

22. R. M. Myers, Z. Larin, and T. Maniatis, *Science* **230,** 1242 (1985).

23. K.-U. Lentes, W. H. Berrettini, M. R. Hoehe, F.-Z. Chung, and E. S. Gershon, *Nucleic Acids Res.* **16,** 23 (1988).

24. P. C. Potter, L. Van Wyk, M. Martin, K. U. Lentes, and E. B. Dowdle, *Clin. Exp. Allergy* **23,** 874 (1992).

25. M. Ohe, M. Munukata, and N. Hizawa, *Am. Rev. Respir. Dis.* **147,** A155 (1993).

26. H. Taguchi, M. Ohe, and N. Hizawa, *A. C. I. News* **89** (*Suppl. 2*), A317 (1994).

27. C. K. McQuitty, C. W. Emala, C. A. Hirshman, and M. A. Levine, *Hum. Genet.* **93,** 225 (1994).

28. S. A. Green, J. Turki, M. Innis, and S. B. Liggett, *Biochemistry* **33,** 9414 (1994).

29. J. Turki, J. Pak, S. A. Green, R.J. Martin, and S. B. Ligget, *J. Clin. Invest.* **95,** 1635 (1995).

30. I. P. Hall, A. Wheatley, P. Wilding, and S. B. Liggett, *Lancet* **345,** 1213 (1995).

31. S. A. Green, G. Cole, M. Jacinto, M. Innis, and S. B Liggett, *J. Biol. Chem.* **268,** 23116 (1993).

32. C. Nuoffer and W. E. Balch, *Annu. Rev. Biochem.* **63,** 949 (1994).

33. A. Hall, *Annu. Rev. Cell Biol.* **10,** 31 (1994).

34. G.L. Pronk and J. L. Bos, *Biochim. Biophys. Acta* **1198,** 131 (1994).

35. E. J. Neer, *Cell* (*Cambridge, Massachusetts*) **80,** 249 (1995).

36. M. A. Levine, P. M. Smallwood, P. T. Moen, L. J. Helman, and T. G. Ahn, *Proc. Natl. Acad. Sci. U.S.A.* **87,** 2329 (1990).

37. J. A. Watson, A. Katz, and M. I. Simon, *J. Biol. Chem.* **269,** 22150 (1994).

38. Y. Fukada, H. Ohguro, T. Saito, T. Yoshizawa, and T. Akino, *J. Biol. Chem.* **264,** 5937 (1989).

39. M. Rahmatullah, R. Ginnan, and J. D. Robishaw, *J. Biol. Chem.* **270,** 2946 (1995).

40. M. Rahmatullah and J. D. Robishaw, *J. Biol. Chem.* **269,** 3574 (1994).

41. L. Birnbaumer, J. Abramowitz, A. Yatani, K. Okabe, R. Mattera, J. Sanford, J. Codina, and A. M. Brown, *Crit. Rev. Biochem. Mol. Biol.* **24,** 225 (1990).

42. G. L. Johnson and N. Dhanasekaran, *Endocr. Rev.* **10,** 317 (1989).

43. M. A. Levine, W. S. Modi, and S. J. Obrien, *Genomics* **11,** 478 (1991).

44. T. Kozasa, H. Itoh, T. Tsukamoto, and Y. Kaziro, *Proc. Natl. Acad. Sci. U.S.A.* **85,** 2081 (1988).
45. P. Bray, A. Carter, C. Simons, V. Guo, C. Puckett, J. Kamholz, A. Spiegel, and M. Nirenberg, *Proc. Natl. Acad. Sci. U.S.A.* **83,** 8893 (1986).
46. Y. Ishikawa, C. Bianchi, B. Nadal-Ginard, and C. J. Homcy, *J. Biol. Chem.* **265,** 8458 (1990).
47. R. H. Kehlenbach, J. Matthey, and W. B. Huttner, *Nature (London)* **372,** 804 (1994).
48. L. Vallar, A. Spada, and G. Giannattasio, *Nature (London)* **330,** 566 (1987).
49. C. A. Landis, S. B. Masters, A. Spada, A. M. Pace, H. R. Bourne, and L. Vallar, *Nature (London)* **340,** 692 (1989).
50. E. Clementi, N. Malgaretti, J. Meldolesi, and R. Taramelli, *Oncogene* **5,** 1059 (1990).
51. J. Lyons, C. A. Landis, and G. Harsh, *Science* **249,** 655 (1990).
52. K. Yoshimoto, H. Iwahana, A. Fukuda, T. Sano, and M. Itakura, *Cancer* **72,** 1386 (1993).
53. C. O'Sullivan, C. M. Barton, S. L. Staddon, C. L. Brown, and N. R. Lemoine, *Mol. Carcinogenesis* **4,** 345 (1991).
54. C. A. Landis, G. Harsh, J. Lyons, R. L. Davis, F. McCormick, and H. R. Bourne, *J. Clin. Endocrinol. Metab.* **71,** 1416 (1990).
55. A. Spada, M. Arosio, D. Bochicchio, N. Bazzoni, L. Vallar, M. Bassetti, and G. Faglia, *J. Clin. Endocrinol. Metab.* **71,** 1421 (1990).
56. C. M. Foster, J. L. Ross, T. Shawker, O. H. Pescovitz, D. L. Loriaux, G. B. Cutler, and F. Comite, *J. Clin. Endocrinol. Metab.* **58,** 1161 (1984).
57. P. P. Feuillan, T. Shawker, S. R. Rose, J. Jones, R. K. Jeevanram, and B. C. Nisula, *J. Clin. Endocrinol. Metab.* **71,** 1596 (1990).
58. L. Cuttler, J. A. Jackson, M. S. Uz-Zafar, L. Levitsky, R. C. Mellinger, and L. A. Frohman, *J. Clin. Endocrinol. Metab.* **68,** 1148 (1989).
59. N. Mauras and R. M. Blizzard, *Acta. Endocrinol.* **256** (*Suppl.*), 207 (1986).
60. E. G. Lever and K. W. Pettingale, *J. Bone Joint Surg.* **65B,** 621 (1983).
61. R. Happle, *Clin. Genet.* **29,** 321 (1986).
62. L. S. Weinstein, A. Shenker, P. V. Gejman, M. J. Merino, E. Friedman, and A. M. Spiegel, *N. Engl. J. Med.* **325,** 1688 (1991).
63. W. F. Schwindinger, C. A. Francomano, and M. A. Levine, *Proc. Natl. Acad. Sci. U.S.A.* **89,** 5152 (1992).
64. A. Shenker, D. E. Sweet, A. M. Spiegel, and L. S. Winstein, *J. Bone Miner. Res.* **7,** S115 (Abstr.) (1992).
65. W. F. Schwindinger, S. Q. Yang, E. P. Miskovsky, A. M. Diehl, and M. A. Levine, *Endocr. Soc. Progr., Abstr.* **75,** 517 (1993).
66. A. Shenker, L. S. Weinstein, A. Moran, O. H. Pescovitz, N. J. Charest, C. M. Boney, J. J. Van Wyk, M. J. Merino, P. P. Feuillan, and A. M. Spiegel, *J. Pediatr.* **123,** 509 (1993).
67. M. A. Levine, W. F. Schwindinger, R. W. Downs, Jr., and A. M. Moses, *in* "The Parathyroids: Basic and Clinical Concepts" (J. P. Bilezikian, R. Marcus, and M. A. Levine, eds.), pp. 781–800. Raven Press, New York, 1994.

68. A. Miric, J. S. Vechio, and M. A. Levine, *J. Clin. Endocrinol. Metab.* **76,** 1560 (1993).

69. W. F. Schwindinger, A. Miric, D. Zimmerman, and M. A. Levine, *J. Biol. Chem.* **269,** 25387 (1994).

70. J. L. Patten, D. R. Johns, D. Valle, C. Eil, P. A. Gruppuso, G. Steele, P. M. Smallwood, and M. A. Levine, *N. Engl. J. Med.* **322,** 1412 (1990).

71. L. S. Weinstein, P. V. Gejman, E. Friedman, T. Kadowaki, R. M. Collins, E. S. Gershon, and A. M. Spiegel, *Proc. Natl. Acad. Sci. U.S.A.* **87,** 8287 (1990).

72. M. E. Luttikhuis, L. C. Wilson, J. V. Leonard, and R. C. Trembath, *Genomics* **21,** 455 (1994).

73. T. Haga, E. M. Ross, H. J. Anderson, and A. G. Gilman, *Proc. Natl. Acad. Sci. U.S.A.* **74,** 2016 (1977).

74. T. Rall and B. A. Harris, *FEBS Lett.* **224,** 365 (1987).

75. K. A. Sullivan, R. T. Miller, S. B. Masters, B. Beiderman, W. Heideman, and H. R. Bourne, *Nature (London)* **330,** 758 (1987).

76. J. M. Nakamoto, E. A. Jones, D. Zimmerman, M. L. Scott, M. A. Donlan, and C. Van Dop, *Clin. Res.* **41,** 40A (1993).

77. T. Iiri, P. Herzmark, J. M. Nakamoto, C. V. Dop, and H. R. Bourne, *Nature (London)* **371,** 164 (1994).

78. P. V. Gejman, L. S. Weinstein, and M. Martinez, *Genomics,* **9,** 781 (1991).

79. C. Waltman, M. A. Levine, F. Schwindinger, and G. S. Wand, *Hum. Genet.* **93,** 477 (1994).

80. I. Marbach, J. Shiloach, and A. Levitzki, *Eur. J. Biochem.* **172,** 239 (1988).

81. C. W. Emala, W. F. Schwindinger, G. S. Wand, and M. A. Levine, *in* "Progress in Nucleic Acid Research and Molecular Biology" (W. E. Cohn and K. Moldave, eds.), Volume 47, pp. 81–107. Academic Press, San Diego, 1994.

82. L. S. Weinstein, P. V. Gejman, P. de Mazancourt, N. American, and A. M. Speigel, *Genomics* **13,** 1319 (1992).

83. L. S. Weinstein, P. V. Gejman, and E. Friedman, *Proc. Natl. Acad. Sci. U.S.A.* **87,** 8287 (1990).

84. W. F. Schwindinger, A. Miric, and M. A. Levine, *74th Annu. Mt. Endocrine Soc.,* Abstr. 3 (1992).

Index